Engineering and Industrial Graphics Handbook

George E. Rowbotham, Editor in Chief
Engineering Services Management Consultant

Engineering and Industrial Graphics Handbook

McGraw-Hill Book Company

New York   St. Louis   San Francisco   Auckland
Bogotá   Hamburg   Johannesburg   London   Madrid
Mexico   Montreal   New Delhi   Panama   Paris
São Paulo   Singapore   Sydney   Tokyo   Toronto

Library of Congress Cataloging in Publication Data

Main entry under title:
Engineering and industrial graphics handbook.

(McGraw-Hill handbook series)
Includes index.
1. Engineering graphics—Handbooks, manuals, etc.
I. Rowbotham, George E.
T353.E615   604.2   81-12399
ISBN 0-07-054080-2   AACR2

1234567890   HDHD   8987654321

ISBN 0-07-054080-2

The editors for this book were Patricia Allen-Browne and Olive H. Collen, the
designer was Mark E. Safran, and the production supervisor was Thomas G.
Kowalczyk. It was set in Aster by University Graphics, Inc.

Printed and bound by Halliday Lithograph.

To progress.

We are living in an extraordinary and challenging era. It is altogether new. The world has never seen anything like it. It is an age marked by exploration into the heavens, the earth, and the sea. Perhaps even more remarkable has been the application of vast scientific and technological discoveries to everyday life.

The ancients saw nothing like it. The moderns have seen nothing like it—until the present generation. The progress of the age has almost outstripped human imagination.

Engineering and industrial graphics, with its endless array of tools, techniques, applications, and innovations, has contributed notably to this tremendous progress.

Contents

Contributors

Regis L. Augusty, National Sales Manager, Microsystems Products, Xerox Corporation, Rochester, New York (Section 12)

Irving A. Dlugatch, Dean of Students, California Western University, Santa Ana, California (Section 11)

Vance D. Dutton, Engineering Supervisor—Design Support, Marion Power Shovel Division of Dresser Industries, Inc., Marion, Ohio (Section 7)

Harold E. Guetzlaff, Senior Designer, John Deere Products Engineering Center, Deere and Company, Waterloo, Iowa (Section 8)

John C. McGlone, Executive Assistant to the President, REDAC Interactive Graphics, Inc., Littleton, Massachusetts (Section 6)

George A. Magnan, Editor, *Design Drafting & Reprographics, Graphics Today,* and *Today's Art,* Ventura, California (Sections 1 and 2)

Robert L. Myers, President, ØMNIMATIØN, San Pedro, California (Section 9)

Arnold C. Noble, Engineering Supervisor, Drafting and Documentation Services, Interstate Electronics Corporation, a subsidiary of Figgie International, Inc., Anaheim, California (Section 5)

George E. Rowbotham, Consultant, author, and engineering management editor of *Design Drafting & Reprographics* (Sections 3, 14, 15, and 16)

Louise G. Schatzman, Senior Systems Engineer, Fluor Engineers and Constructors, Inc., Irvine, California (Section 13)

Gilbert A. Thomas, Attorney, Thomas and Virden, A Professional Law Corporation, Tustin, California (Section 10)

Gary Whitmire, Manager of Engineering Services, Teledyne, McCormic and Selph, Hollister, California (Section 4)

Preface

Graphics is a highly effective communications vehicle. Graphics communication encompasses a diversity of disciplines, including design methodology, engineering detail and assembly drawings, technical illustrations, interactive graphics, computer-aided design and drafting, printed-wiring drawings, schematics, logic diagrams, undimensioned drawings, drafting, patent drawings, mathematical graphics, graphics presentations, and specialized drawings—coupled with supportive documentation storage and retrieval and reproduction methods.

Recent years have seen drastic changes in graphics disciplines. Keeping up with these changes is imperative for people in industry and in education. Managers, engineers, and drafters who have developed the ability to visualize geometric configurations and to "think" graphically have a decided advantage in creating effective tools for achieving their objectives. Educators are seeing how graphics can be a useful tool in their work. Students reared on television are improving their verbal skills through the use of graphics and animation.

This Handbook provides the reader with the latest information on graphics technology. It is not a book of instructions on the rudiments of drafting and illustration; such elementary detail is easily found in other texts. Rather, the state of the art is set forth by a group of experts representing the various graphics disciplines. Their contributions, and the topics covered in this Handbook, are touched on briefly in the following paragraphs.

First, two sections of the Handbook serve as a practical working guide and reference—for both managers and technicians—for preparing and handling all types of graphics employed in creating visual aids and technical illustrations used in industry. In another section, specialized drawings used in industry are covered in detail. The responsibilities of all the participants involved in creating these drawings are explained, and the various styles and types of drawings are thoroughly discussed.

The cost-conscious, and efficiency-conscious, graphics professional will find a variety of up-to-date information. For example, preparing a design/drafting manual, developing a checker's guide, and establishing document numbering systems, as well as the economics of preparing layout drawings, are all covered in the Handbook.

Proper dimensioning and tolerancing make it possible to achieve 100 percent trouble-free interchangeability when parts are designed, manufactured, inspected, and assembled at widely disparate geographic locations. Costs as well as rejection rates are reduced, and tolerance accumulations are eliminated, with close attention to dimensioning and tolerancing. Obviously, graphics professionals need to have a clear understanding of this proven technology.

Knowledge of the international (SI) metric system is imperative, for many industries are in the process of making the transition to this system. It is simply sound business to be familiar with the graphics involved in presenting patent drawings. In many areas of engineering, mathematical solutions may be possible, but they may not be readily accessible; graphics methods may be the solution.

With the maturing of electronics, close attention should be focused on electronic drafting and its highly specialized contributions to graphics. The importance of the relationship between the designer and the computer is well illustrated by the increasing complexity of printed circuit boards. The section on printed-wiring drawings presents both an overview of this technology and an example of the use of a typical computer-aided design system.

Graphics transforms the engineer's rough sketch into a polished layout and translates words into graphic details. Graphics puts life into ideas and is, indeed, the engineer's primary link between the concept and its realization. Until recently drafting has depended largely upon the manual skills of human beings; in fact, drawings have been prepared in virtually the same way for more than four thousand years. Today, interactive graphics and computer-aided design systems are revolutionizing design, drafting, and manufacturing processes. Since the 1970s these systems have gained credibility as cost-effective tools in a wide variety of applications. Inexpensive, reliable minicomputers have made it possible for the technology to be within the reach of most engineering and manufacturing organizations. Interactive graphics systems provide an accurate and accessible database that is a key factor in increasing productivity throughout the design and manufacturing cycle.

The tremendous increase in documentation—the glut of paper in industry—has been addressed by micrographics and reprographics. Substantial space savings can be achieved by the use of micrographics, as can other significant benefits such as the establishing of a disaster file, prolonged life of originals, lower mailing costs, and faster handling. Related to this advance are developments in xerography, until recently the missing link in a total micrographics system.

The paper explosion has also greatly increased the problems involved in retrieving information. The systems that have been developed combine a simple file structure, economical computer assistance, flexible applications, and updatable micrographics with sophisticated indexing and abstracting.

The preparation of this Handbook has presented many challenges. The knowledgeable people who contributed sections to it are to be commended for having created a useful and practical work for professionals in graphics and industry, a comprehensive reference volume providing quick and reliable answers on engineering and industrial graphics problems. Credit would not be complete without mentioning the publisher. McGraw-Hill Book Company provided guidance and has been instrumental in recognizing the need for a handbook of this type.

George E. Rowbotham
Santa Ana, California

Technical Illustrations

George A. Magnan

Although orthographic engineering drawings—those which depict each side of an object in two dimensions (i.e., in a plane)—remain the chief means of conveying technical information in industry today, these drawings are now being supported and in many cases replaced by what we call technical illustrations. For assembly, installation, and other types of drawings that do not require the kinds of detailed instructions needed in fabrication—for drawings which, when expressed orthographically with several cross-sections, auxiliary views, and other detailed delineations, can become very complex—three-dimensional depictions are much simpler and more easily understood.

USES IN INDUSTRY

As easily grasped visuals, technical illustrations are a vital link between design and manufacturing. They are also the mainstay of every type of in-plant manual, handbook, report, proposal, catalog, and other technical publication for repair, training, operation, and the like, as well as consumer literature which shows assembly and maintenance of a variety of products ranging from toasters to trucks. The U.S. government requires that illustrated technical literature accompany all hardware procured by military or civilian agencies, and that such illustrations conform to detailed military specifications (MIL specs). Technical illustrations are also used for other purposes, such as visual aids to help assembly-line workers, technical and instructional audiovisual presentations, and promotional literature.

TYPES OF TECHNICAL ILLUSTRATIONS

Although the term *technical illustration* is often interpreted to mean a three-dimensional drawing, the definition is a broader one. The forms and methods of technical illustration include orthographic drawings, diagrams, charts, and other graphics. These illustrations are made clear and useful by the application of such art principles as toning, shading, art design and composition, the use of colors, and a variety of art techniques.

The principal types of technical illustrations are:

Orthographic drawings   Two-dimensional engineering projections

Diagrams and charts   Schematics, block diagrams, flowcharts, system diagrams, wiring diagrams, and graphs

Axonometric projections   Parallel-plane drawings showing three dimensions

Perspectives   Three-dimensional drawings whose picture planes converge to a vanishing point

Photographs   Continuous-tone camera pictures modified by art techniques to convey technical information

Orthographic Illustrations

Very often an orthographic drawing will communicate a clearer message for a given purpose—if the drawing is modified or redrawn so as to eliminate details and highlight the concept in mind—than any perspective type of delineation. Great time savings can also be realized when such flat views are already available from engineering files in the form of diazo checkprints or the like. These prints may be adapted, reproduced, or traced for use in basic layouts that do not require the use of perspective. Examples of orthographic illustrations are shown in Fig. 1-1.

Diagrams and Schematics

Schematic diagrams, which show the functions, path of flow, and interconnections among components of a system, and which also identify terminal points, often use standard symbols to designate commonly understood units, such as in a wiring diagram. Many schematics are block diagrams that show the relationships among subsystems within a major system. Some examples of the uses for schematic diagrams are in electrical and electronic systems and wiring; industrial and architectural piping; systems for flow of air, gases, hydraulic fluid, and oil; and aircraft controls. See Figs. 1-2 through 1-5.

Most schematics are drawn as pencil tracings, although some are generated in ink layout form by plotters from computer-stored data under software programs. Incorporated in the engineering drawing system are prints made from these drawings, which are used for reference in shop assembly and installation. For publication in handbooks or manuals, however, many schematics must be retraced in ink to meet

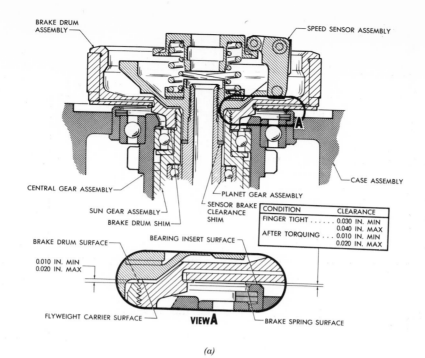

BRAKE DRUM ASSEMBLY

SPEED SENSOR ASSEMBLY

CASE ASSEMBLY

CENTRAL GEAR ASSEMBLY

PLANET GEAR ASSEMBLY

SUN GEAR ASSEMBLY

SENSOR BRAKE CLEARANCE SHIM

BRAKE DRUM SHIM

CONDITION	CLEARANCE
FINGER TIGHT	0.030 IN. MIN 0.040 IN. MAX
AFTER TORQUING . . .	0.010 IN. MIN 0.020 IN. MAX

BRAKE DRUM SURFACE

BEARING INSERT SURFACE

0.010 IN. MIN
0.020 IN. MAX

FLYWEIGHT CARRIER SURFACE

VIEW A

BRAKE SPRING SURFACE

(a)

a. LINKAGE LATCHED, HOOKS CLOSED

TIE LINK

HOOK LATCH

BUMPER

PAWL LOCK LINK

PAWL RETURN SPRING

LINK

RELEASE CABLE

HOOK SPRING

RELEASE LINK

BELLCRANK

RELEASE PLUNGER

AFT LINK

RELEASE ASSEMBLY

b. LINKAGE RELEASED, HOOKS OPEN

d. MANUAL RELEASE

c. LINKAGE LATCHED, HOOKS OPEN

WING RACK - AERO 65A-1 - INTERNAL DETAILS (SHEET 2 OF 2)

(b)

FIG. 1-1 Typical orthographic illustrations for a technical manual depicting conditions that would be more difficult, complex, and time-consuming to show in three-dimensional views. (a—Northrop Norair, artist D. H. Boyd; b—Lockheed-California, artist M. Haro.)

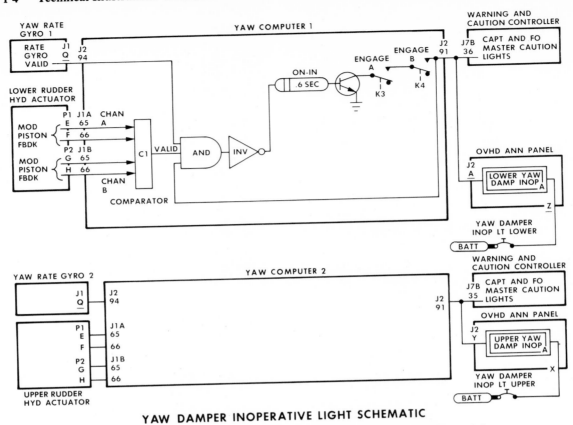

FIG. 1-2 Schematic drawing of aircraft system. (*McDonnell Douglas, Product Support Training Department.*)

publication requirements, with improved layout to simplify lines of flow, heavier line-work so the drawing will print well at a reduced scale, and usually the replacement of small hand lettering with larger, heavier mechanical lettering or typed copy. Because schematics can be very long, they are often shortened so as to avoid bulky fold-outs in the final printed book. When possible, an overlong schematic should be divided at strategic breaks, where there are the fewest connector lines, into two or more elements keyed to the overall diagram. Where many such divided sections make continuity a problem to decipher, a simplified master diagram, or "locator," of major elements can be added as an aid in visualizing the overall diagram.

Wherever possible, schematics, diagrams, and other artwork should not be laid out vertically; that is, the reader should not need to turn the book to read them.

Drawing Diagrams for Reproduction

Although some pencil-on-vellum schematics done for engineering use can be adapted and reproduced satisfactorily for manuals and handbooks, these reproductions often fail to meet contractual standards for legibility, and their general use in publications is not recommended. Good resolution and readability are extremely important, and linework can break or disappear when an image is reduced to one-fourth or one-eighth scale, as is often required in order to fit a diagram into a publication format. The best practice is to adapt diagrams for the necessary offset printing scale by rearranging elements for better layout, and then tracing this with technical pens to yield suitably heavy and uniform linework. Translucent vellum, tracing cloth, and polyester drafting film are the surfaces most often used for tracing layouts over engineering prints.

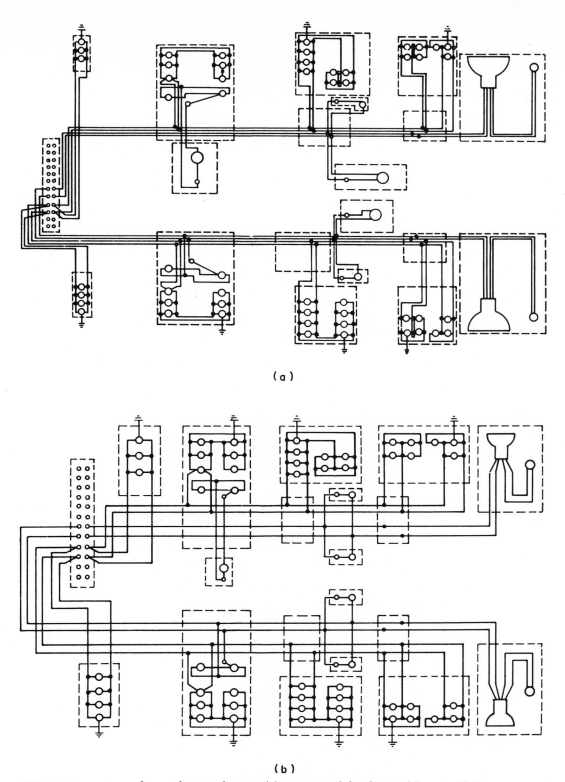

FIG. 1-3 Comparison of two schematic layouts. (*a*) A congested, hard-to-read layout with lines and terminals close together; (*b*) a correctly laid-out, easy-to-read version of the same schematic with clear separations between elements. *(MIL-M-38784.)*

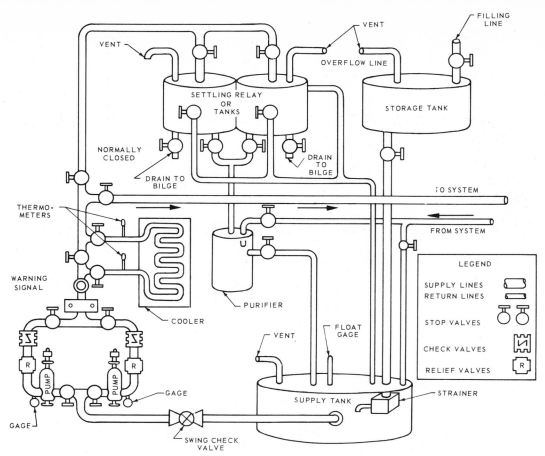

FIG. 1-4 Showing cylindrical tanks in three dimensions adds clarity to this schematic diagram of a piping system. *(NAVPERS 10470.)*

Because diagrams are often filled with repetitive symbols, several methods may be used to avoid copying each symbol by hand. The most popular method is to use symbols and repetitive nomenclature that have been preprinted on an adhesive-backed clear film. These are cut out and applied to the drawing, saving a great deal of time and effort on diagrams in which dozens or even hundreds of identical symbols may be involved. Standard symbols come in sheet form from many suppliers; special symbols, logotypes, and the like may be ordered from a manufacturer or reproduced onto adhesive-backed materials in a company's own reproduction shop. Other symbols may be readily traced using a technical pen and plastic or metal templates.

Lettering can be drawn using Leroy, Wrico, or other mechanical template devices; it can also be typed on sticky-backed matte films and placed on a drawing, or typed directly onto a tracing in an open-ended reproduction typewriter.

Diagrams and other artwork that incorporate appliqués, tapes, or other stick-ons and which are intended for offset publication reproduction should not be rolled up for storage. They should instead be taped to a stiff white artboard, covered with protective tissue and paper flap overlays, and stored flat.

Many companies today are schematizing wiring, electronic, and similar diagrams that are in high-volume use through computer graphics techniques, and some of these drawings require little modification for use in handbooks. There are so many variables in the broad variety of general diagram graphics, however, that progress in automating all these drawings is unlikely to occur.

In some large corporations much thought has gone into methods for avoiding duplication of effort in making schematics twice—once for engineering reference and then

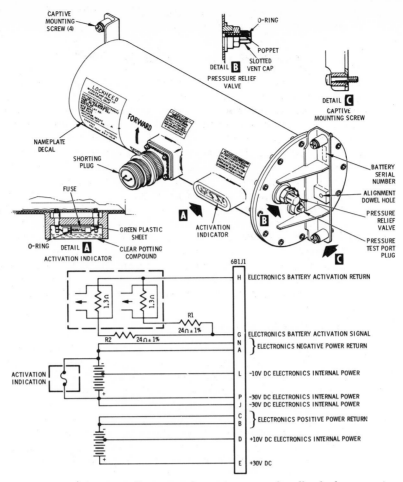

FIG. 1-5 Combination illustration for maintenance handbook shows equipment in isometric, with orthographic sections and wiring schematic. *(Lockheed Missiles and Space Company, artist R. Wagner.)*

again for handbook reproduction. Such methods include (1) upgrading graphic quality and line resolution in computer plots for optimum flexibility in all applications; (2) drafting pencil vellums that serve all purposes by better initial planning and layout, including heavier lines and larger lettering; (3) combining art paste-up and inking techniques on opaque paper, and then microfilming the diagram. The microfilm may be used to reproduce second originals on workable polyester film, wash-off film, or some other translucent base from which prints can be run off, or for the production of scaled photoprints as original copy for handbook publication.

For preparing diagrams and other large drawings to meet microfilming and publication requirements, see American National Standards Institute (ANSI) standards Y32.1-1959, Illustrations for Publication and Projection, and Y14.2-1957, Line Conventions, Sectioning and Lettering, plus "Modern Drafting Techniques for Quality Microreproduction," by Carl E. Nelson, issued by the National Micrographics Association, Suite 1101, 8728 Colesville Rd., Silver Spring, MD 20910.

Charts and Graphs

Facets of scientific, financial, and other types of data and their relationships can be quickly grasped in graphic form as bars or curves, giving immediate insight into complex data interactions. Any graphical means to accomplish this is broadly defined as a chart, whereas the types of charts generally used by engineers and scientists in plotting scientific data are considered graphs.

Technical Engineering Charts and Graphs

Technical charts and graphs are used to record the progress of laboratory experiments, to express theoretical mathematical data graphically, to analyze design alternatives, and the like. Most graphs are based on a rectangular grid system, although circular grid systems may be used for special map charting applications. All points plotted on a graph must be related to known values, which are placed in increments along two primary reference lines that are perpendicular to each other and which intersect. These are called coordinate axes, and the auxiliary reference lines or coordinates that complete the framework carry numerical values proportionate to their distance from the coordinate axes. Within this framework any given point can be located.

Although mathematical graphs are commonly plotted on grids whose coordinate axes intersect to form four parts, or quadrants, display graphs show values on one quadrant only in order to achieve a simpler, more easily understood presentation.

Two types of coordinate grids for graphs that are commonly used are the rectilinear chart and the logarithmic chart. Rectilinear charts have equal spaces between lines in arithmetic progression and may be in the form of either a square grid or a rectangular grid having closer divisions along one coordinate than the other.

Types of Rectilinear Charts

The rectilinear charts most often used are the time-series graph, the profile graph, and the multiple-curve graph. In a time series, the values of products, dollars, or some other subject are scaled along the vertical axis while the time values of days, months, years, and so forth are plotted along the horizontal axis. A profile graph is a plotted curve with areas to the base line rendered solid or shaded to represent a quantity rather than a linear trend. A multiple-curve graph employs several curves plotted within the same frame of reference in order to compare different aspects of a common situation, such as the longevity of workers in different occupations or the sales of different models of the same basic product (Fig. 1-6a and b).

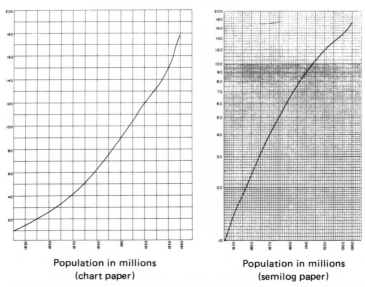

The growth of population of the United States in millions (1830-1960)

Population in millions
(chart paper)

Population in millions
(semilog paper)

FIG. 1-6 (a) Handbook illustration combines orthographic views and dimensions with isometrics. (*Rockwell International, Los Angeles Division.*)

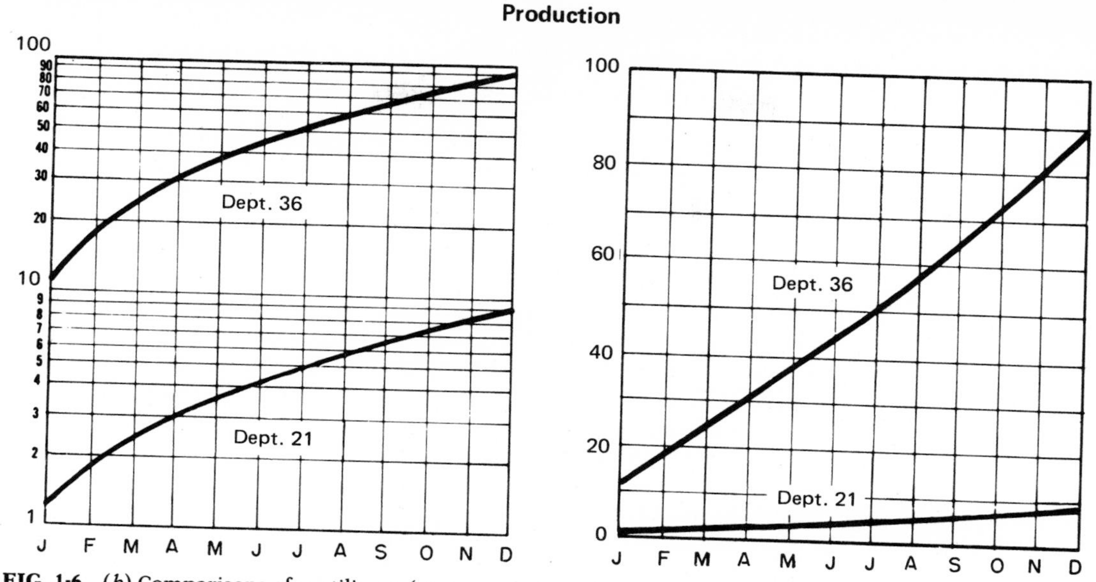

Production

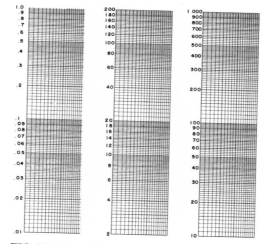

FIG. 1-6 (*b*) Comparisons of rectilinear (amount-of-change) chart with semilog (rate-of-change) chart. Note that congestion of fine lines in semilog graph paper makes unclear the printed reproduction in Fig. 1-6*a*, but the retraced and simplified version in *b* reads properly.

FIG. 1-7 Logarithmic scales.

Logarithmic Charts

Logarithmic values are based on exponential progression, of units multiplied by themselves, rather than on arithmetical progression, as is the case with rectilinear charts (Fig. 1-7). There are two basic types of logarithmic graphs. One is the semilogarithmic, known as the semilog, which has equally spaced vertical grid lines but whose horizontals are unevenly laid out in logarithmic ratios. These graphs are known as rate-of-change charts or ratio charts.

In the other type, the log-log graph, both the horizontal and vertical lines are laid out on a logarithmic scale. The log-log graph is most often used for solving mathematical problems (Fig. 1-8).

Nontechnical and Display Charts

Most charts that are not specifically intended for recording engineering or scientific data fall into the broad category of charts, which are intended to tell a story by the

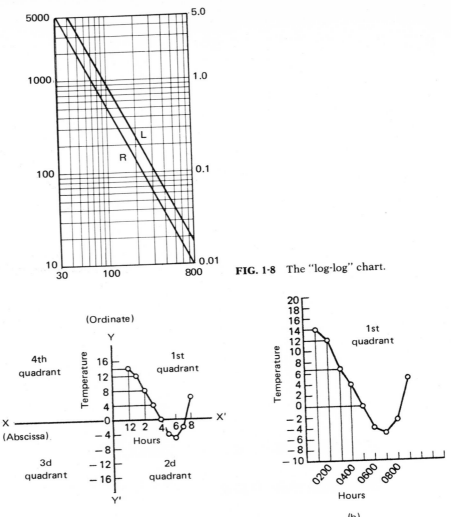

FIG. 1-8 The "log-log" chart.

FIG. 1-9 (*a*) Technical or mathematical chart versus (*b*) display chart.

most effective graphic means in a display or presentation (Fig. 1-9*a* and *b*). There are many types of charts that can be used to depict the same figures, trends, conditions, and so forth, and these types must be carefully reviewed in every situation calling for a chart in order to select the one which carries the desired message with the greatest clarity and impact (Fig. 1-10). The basic types of display charts are outlined as follows.

Bar Charts Bar charts may be drawn horizontally or vertically; the vertical type is often called a column chart. There are two major types of bar charts. The first is the 100 percent bar chart, in which a whole bar or column is divided by shading or colors into proportionate segments showing various percentages of a total. By using a series of segmented 100 percent bars, comparisons of percentages can be visually apparent (Figs. 1-11 and 1-12). The second type is the multiple-bar chart, or multiple-column chart, in which the length of a bar along a grid scale denotes its value or quantity for comparison with similar bars (Figs. 1-13 and 1-14). There are many variations of bar and column charts, such as connected-column, floating-column, deviation-column, sub-divided-bar, deviation-bar (Fig. 1-15), and sliding-bar charts.

Curve, or Line, Charts The curve chart, sometimes known as a line chart, is best suited to showing variables along a time span. Along with the bar chart, it is most pop-

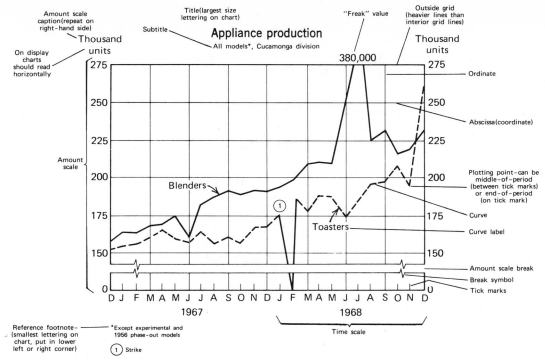

FIG. 1-10 Display chart—structure and nomenclature.

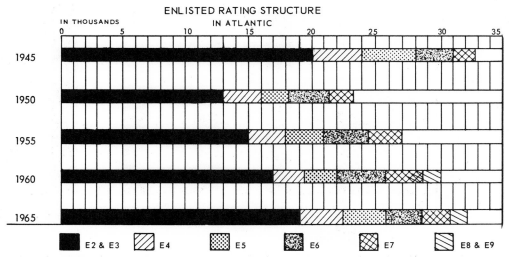

FIG. 1-11 Multiple 100 percent bar chart.

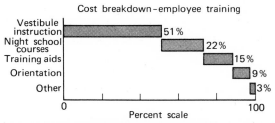

FIG. 1-12 Segmented 100 percent bar chart.

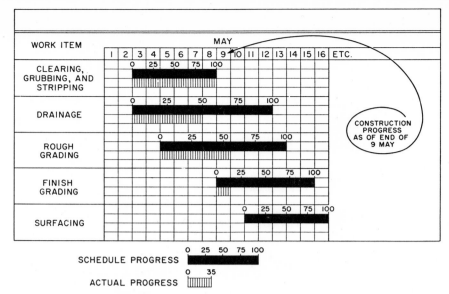

SCHEDULE PROGRESS

ACTUAL PROGRESS

FIG. 1-13 Multiple bar chart, as used for scheduling a construction project.

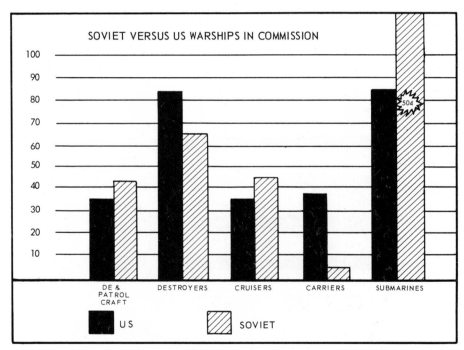

FIG. 1-14 Multiple column chart. *(NAVPERS 10470.)*

ularly used to express statistical situations. The vertical ruling represents in increments quantitative numbers or percentages, while the horizontal scale refers to time. The data is plotted as points along the coordinate lines, which are then connected as straight lines or interpreted as smooth curves. The types of line or curve charts in greatest use are:

- **Cumulative line chart** Each plotted point represents an addition to total previous quantities (Fig. 1-16).

- **Step, or staircase, chart** Used instead of a line chart when quantities fluctuate so greatly that a line progression would be incoherent (Fig. 1-17).

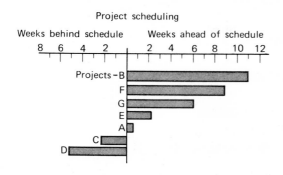

Project scheduling

Weeks behind schedule | Weeks ahead of schedule

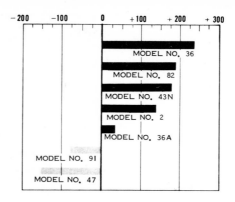

FIG. 1-15 Typical deviation bar charts.

DEFECTIVE UNITS–CUMULATIVE REJECTS

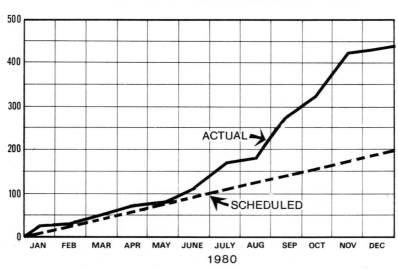

1980

FIG. 1-16 Cumulative line chart.

Air Express shipments

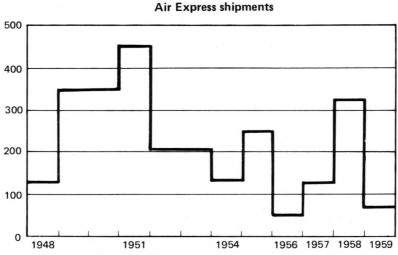

FIG. 1-17 Step or staircase chart.

1-13

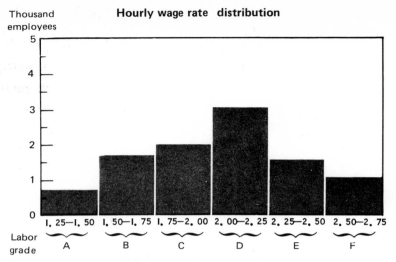

FIG. 1-18   Frequency chart, or histogram.

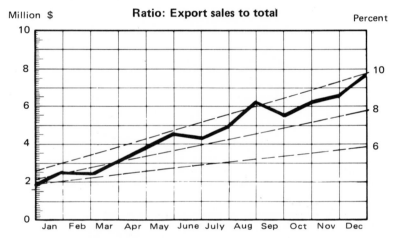

FIG. 1-19   Supplemental scale chart.

- **Frequency chart (histogram)**   Often use to express grouped quantities called *class intervals*, rather than a continuous time span (Fig. 1-18).

- **Supplemental scale chart**   Shows an actual situation against comparative ratios (Fig. 1-19).

- **Layer curve, or surface, chart**   Shown with curve areas and strata filled in with different colors or shadings, often as a means of graphically dramatizing trends, results, and so forth, over a time period as shown in a curve chart (Figs. 1-20 and 1-21).

- **Combined bar-and-curve chart**   Combines the quantities shown on a bar or column chart with a line or curve so as to present an effective graphical comparison (Fig. 1-22).

- **Index chart**   A useful device for comparing trends among items that may seem to have little in common, related to a base year. An example is the Cost-of-Living Index, which converts to a common denominator the prices of such dissimilar items as rent, food, and clothing. Given the amounts, they are converted to an index chart by selecting a base year or period; the given amount is divided by the value of the base year and the answer multiplied by 100 (Fig. 1-23*a* and *b*).

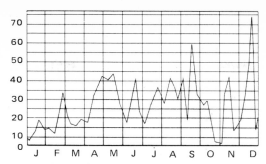

FIG. 1-20 Surface curve chart.

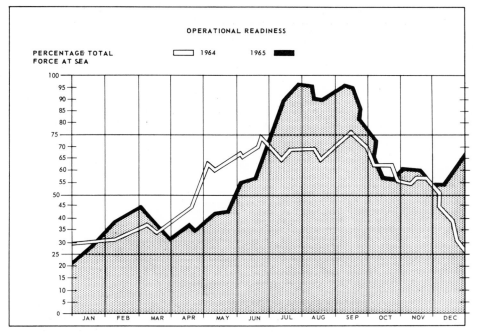

FIG. 1-21 Multiple curve chart. *(NAVPERS 10470.)*

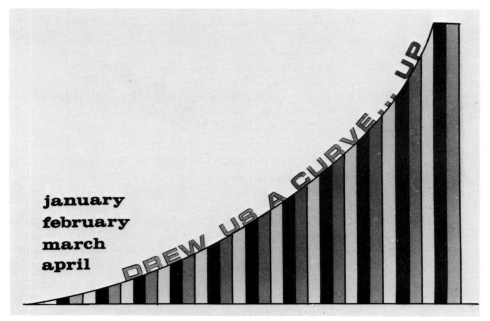

FIG. 1-22 Combined bar-and-curve chart. *(IAM magazine.)*

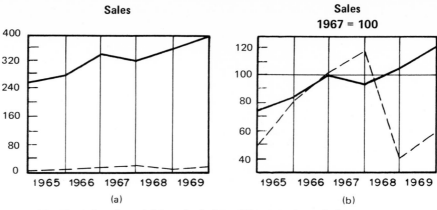

FIG. 1-23   The index chart. (*a*) Standard chart; (*b*) conversion to index.

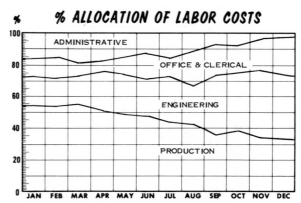

FIG. 1-24   The 100 percent surface chart.

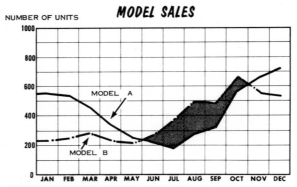

FIG. 1-25   Shaded zone chart.

Miscellaneous Charts   Many other types of charts are used to dramatize and accent trends for statistical display purposes, often combining pictures and symbols with grids and other charting forms for greater effect. The most important and frequently used charts in this category are:

- **Surface chart**   Expresses data by whole areas filled in with shading or color to become solids that indicate mass or volume. The 100 percent surface chart (Fig. 1-24) shows several curve trends within a total 100 percent area (the curve chart counterpart of the 100 percent bar chart). The shaded zone chart (Fig. 1-25) employs shading within an area between the points at which trend lines cross each other, to emphasize the reversal of trends and the duration of changes. A pictorial surface chart combines a solid-area curve with illustrations or photographs as an effective means of telling statistical a story.

- **Pie chart** Consists of a circle segmented around the center like the slices of a pie. A significant segment may be stressed by coloring it or by pulling it out a little from the rest of the pie. This compelling device to show proportions at a glance may also be combined with bars or columns (Figs. 1-26, 1-27, and 1-28).

- **Pictograph** An abbreviation of "pictorial graph," a graph in which products or other items trended in bar or curve form are identified by pictures or graphic symbols for greater attention value (Fig. 1-29).

Distribution of the "O" and "M" dollar

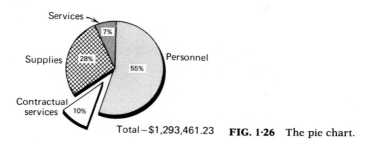

Total—$1,293,461.23   **FIG. 1-26**   The pie chart.

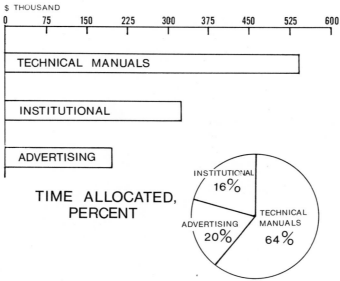

FIG. 1-27 Pie chart combined with bars.

Maintenance activity

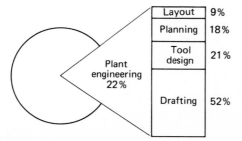

FIG. 1-28 Combination column and pie chart.

• **Pictogram** Besides illustrating the subject of the graph, a pictorial symbol is used repetitively or in multiple groups to express quantities (Figs. 1-30 and 1-31).

Nonstatistical Charts Nonstatistical charts comprise the many types of charts that are used neither for scientific purposes nor for the display or promotion of statis-

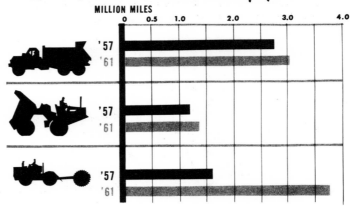

FIG. 1-29 Pictograph.

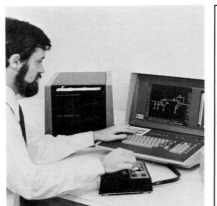
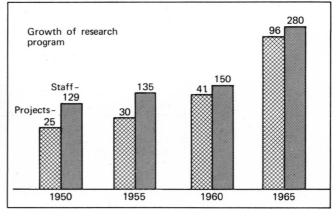

FIG. 1-30 The use of illustrations or photographs with charts adds interest and impact to an otherwise dry presentation of charting statistics.

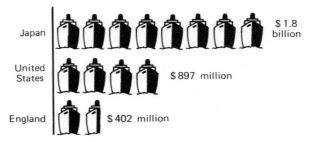

FIG. 1-31 Pictogram.

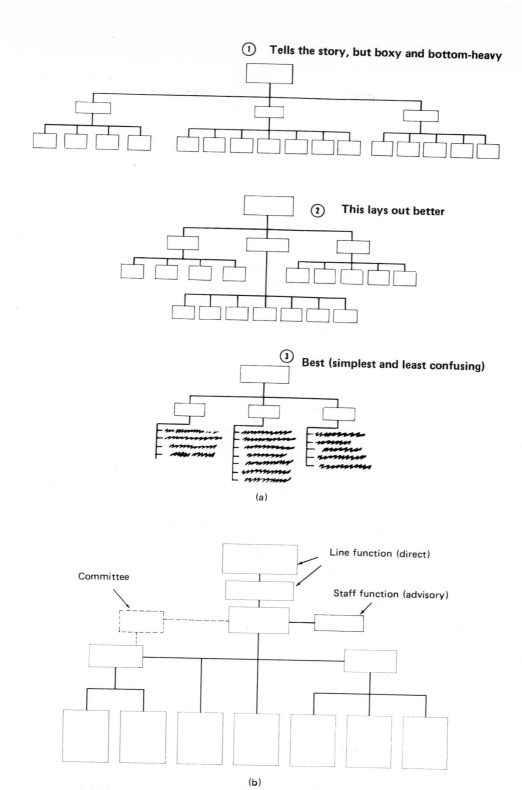

FIG. 1-32 (*a*) Organization chart layout. (*b*) Organization chart.

tical concepts, but which function instead as graphic aids in day-to-day work situations. Some of these charts are:

- **Organization chart**   Depicts the distribution of functions, personnel, and executive authority within an organization. Organization charts define the two main types of organizations: (1) line-type, as exemplified by the military services, in which authority is structured in a pyramid from bottom to top, and (2) staff-type, as in many research institutions in which work is cooperative and project-oriented rather than being centered on levels of management. When names and functions are assigned to connected boxes, the relationships of each element within an organizational structure can be seen at a glance. Direct connections are shown with solid lines, and staff and advisory functions are indicated with dotted boxes and lines (Fig. 1-32*a* and *b* and Fig. 1-33).

- **Flowchart**   A flowchart, which shows the consecutive manufacturing steps and work flow involved in producing a commodity, can graphically illustrate the production process to workers and planners. Assembly and other processes are also often shown in flowchart form, as visuals that can assist planners in simplifying and improving production (Fig. 1-34).

- **Scheduling chart**   Sometimes called Gantt charts or time-phased schedules, scheduling charts are usually large boards using colored pins, tapes, pegs, and so forth, to track planned versus actual progress in meeting project work schedules, often on a day-to-day basis.

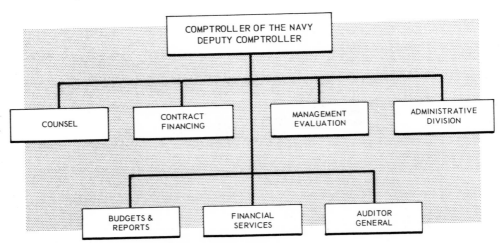

FIG. 1-33   Typical organization chart, with shading.

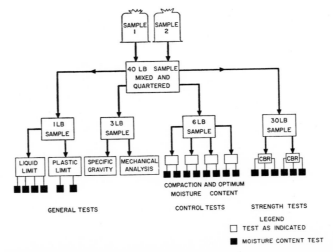

FIG. 1-34   Process flowchart.

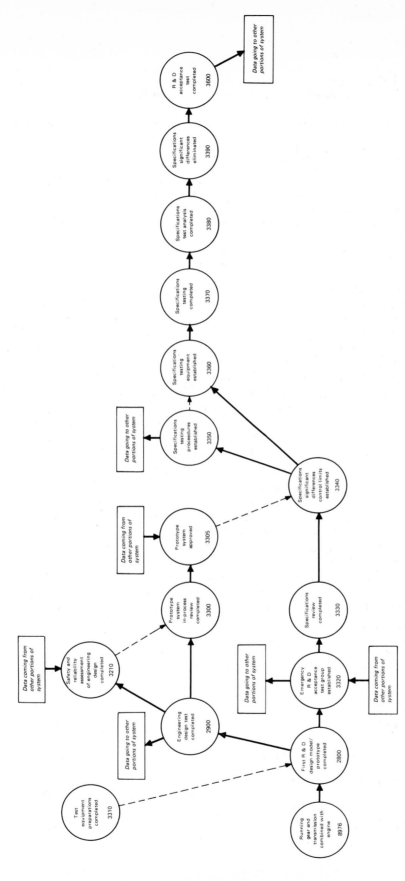

FIG. 1-35 Portion of a typical PERT network chart.

- **PERT chart**   An acronym for Program Evaluation Review Technique, PERT represents a sophisticated flowchart plan that shows the sequence, interdependencies, interrelationships, and estimated times of planned project activities and events, with a critical path for accomplishing program objectives. PERTs are a way to present graphically all factors of a project encompassing the variables of time, resources, and performance, linking all activities within the format of the whole, in a network that may be used as a valuable analysis, planning, and scheduling tool (Fig. 1-35).

Guidelines for Charting

In adapting or developing charts and graphs for presentation, display, or publication, the available text matter and other copy should be studied with care to provide answers to these key questions: What are the important trends, concepts, or facts to be brought out in the chart? What is the simplest, clearest type of chart that can accomplish this with the strongest visual effect? What art methods can do the most successful job within the budget, allowing for deadlines and other constraints?

Suggestions to keep in mind when planning charts are:

- In making a speech or presentation, one must present one idea or situation at a time; the same is true of display charts. Analyze the data by breaking them down into separate elements; if this reveals several sets of statistical facts, present these facts in several charts.
- Do not present special technical charts such as logarithmic or triangular in publications or presentations whose audience will be nontechnical people. Translate chart forms requiring special knowledge into display charts that anyone can understand.
- Simplify multiple-curve comparisons to make differences highly visible by magnifying grid scales. If too many curves appear for clear viewing, consider dividing them into two charts, perhaps keyed to an average or median curve to establish a frame of reference.
- Simply chart text and exclude lengthy notes. Wherever possible, use round numbers.
- Use large, uniform, easy-to-read gothic (sans serif) lettering.
- Do not make charts that cannot be reproduced. Although the use of color is preferred, if the chart is to be reproduced in black and white, then the different colors of curves or bars should be applied in patterns (cross-hatching, dots, dashes, etc.) rather than solid colors so that the different areas can be identified in both the display and the black-and-white printed versions. *Note:* Avoid light blue as a chart color; it will not reproduce.
- If a chart is to reflect new data additions or changes, make it as a base map on artboard and apply an overlay of clear acetate on which colored tapes, adhesive color sheets, etc., may be added or removed without damaging the base.
- Graphs drawn on standard orange or green graph paper with a finely graduated grid pattern should never be used for reproduction: the colored lines will print black with blotchy, unreadable results. Always trace such charts in ink, tracing only the major grid coordinates, for use in printed literature.

PICTORIAL DRAWING[1]

Pictorial Terminology

The following definitions of terms are commonly accepted for pictorial drawings.

axonometric projection   One of several forms of one-plane projection giving the pictorial effect of perspective with the possibility of measuring the principal lines directly.

bird's-eye view   A view of an object as seen from above.

[1]Section on "Pictorial Drawing," from "Pictorial Terminology" through "Line Conventions," including illustrations, excerpted with permission from *General Motors Drafting Standards Manual*, General Motors Corp., Detroit, Mich.

center of vision (CV)   The point at which each imaginary line of sight to any point of an object intersects the plane of projection.

cutaway   The visual removal of part of an outer surface of an object in order to show underlying mechanisms or parts more clearly.

dimetric   An axonometric projection in which two sides and two axes make equal angles with the plane of projection. The third face and axis make a different angle.

dummy   A rough outline used to plan the arrangement of a drawing, indicating the desired positions or locations of items which will later be added or drawn in further detail.

explode   To pictorially displace parts of an assembly for improved viewing clarity and understanding.

eyeball   To draw an object by the use of judgment, or by eye, rather than by using a grid or similar aid.

grease-rack view   A view of an object as seen from below.

grid   A plotted or predetermined perspective-lined graph, having an established structure of measuring increments, that is used to create a three-dimensional drawing.

horizon line (HL), or horizon   An imaginary line along which horizontal vanishing points are located.

isometric   An axonometric projection in which the three principal faces or axes of an object are equally inclined to the plane of projection and make equal angles of 120° with each other.

keyline   The process of combining and arranging views, notes, and other materials so as to produce a finished drawing.

leader   A line used to connect an object to a note, symbol, number, or caption.

oblique drawing   A method of drawing in which one side of an object is parallel to the plane of projection and the lines of projection are not at right angles to the plane of projection.

overlay   A transparent sheet that is placed over a drawing for protection. An overlay may also be used for adding notes or other information; it may be a temporary addition or a permanent part of the drawing.

perspective   A method of drawing that depicts a three dimensional object as it would appear to the eye.

phantom view   A method of drawing that uses a transparent or see-through effect to expose inner structure or components.

picture plane (PP)   The plane of projection or drawing surface on which an object is drawn or viewed.

plotting   The process of locating points on the planes of a grid.

register marks   Predetermined points or definite locations on a drawing that are used for the purpose of marking the exact location and position of overlay material.

rendering   The art of shading a line drawing so as to add depth, shape, and other definition.

scissors drawing   A method of creating a new drawing by cutting and rearranging views.

sighting point (SP)   The position of the observer's eye relative to the perspective drawing.

station point   Same as *sighting point*.

symbols   Symbols are used on drawings to express commonly used words, hardware, processes, and procedures. Symbols should not be used on drawings that will be used by persons unfamiliar with symbol terminology. Examples are shown in Figs. 1-36 and 1-37.

SYMBOLS			
SYMBOL	REPRESENTS	SYMBOL	REPRESENTS
△2	Torque	◼	Existing part or part of an assembly
1 B	Reference designation	◇C ◇	Process material

FIG. 1-36   Examples of symbols.

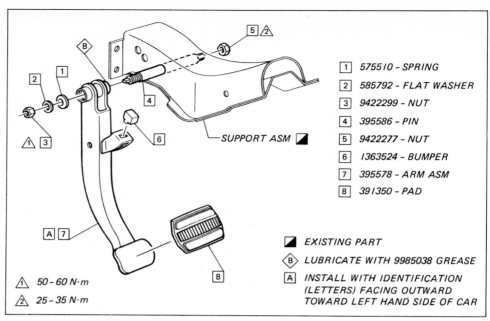

1	575510 – SPRING
2	585792 – FLAT WASHER
3	9422299 – NUT
4	395586 – PIN
5	9422277 – NUT
6	1363524 – BUMPER
7	395578 – ARM ASM
8	391350 – PAD

◢ EXISTING PART

⟨B⟩ LUBRICATE WITH 9985038 GREASE

[A] INSTALL WITH IDENTIFICATION (LETTERS) FACING OUTWARD TOWARD LEFT HAND SIDE OF CAR

⚠1 50 – 60 N·m

⚠2 25 – 35 N·m

FIG. 1-37   Use of symbols on pictorial drawing.

trimetric   An axonometric projection in which all three faces and axes make different angles with the plane of projection.

vanishing point (VP)   A point on the horizon at which a group of receding parallel lines meet when represented in linear perspective.

worm's-eye view   A view of an object as seen from below.

Abbreviations

Abbreviations that are unique to pictorial drawings are listed below; otherwise the rules for use of abbreviations in pictorial drawings are the same as those specified for orthographic drawings (Fig. 1-38).

Center of vision	CV	Picture plane	PP
Horizon line	HL	Right vanishing point	RVP
Left vanishing point	LVP	Sighting point	SP
Measuring point	MP	Vanishing point	VP

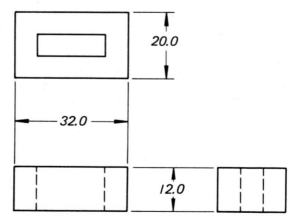

FIG. 1-38   Orthographic drawing.

Axonometric Projections

Several forms of one-plane projection have been devised for drawing an object in such a way that three of its faces are visible. Axonometric drawings combine the pictorial effect of perspective drawing with the advantage of measuring an object's lines directly.

There are three types of axonometric drawings. They are known as:

Isometric Three axes making equal angles with a plane (Figs. 1-39 and 1-40).

Dimetric Two of the three axes making equal angles with a plane (Fig. 1-41).

Trimetric Three axes making unequal angles with a plane (Fig. 1-42).

Isometric In the isometric drawing method, the three principal faces or axes of an object are equally inclined to the plane of projection and also make equal angles with each other. Lines parallel to these axes appear in their true lengths. Isometric is not a true pictorial representation of an object; it is, however, a convenient method of making a pictorial drawing to scale and is the most extensively used of the axonometric

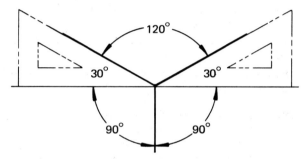

FIG. 1-39 Isometric axes.

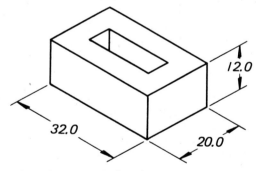

FIG. 1-40 Isometric drawing.

Same scale on all axes

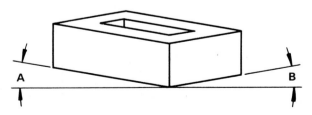

Angles **A** and **B** equal

FIG. 1-41 Dimetric drawing.

Different scale on each axis

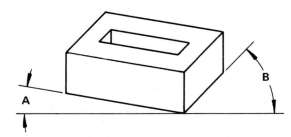

Angles **A** and **B** not equal

FIG. 1-42   Trimetric drawing.

drawing methods. To make an isometric drawing of the block shown in Fig. 1-38, use the following procedure:

1. Start with a point representing the front corner of the block and from that point draw three isometric axes 120° apart, drawing one axis vertically and the other with a 30° triangle as shown in Fig. 1-39.

2. Point off the length, width, and thickness of the block along these three lines as indicated in Fig. 1-40.

3. Draw lines to these points parallel to the axes, thereby completing a figure in which all dimensions are to scale along their respective planes.

Dimetric   In the dimetric drawing method, two sides and two axes make equal angles with the plane of projection. The third face and axis make a different angle. The two angles that are equal may vary from 0° to 45°, with the exception of 30°. The same scale is used along the two equal axes (Fig. 1-41).

Trimetric   Drawings in which all three faces and axes make different angles with the plane of projection are known as trimetric. The angles are variable but not equal, and none is 0°, and the sum of the two angles is less than 90°. Trimetric, although appearing more like perspective, necessitates the use of three different measuring guides or scales. This requirement makes trimetric more time-consuming than other axonometric methods and therefore the least practical for extensive use (Fig. 1-42).

Oblique Projection

A projected view in which the lines of sight are parallel to each other but inclined to the plane of projection is known as oblique. The principal face or long edge of an irregular contour is parallel to the plane of projection, thus making the principal face and faces parallel to it appear in true shape. Two of the axes are always at right angles to each other, while the third axis may be at any angle to the horizontal. To make an oblique drawing of the rectangular object shown in Fig. 1-38, start with a point representing a front corner and draw three axes: one vertical, one horizontal, and one at an angle. Height, width, and depth on these axes are measured in full scale (Fig. 1-43).

Full scale on all three axes

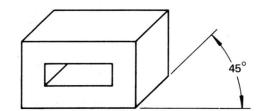

FIG. 1-43   Oblique drawing—cavalier.

Half scale on one axis

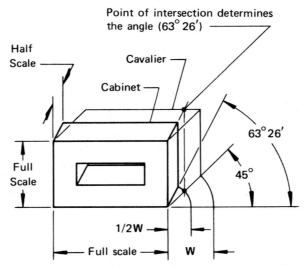

FIG. 1-44 Oblique drawing—cabinet.

There are two types of oblique projection. They are known as:

Cavalier When the projected lines make an angle of 45° (Fig. 1-43).

Cabinet To overcome the appearance of excessive thickness in cavalier projection, cabinet projection may be used. Full scale is used on the two frontal axes. One-half scale is used on the third axis. Measure and develop this projection as demonstrated in Fig. 1-44.

Perspective Drawing

Perspective is a method of drawing that depicts a three-dimensional object on a flat plane in a manner as it appears to the eye. The following is intended to present the basic principles of perspective drawing as used in engineering and design drafting.

Basic Elements

The basic elements of perspective drawing as shown in Fig. 1-45 are:

- **The object** That which is being viewed.
- **The observer's eye** The position of the observer's eye is called the *sighting point* (SP).
- **Plane of projection** A drawing surface or a plane on which a likeness of the object being viewed is produced. In perspective drawing this is known as the *picture plane* (PP). Because the drawing on the picture plane is both the purpose and the end result of perspective construction, the location of the picture plane plays a very important part in perspective drawing. The entire drawing should be planned around it.
- **Imaginary lines of sight** As the imaginary lines of sight (to all points on an object) pierce the plane of projection, they produce imaginary intersection points which when connected together create a perspective drawing.
- **Center of vision (CV)** A point on the horizon line opposite the observer's eye, or sometimes, the point of sight or center of the picture.
- **Vanishing point (VP)** The point at which a group of receding parallel lines meet when represented in linear perspective.
- **Horizon line (HL)** An imaginary line representing the eye level of the observer. The center of vision and the right and left vanishing point fall on this line.

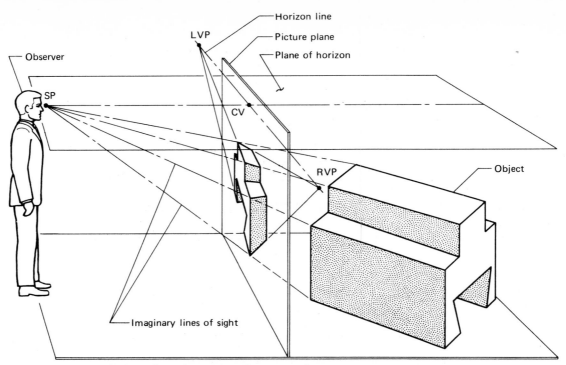

FIG. 1-45 Observer looking through a picture plane at an object.

Types of Perspective Drawings

There are three types of perspective drawings. They are known as:

Parallel One vanishing point (Fig. 1-46).

Angular Two vanishing points (Fig. 1-47).

Oblique Three vanishing points (Fig. 1-48).

In industry these are normally referred to as one-point, two-point, and three-point perspective.

Parallel (One-Point) Perspective Parallel perspective is a method in which the vertical and horizontal axes of the object being viewed are parallel to the picture plane and the third axis appears at a right angle to the picture plane (Fig. 1-49).

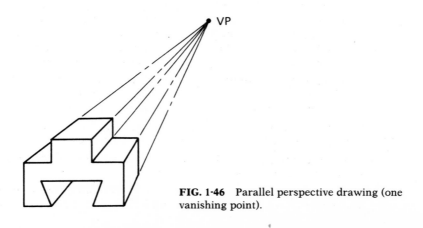

FIG. 1-46 Parallel perspective drawing (one vanishing point).

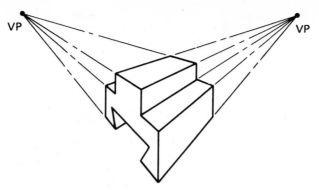

FIG. 1-47 Angular perspective drawing—two vanishing points.

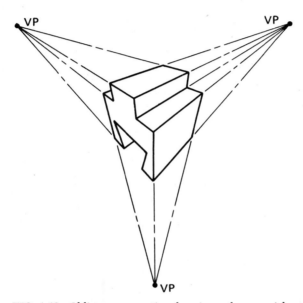

FIG. 1-48 Oblique perspective drawing—three vanishing points.

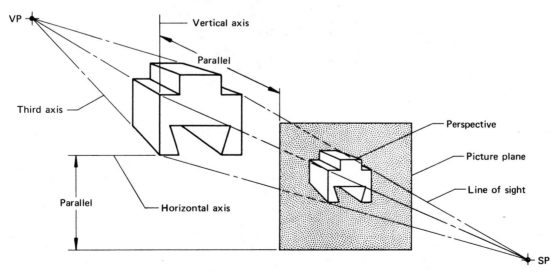

FIG. 1-49 Parallel (one-point) perspective.

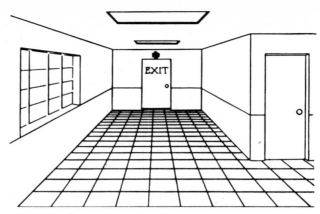

FIG. 1-50 One point perspective.

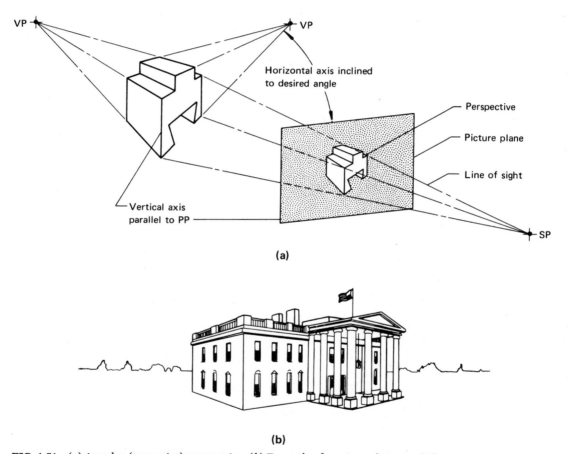

(a)

(b)

FIG. 1-51 (*a*) Angular (two-point) perspective. (*b*) Example of use in architectural illustration.

One-point perspective is not extensively used. It has particular value for illustrating the interiors of architectural structures or automotive interiors. In Fig. 1-50 a room interior is illustrated in this manner.

Angular (Two-Point) Perspective Two-point perspective is a method in which one axis of an object, usually the vertical axis, is parallel to the picture plane and the other two axes are inclined to the picture plane (Fig. 1-51*a*).

Two-point perspective is used quite extensively for architectural and product illustration (Fig. 1-51*b*).

Oblique (Three-Point) Perspective Three-point perspective is a method in which all three principal axes of an object are oblique to the picture plane (Fig. 1-52a).

Three-point perspective introduces a third point above or below the horizon line. Because of the lengthy process involved in projecting, three-point perspective is not used extensively for technical illustration. An example of oblique perspective is illustrated in Fig. 1-52b.

Sighting Point Location

In order to avoid undue distortion in a perspective drawing, the sighting point should be so located that the lines of sight encompassing the entire object form a cone having an apex angle of 30° or less (Fig. 1-53). When the object is close to the horizon or at eye level, as in architectural work, a greater angle may occasionally be used. Pleasing results may also be obtained if the sighting point is located centrally in front of the

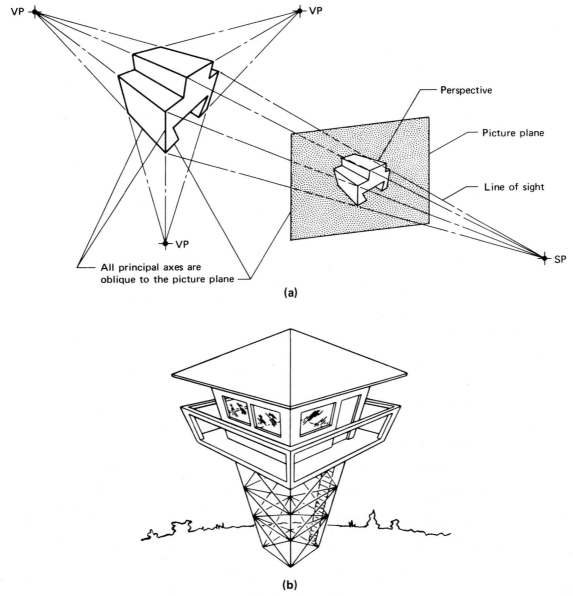

FIG. 1-52 (a) Oblique (three-point) perspective. (b) Example of use in illustration.

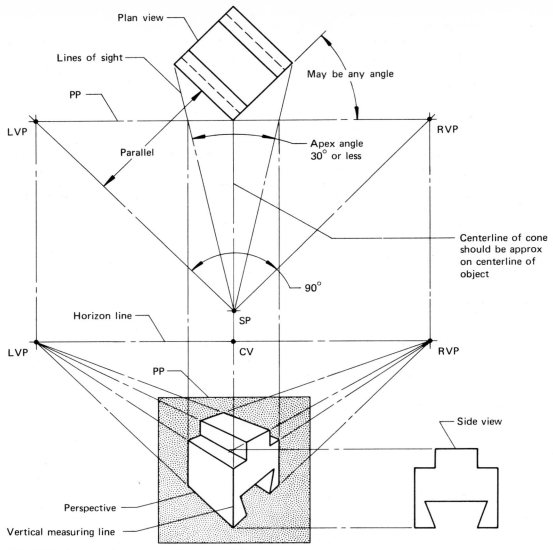

FIG. 1-53   Constructing a two-point perspective.

object and at a height that will show the desired amount of horizontal surface (also shown in Fig. 1-53).

Horizon Lines

Figure 1-54 illustrates different effects produced by repositioning an object with respect to the horizon.

Picture Plane Location

Normally the picture plane is located between the object and the observer (Fig. 1-55). The location of the picture plane in relation to an object determines the apparent size of the object. The closer the picture plane is located to the object, the larger the object will appear. The farther away the picture plane is located from the object, the smaller the object will appear. The size will change, but proportions remain the same.

Construction of Circles and Curves in Perspective

Circles and curves may be constructed in perspective as illustrated in Fig. 1-56. From orthographic projections of the subject in plan or side views, plot and label the desired points to establish the curved line or shape. Project these points to the picture plane and connect them with a French curve or an ellipse template. The more reference points used, the more accurate and smooth the curve will be.

Multiview Method in Perspective

The multiview method is another technique that is useful in perspective drawing. By using the ordinary method of multiview projection, as in orthographic drawing, a perspective of any object can be constructed (Fig. 1-57).

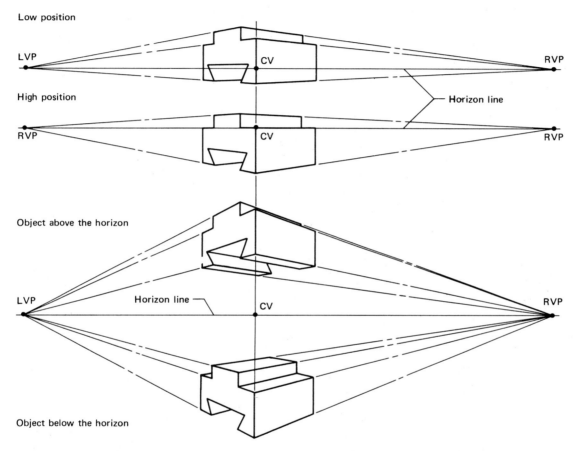

FIG. 1-54   Horizon lines.

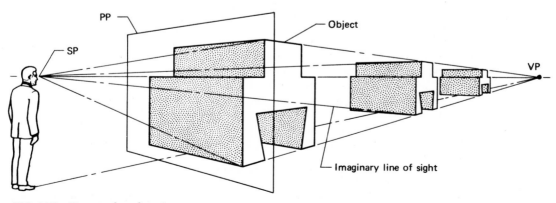

FIG. 1-55   Picture plane location.

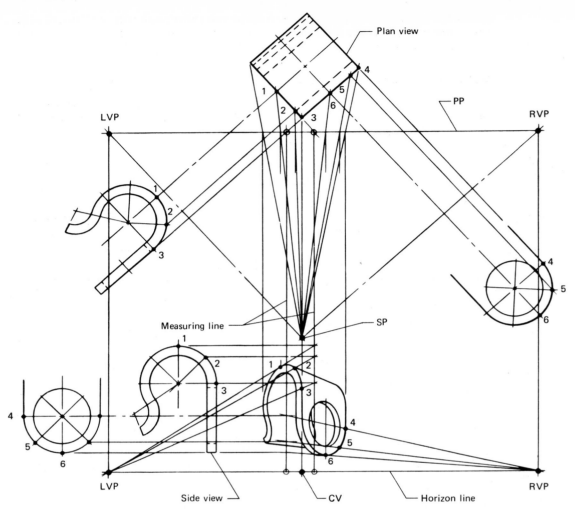

FIG. 1-56   Construction of curved lines.

The multiview method is achieved in the following manner. Position the plan view, side view, picture plane, and station point on the drawing surface, and then label all the points that are to be projected. With the front-view location of the object and the drawing surface used as the picture plane, the perspective can then be constructed. Project the corresponding intersection points from the plan-view and side-view picture planes into the front-view plane. Connect the points, thus forming the perspective. The use of vanishing points is not needed with this method. However, they may be established by extending the converging lines of the object until they meet.

Grid Projection

Grid projection is a graph-oriented method of making accurate perspective drawings of complex mechanisms or installations. It is normally used to develop master perspective layouts from which numerous coordinated drawings can then be generated. These coordinated drawings usually feature related components and may require simultaneous drawing by several illustrators. To achieve this flexibility, and at the same time maintain accuracy and true relationships among components, some type of common drawing standard is required. One standard device frequently used is called a perspective grid (Fig. 1-58).

The perspective grid resembles a cube shape drawn in perspective with its three visible planes subdivided vertically and horizontally by lines to create multiple measuring increments. These lines have been located by projection utilizing vanishing points. The lines will, if extended, lead to the vanishing point, although the vanishing point itself is not usually shown on the grid (Fig. 1-58).

Grid Advantages   Some of the advantages provided by the grid are:

1. Several illustrators can draw related components independent of one another and be assured that the drawings will match with respect to size, angle, and perspective.

2. Drawing detailed parts of a component can be accomplished even though the total design has not been resolved.

3. The grid eliminates the tedious effort of establishing and projecting from the vanishing, measuring, and sighting points for each individual part.

4. A grid makes it unnecessary to work with points beyond the drawing area, as may often occur with vanishing points.

5. Using a grid makes it possible to rapidly locate the relative positions of parts at some future date for additions or revisions.

All these factors are predetermined and incorporated in the preparation of the grid.

Grid Types and Variations   There are many varied applications of grids. Aircraft, architecture, and automotive illustration are but a few of these uses. There are also several types of grids to suit specific needs. The cylindrical grid, used for drawing an aircraft fuselage, and the centerplane grid, which is used for perspective centerline

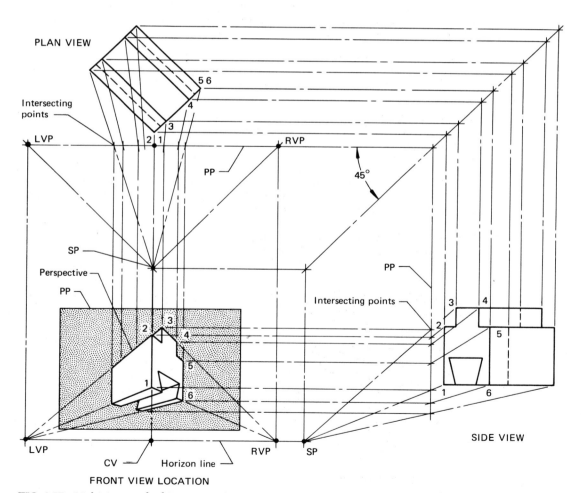

FIG. 1-57   Multiview method in perspective.

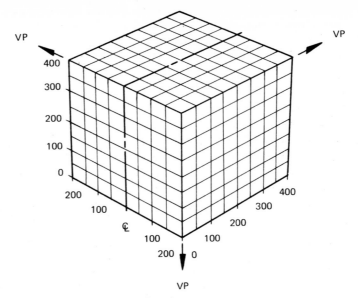

FIG. 1-58 Perspective grid.

sections, are prime examples of special applications. The cube grid, however, is the most widely used for general-purpose illustration. It is produced in one of two basic variations: as an exterior grid and as an interior grid.

- **Exterior grid** When the three adjacent exterior planes of the cube are developed, the resultant image is referred to as an exterior grid. In this variation, points are projected from the top plane downward and from the picture planes away from the observer (Fig. 1-59).

- **Interior grid** When the three adjacent interior planes of the cube are exposed and developed, the resultant image is referred to as an interior grid. In this variation, points are projected from the base plane upward and from the picture planes toward the observer (Fig. 1-59).

The choice and use of either the exterior grid or the interior grid is a matter of individual preference. Each will produce the same results, and the ability to match drawings from either or both grids is still retained.

Exterior and Interior Grid Variations Two further variations of both the exterior and interior grids are known as the bird's-eye and worm's-eye grids. These effects are

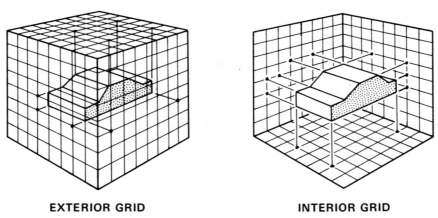

EXTERIOR GRID   INTERIOR GRID

FIG. 1-59 Cube grid variations.

achieved by rotating the vertical plane of the grid about the horizon line during the development stage.

Objects drawn in the bird's-eye grid appear as if they were being viewed from above the horizon line, as seen in Fig. 1-60. Objects drawn in the worm's-eye grid appear as if they were being viewed from below the horizon line, as seen in Fig. 1-60 (a drawing projected in this grid may also be referred to as a *grease-rack view*).

Grid Development

A grid is developed on a stable, preferably transparent material such as drafting film. The size of the grid used depends on the size and scale of the components to be drawn. A grid may be drawn in either pencil or ink using the same methods and principles defined for perspective drawings. The object in this application, however, is to draw the three adjacent planes of the grid, rather than a specific part.

Grid Increments   The three surfaces or planes of the grid are divided into multiple, vertical, and horizontal increments to aid in plotting. These increments usually represent 100-mm lines in each plane and serve the same locating and measuring functions as do station lines in orthographic drawing (Fig. 1-61). Each increment is proportionately foreshortened as it recedes from the picture plane, thus creating the perspective illusion. A grid can also be subdivided into smaller increments when necessary.

Horizontal increments are used to determine fore-and-aft measurements in the plotting process and are evolved in the following three basic steps:

1. Subdivide line AB in the plan view of the grid to the desired number of equal increments as shown in Fig. 1-62.

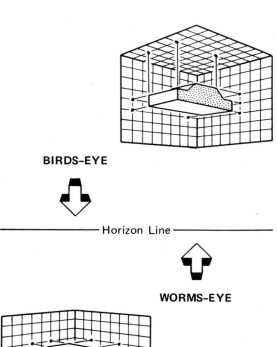

BIRDS-EYE

Horizon Line

WORMS-EYE

FIG. 1-60   Grid variations.

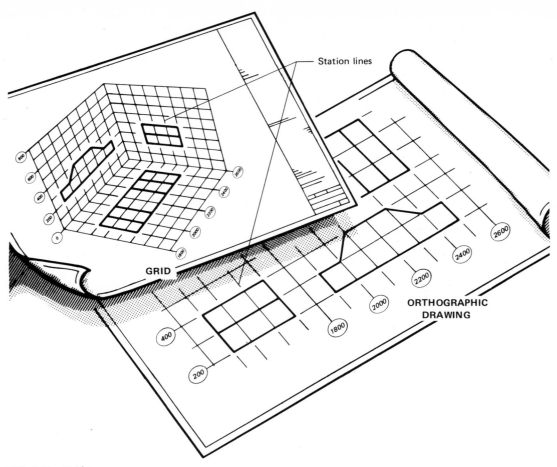

FIG. 1-61 Grid increments.

2. Project the equal increments from line AB in the plan view to the picture plane using sighting lines from the sighting point. This function foreshortens each increment of line AB as it recedes from the picture plane.

3. Subdivide the side plane of the grid by constructing perpendicular lines from the foreshortened increments located on the picture plane.

These three steps are for two-point grids. When three-point grids are used, steps (1) and (2) remain the same, but step (3) must be modified and step (4) added as follows:

3a. Subdivide line A′B′ of the side plane of the grid by constructing perpendiculars from the foreshortened increments on the picture plane.

4. Subdivide the side plane of the grid by projecting lines from the lower vanishing point through the newly established points on line A′B′ of the grid.

Repeat these same steps to subdivide the rear plane of the grid using line BC of the plan view. These steps complete the horizontal measuring increments of the side plane and the lateral measuring increments of the rear plane.

Vertical increments are used to measure vertical lengths in the plotting and projection process. These increments can be evolved by numerous methods. Many are too complex and time-consuming to be practical for common use. One simple method considered both practical and effective involves the following steps:

1. Measure the foreshortened increment marked X immediately adjacent to the center of vision on the picture plane at point D as shown in Fig. 1-63.

2. Lay off increment X on line B′F′ of the grid.

3. Subdivide the side plane of the grid vertically by projecting lines from the right vanishing point (RVP) through the newly established points on line B'F'.

4. Repeat step (3) using the left vanishing point (LVP) and complete the vertical increments of the rear plane.

Base-plane increments are used to determine fore-and-aft measurements, as well as lateral measurements, from the centerlines of the grid. These increments are generated by the following steps:

1. Project lines from the LVP through the previously established points on line A'B' of the base plane. This step establishes the fore-and-aft increments of the base plane (Fig. 1-63).

2. Project lines from the RVP through the previously established points on line B'C' of the base plane. This step establishes the lateral increments of the base plane.

Expanding the Grid   Objects of an oblong nature do not project favorably in a cube-shaped grid and therefore require an elongated grid. An elongated grid can be developed by expanding the individual planes of the basic grid in the direction desired. When expanding a grid, it is wise to add units to both the front and the rear of the existing grid. This practice maintains a comparable center of vision in the expanded grid. The example described here will add one unit to the front and three to the rear of the grid.

Side plane: Expanding the side plane of the grid is accomplished by the three steps shown in Fig. 1-64.

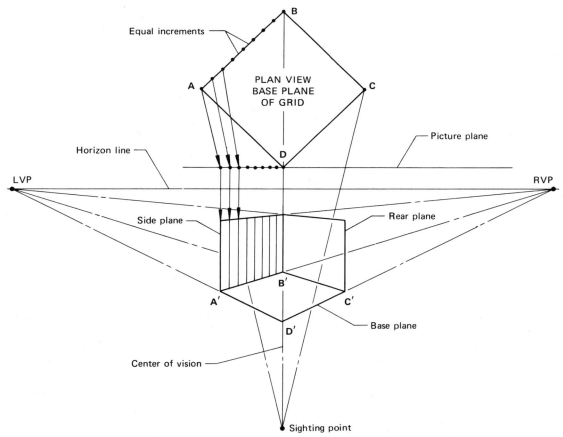

FIG. 1-62   Horizontal increments.

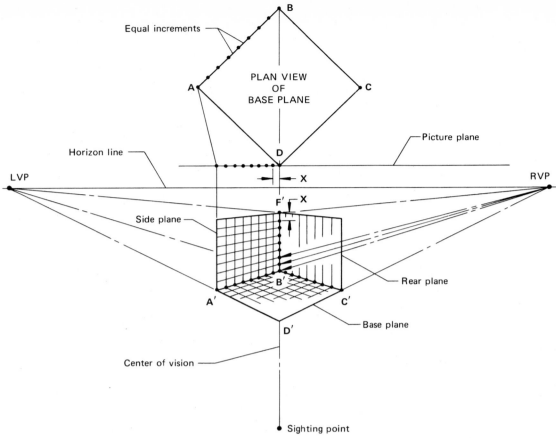

FIG. 1-63   Vertical increments.

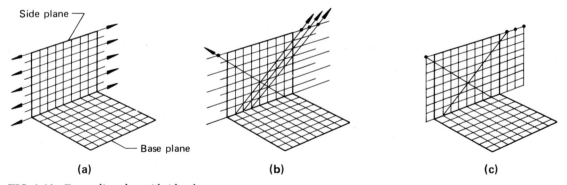

(a)   **(b)**   **(c)**

FIG. 1-64   Expanding the grid side plane.

1. Extend the horizontal lines of the side plane forward and rearward to the approximate new length desired.

2. Construct diagonals through the intersections of the existing horizontal and vertical lines of the side plane. Continue the diagonals through the newly extended horizontal lines.

3. Draw vertical lines through the new intersections created by the extended diagonal and horizontal lines.

Base plane: Expanding the base plane of the grid is accomplished by the three steps shown in Fig. 1-65.

1. Extend the longitudinal lines of the base plane to the same length as the expanded side plane.

2. Construct diagonals through the intersections of the existing longitudinal and lateral lines of the base plane. Continue the diagonals through the newly extended longitudinal lines.

3. Locate the extended intersecting points at the base of the side plane. Draw lateral lines from these points to the new intersections created by the extension of the diagonal and longitudinal lines.

The expansion of the side plane and the base plane provides new horizontal and lateral measuring increments in a fore-and-aft direction.

Rear plane: Completion of the expanded grid requires the development of the rear plane in its new location. The rear plane outline is developed as follows:

1. Establish the top plane of the unexpanded grid, defined by letters ABCD in Fig. 1-66a.

   a. Project a line from the LVP through point A to the approximate width of the grid.

   b. Project a line from the LVP through point B to the approximate width of the grid.

   c. Project a perpendicular line from point D' to intersect the line projected from the LVP through point A. This establishes point D.

   d. Project a line from the RVP to point D and establish point C at the intersection of the line from the LVP through point B.

2. Draw a diagonal line through points A and C and a parallel line from point E. The intersection of the parallel line and the extended line DC establishes point G of the

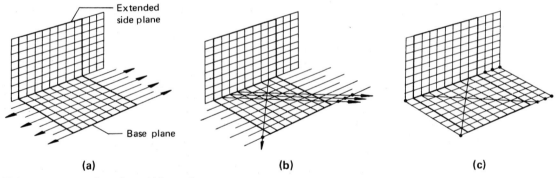

(a)   (b)   (c)

FIG. 1-65 Expanding the grid base plane.

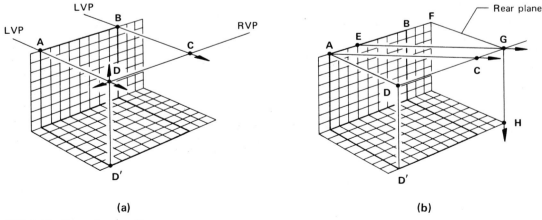

(a)   (b)

FIG. 1-66 Rear plane outline.

rear plane. *Note:* AE is equal to the number of units added to the side plane of the grid, as shown in Fig. 1-66*b*. Draw lines FG and GH.

The rear plane detail as shown in Fig. 1-67 is obtained as follows:

1. Draw perpendiculars from the intersecting points on line JH. This step provides the new lateral measuring increments (Fig. 1-67*a*).
2. Draw diagonal lines JG and FH (Fig. 1-67*b*).
3. Draw lateral lines through the intersections of the vertical lines and the diagonals FH and JG (Fig. 1-67*c*).

The development of the rear plane provides the vertical and lateral measuring increments for the rear plane and completes the expansion of the grid.

Line Conventions

The various widths of lines used for orthographic drawings have symbolic meanings. These meanings also apply to pictorial drawings in the use of object lines, dimension lines, centerlines, leaders, cross-section lines, breaklines, and cutting-plane lines.

Line Depth Illusion   Line thickness in a pictorical drawing has an illusionary meaning. A pictorial drawing created with lines only—that is, without shading—must rely on converging lines and subtle degrees of line weight to appear three-dimensional (Fig. 1-68).

Line Weight or Thickness   Line thickness or weight is the width of the line as it is actually drawn. The weight of the lines in any drawing depends basically on the size of the drawing and its intended use. If the drawing is to be reduced in size for final reproduction, the line weight should be drawn accordingly to offset loss of line delineation in reduction. Three basic types of lines are required to create a good linear pictorial drawing: heavy, medium, and fine.

Heavy Lines   Heavy lines are used to emphasize a part in a pictorial drawing. Also, an illusion of depth is created when certain lines are heavier than others. Similarly, a heavy, tapered line is used to illustrate depth on a curved surface (see Fig. 1-68).

Medium Lines   Medium lines are used as secondary emphasizing lines in the main object of a pictorial, as shown in Fig. 1-68. They are also used as depth lines in an unemphasized or background object, as shown in Fig. 1-69.

Fine Lines   Fine lines are used as front edge lines on the surface of an object nearest to the light source. This gives a feeling of light on this edge, which further enhances the illusion of perspective. Another common use of fine lines is in subduing or de-emphasizing an object that is of secondary importance in a drawing. Background

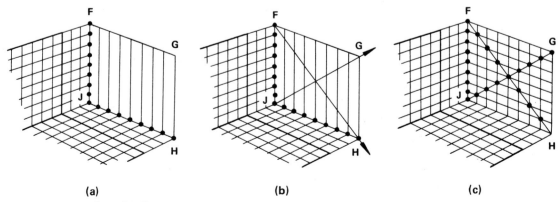

(a)   (b)   (c)

FIG. 1-67   Rear plane detail.

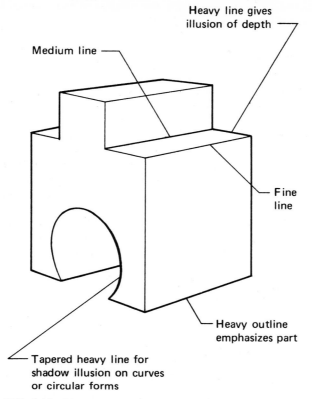

Heavy line gives illusion of depth —

Medium line —

Fine line

Heavy outline emphasizes part

Tapered heavy line for shadow illusion on curves or circular forms

FIG. 1-68   Line usage.

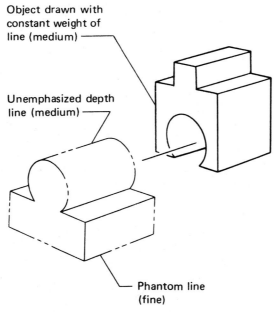

Object drawn with constant weight of line (medium) —

Unemphasized depth line (medium) —

Phantom line (fine)

FIG. 1-69   Line usage.

parts, for instance, are illustrated this way (Fig. 1-69). Phantom lines or fine lines are also used to illustrate background parts.

Pictorial Sectioning

The rules for sectioning of orthographic drawings apply as well to pictorial drawings. In addition, examples and definitions representing the best approach to drawing pictorial sections are presented here.

Cutting-Plane Lines   The orthographic definition of cutting-plane lines also applies to pictorial sections, except that the direction arrows in a pictorial section should be drawn parallel to the lines of perspective (Fig. 1-70). The illustrator should be careful in the placement of the cutting-plane line in order to avoid confusion of line direction. Simplicity should be the rule in determining the placement.

Development of Sections   The use of an orthographic section is the acceptable practice in pictorial drawing (Fig. 1-71a). If clarity is lacking, a pictorial section may be developed from an orthographic section simply by adding vanishing points so as to give the illusion of three dimensions (Fig. 1-71b). The orthographic method uses readily available information taken from a design or detail orthographic drawing, thus making it unnecessary for an illustrator to develop the section.

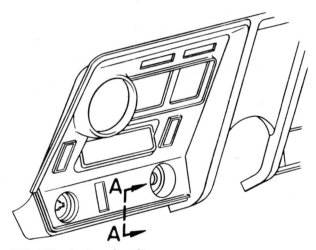

FIG. 1-70   Cutting-plane lines.

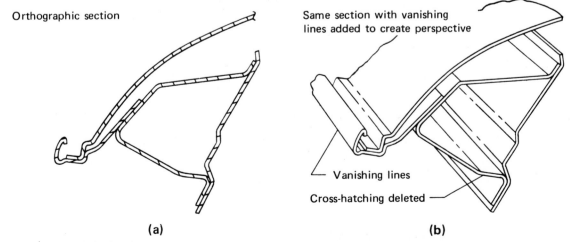

Orthographic section

Same section with vanishing lines added to create perspective

Vanishing lines

Cross-hatching deleted

(a)

(b)

FIG. 1-71   Section development.

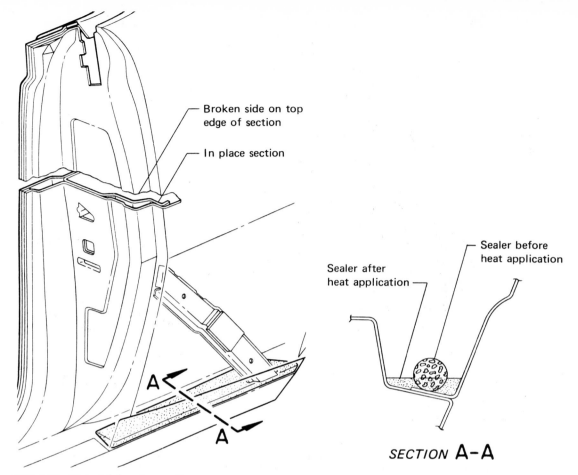

Broken side on top
edge of section

In place section

Sealer before
heat application

Sealer after
heat application

SECTION **A-A**

FIG. 1-72   Pictorial drawing section.

An illustrator should use the following guidelines in the development of any pictorial section:

1. Keep the drawing simple. Use two or more simple sections in place of one complicated section.

2. Avoid excessive detail.

3. Exaggerate metal thicknesses so as to avoid the "plugging" of lines in reduction and reproduction.

4. Use cross-hatching for material delineation only when necessary for the definition of parts.

5. Avoid tone shading, as it requires complicated and expensive reproduction methods.

The development of a pictorial section requires that the section be visualized in its perspective view. The following examples may be used as a guide to the proper methods of making the various types of sections that may be required.

An illustration of a door pillar with a pictorial section cut "in place" is shown in Fig. 1-72. Note that the absence of cross-hatching and the exaggeration of metal thickness increase clarity. The broken side of the section should not be extremely ragged and should follow the contour of the outer surface. Section A-A is an orthographic section portraying a sealant before and after heat application. The section cutting-plane line is placed simply and clearly in an open area.

Figure 1-73 shows an "in-place" section with the sheet metal slightly displaced to more clearly define a sealant application for body welding. Section A-A illustrates the

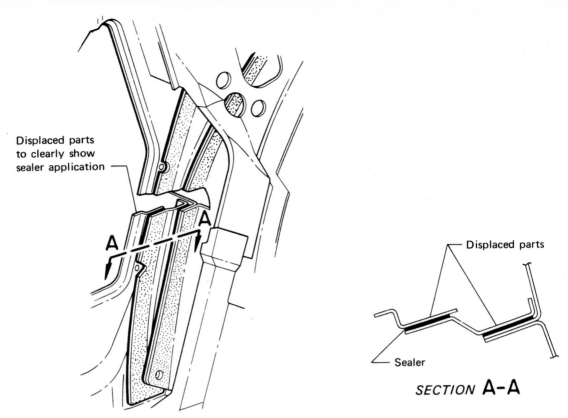

Displaced parts
to clearly show
sealer application

Displaced parts

Sealer

SECTION **A-A**

FIG. 1-73   Pictorial drawing section.

actual sealant relationship to the sheet metal before welding. After welding, the sheet metal will be in a metal-to-metal condition.

Exploded Section   Examples of exploded sections relative to assembly buildups are shown in Figs. 1-74 and 1-75. These types of sections are widely used in the assembly of moldings and other trim items.

A section cutting-plane line extended to include parts exploded in assembly sequence is shown in Fig. 1-76a. The orthographic section depicts the application of sealant in areas that are inaccessible after assembly.

Hidden Part   An example of a section used to show a hidden part in its assembled position is shown in Fig. 1-76b. This type of section must be carefully constructed,

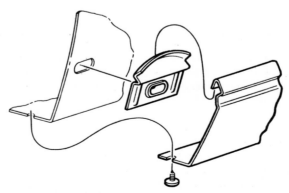

FIG. 1-74   Exploded section.

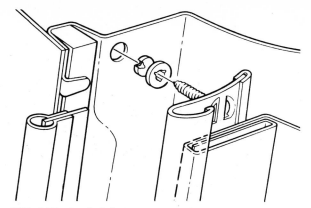

FIG. 1-75 Exploded section.

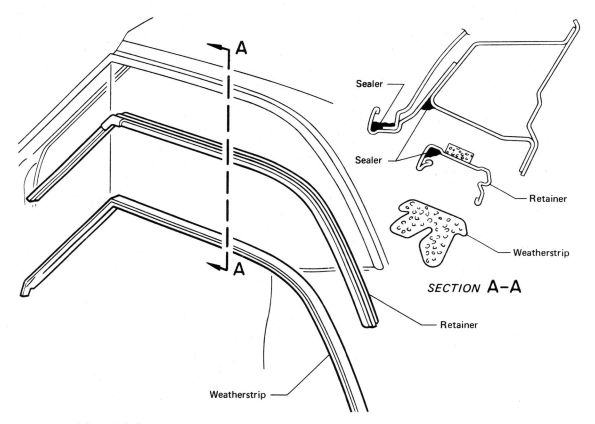

FIG. 1-76 (*a*) Exploded section.

considering the close relationship of hidden lines and phantom lines; the phantom lines suggest the completed shape.

Fig. 1-77 shows a "telltale" section and a "hidden" section. Both are valuable in the illustration of metal relationships wherever the illustrator wishes to show the outside piece in its entirety.

Cross-Hatching in Pictorial Sections An example of a half section through a rubber bushing is shown in Fig. 1-78. Because it is not general practice to section standard parts, the bolt is not sectioned. The sectioning through the bushing is clarified by the opposing direction of the cross-hatching in adjoining parts. In Fig. 1-78a, the parts are crosshatched with a symbolic pattern to illustrate the nature of the material of which

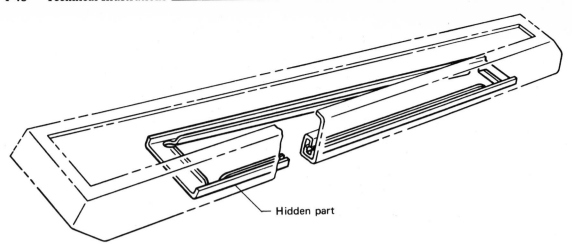

Hidden part

FIG. 1-76 (*b*) Hidden part in sectional view.

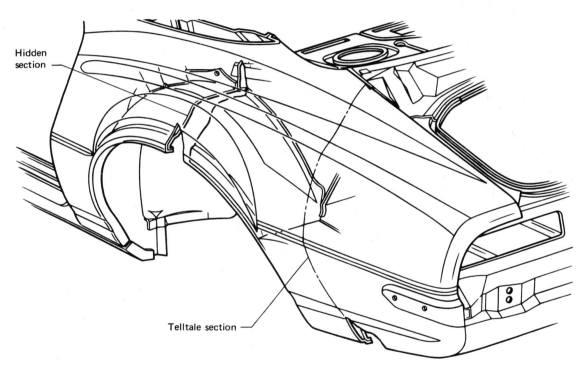

Hidden section

Telltale section

FIG. 1-77 Hidden and telltale sections.

the part shown is made. If symbolic cross-hatching complicates the illustration and causes confusion, general-purpose cross-hatching may be used as shown in Fig. 1-78*b*.

Peeled Section A method for illustrating material that has been "peeled back" to expose a section, for instance, through a seat back, is shown in Fig. 1-79. This typical example of a very complex section requires a detailed knowledge of seat construction. Such sections require artistic judgment on the part of the illustrator and checker.

Complicated assemblies are often difficult to section in their entirety. Figures 1-80 and 1-81 show an engine that has been sectioned through the left front cylinder and the water pump assembly. These views illustrate how additional clarity can be attained by "rendering" the same view with additional artwork. Figure 1-80 shows the section without shading. The section is readable, but it requires study. In Fig. 1-81 an illusion of depth has been created by the addition of stipple shading. These views represent a

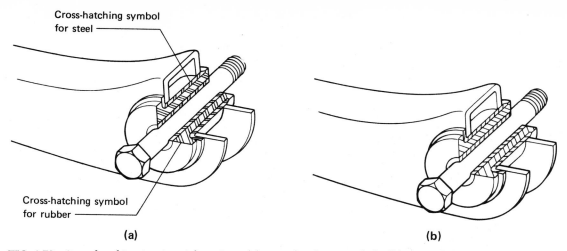

FIG. 1-78 Cross-hatching in pictorial sections. (*a*) Cross-hatching symbols. (*b*) General-purpose cross-hatching.

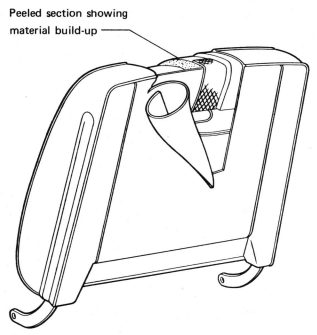

FIG. 1-79 Peeled section.

different approach to the same problem and require technical skill for successful completion. If the available reproduction equipment can produce only low-cost black-and-white copy, without halftones, it may be practical to produce a sectional view in stippled shading for improved clarity, even though the application of this type of shading is complicated and should be avoided. If halftone reproductions can be made, the section should be rendered in shaded tones to make it more readable.

Photographs

Indispensable as visual aids for almost every graphic application, photographs serve the primary purpose of verifying and authenticating text information as well as clearly depicting objects and the steps and procedures involved in technical operations. In technical, training, repair, and other types of technical publications, photographs often

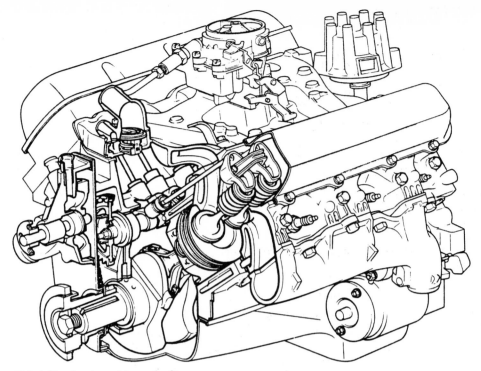

FIG. 1-80 Section without shading.

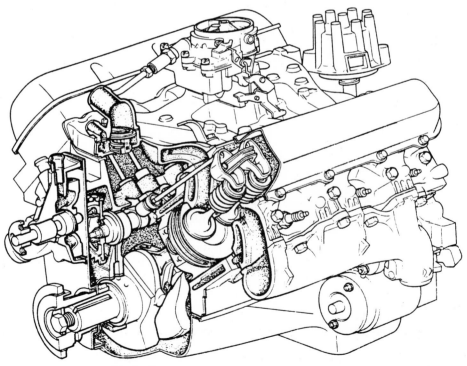

FIG. 1-81 Section with shading.

may be preferable to drawings, taking into consideration such factors as suitability, time, availability, and cost. Photographs also play an important role in presenting copy in films, slide shows, and overhead projection, as well as in product displays, charts, and other communication media, including sales promotion, advertisements, and the like.

Art or Photograph?   The decision on whether to use art or a photograph in any given situation is usually arrived at through consultation among the writer, the editor, and the art director or graphic designer whose responsibility it is to design the total format of the book. It is important to consider such factors as:

- **Production deadline**   If time is short, it is often best to shoot photographs, which can be processed and delivered on a day's notice, rather than try to produce hand-drawn art. In many circumstances, however, an illustrator can produce drawings in less time and at a lower cost than a photographer can produce finished prints. A photographer must be contacted, perhaps sent to a field site, and given time to set up equipment, shoot the photographs, and process the film.

- **Photo retouching**   In technical manuals in which the photographs must show details sharply and clearly, photo retouching by an artist may be necessary, adding time and cost.

FIG. 1-82 Photos shot through special process screens can simulate etchings or other art effects for advertisements, sales booklets, and the like. *(Drafting and Repro Digest.)*

- **Platemaking**   In order to print a photograph in its continuous tones, it must be photocopied for platemaking through a halftone screen, which breaks up the image into a series of tiny dots. Solid linework that passes through these tones, such as callout names and arrows, are put on a separate art overlay. This is then shot separately (without a screen) and "burned" onto a combination plate combining the screened halftone image with the line art. The combination plate costs more than the halftone, which in turn is more expensive than line art alone.

    For promotional literature, a photograph may be shot through one of many special process screens so as to impart a hand-drawn look simulating an etching or engraving (Fig. 1-82).

Technical Illustration Applications

Technical illustrations may be classified in terms of distinct formats, each designed to best depict a given situation. These formats are as follows:

- **Assembly illustration**   All parts of a unit as they fit together (Fig. 1-83).
- **Installation illustration**   System components as they are to be connected, for example, hydraulic lines or valves on a car body (Fig. 1-84*a*, *b*, and *c*).
- **Exploded view**   Shows each unit of an assembly separately as the parts would be removed along the axis of disassembly (Fig. 1-85). Each part is identified in a callout by its name or a number. Exploded views are used most often to illustrate the parts catalogs used by retailers, mechanics, assemblers, etc., and in the assembly, service, and repair instruction sheets that accompany certain products.

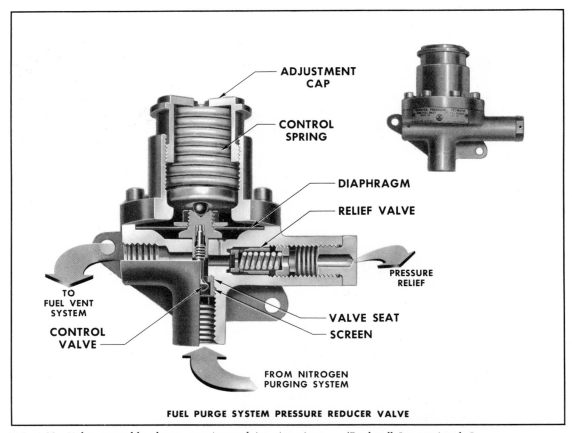

FIG. 1-83   Valve assembly shows exterior and interior views.   (*Rockwell International, Los Angeles Division.*)

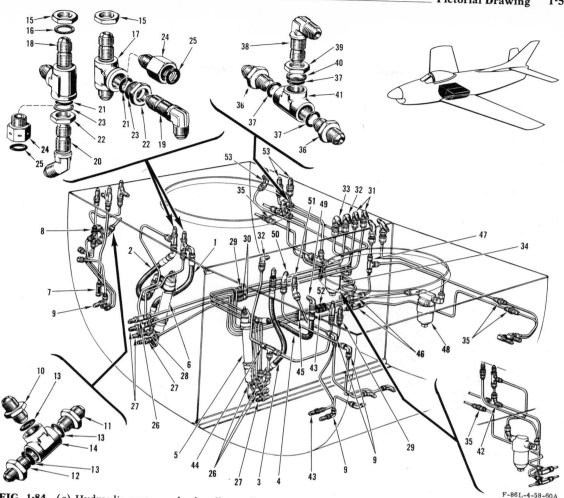

FIG. 1-84 (a) Hydraulic system wheel-well installation shows blown-up views of key fittings. (*Rockwell International, Los Angeles Division.*)

F-86L-4-58-60A

- **Cutaway view** Shows a structure or assembly with portions removed to reveal inside parts, contours, and relationships. Cutaways are often used in instruction manuals and in audiovisual training programs to show the inner structure of such things as engines, pumps, and geological strata, as well as in biological and medical illustration. Cutaways are often prepared as tonal renderings with lines and areas coded in black-and-white line patterns, or in color, to show different materials or the flow of various fluids, gases, etc. (Fig. 1-86a and b).

- **Phantom view** In illustration, as in orthographic engineering drawings, a "phantom" is a part of a drawing that depicts parts, structures, etc. in light line, or, following the drafting convention of a long line followed by two dash lines, as a background for orienting the position of subject units that are drawn to stand out in heavy line. In a continuous-tone illustration, the key units are depicted in strong colors or values against the subdued tones of the phantom area (Fig. 1-87a and b).

- **Application illustration** Shows a product or device functioning in the situation or environment in which it was designed to operate, for example, a space vehicle landing on the moon (Fig. 1-88). An advertising application might be a picture of a farmer using a tractor in a field, or a kitchen appliance in use. In military terminology, an illustration of an artillery system shown deployed in actual battlefield conditions, for example, would be called a *mission study*.

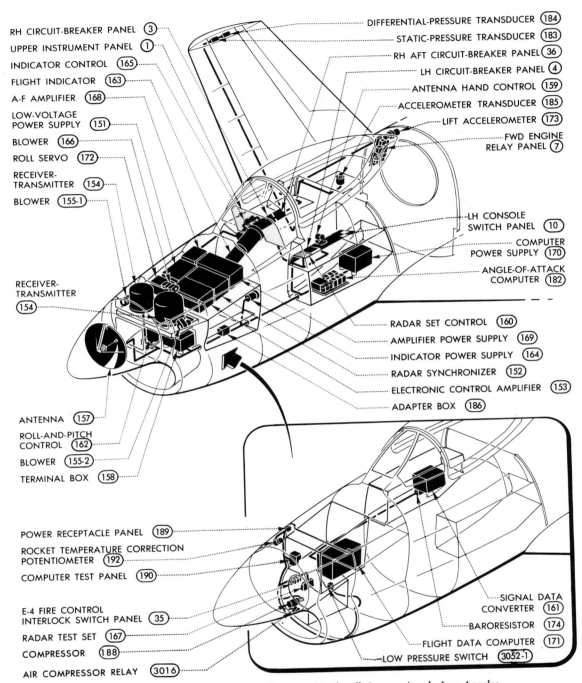

RH CIRCUIT-BREAKER PANEL (3)
UPPER INSTRUMENT PANEL (1)
INDICATOR CONTROL (165)
FLIGHT INDICATOR (163)
A-F AMPLIFIER (168)
LOW-VOLTAGE POWER SUPPLY (151)
BLOWER (166)
ROLL SERVO (172)
RECEIVER-TRANSMITTER (154)
BLOWER (155-1)

RECEIVER-TRANSMITTER (154)

ANTENNA (157)
ROLL-AND-PITCH CONTROL (162)
BLOWER (155-2)
TERMINAL BOX (158)

DIFFERENTIAL-PRESSURE TRANSDUCER (184)
STATIC-PRESSURE TRANSDUCER (183)
RH AFT CIRCUIT-BREAKER PANEL (36)
LH CIRCUIT-BREAKER PANEL (4)
ANTENNA HAND CONTROL (159)
ACCELEROMETER TRANSDUCER (185)
LIFT ACCELEROMETER (173)
FWD ENGINE RELAY PANEL (7)

LH CONSOLE SWITCH PANEL (10)
COMPUTER POWER SUPPLY (170)
ANGLE-OF-ATTACK COMPUTER (182)

RADAR SET CONTROL (160)
AMPLIFIER POWER SUPPLY (169)
INDICATOR POWER SUPPLY (164)
RADAR SYNCHRONIZER (152)
ELECTRONIC CONTROL AMPLIFIER (153)
ADAPTER BOX (186)

POWER RECEPTACLE PANEL (189)
ROCKET TEMPERATURE CORRECTION POTENTIOMETER (192)
COMPUTER TEST PANEL (190)

E-4 FIRE CONTROL INTERLOCK SWITCH PANEL (35)
RADAR TEST SET (167)
COMPRESSOR (188)
AIR COMPRESSOR RELAY (3016)

SIGNAL DATA CONVERTER (161)
BARORESISTOR (174)
FLIGHT DATA COMPUTER (171)
LOW PRESSURE SWITCH (3052-1)

FIG. 1-84 (b) Aircraft fire-control-system installation. (*Rockwell International, Los Angeles Division.*)

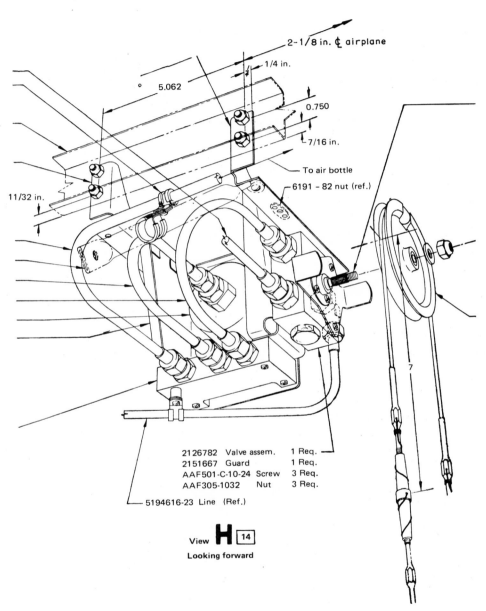

2-1/8 in. ₵ airplane

1/4 in.

5.062

0.750

7/16 in.

To air bottle

6191 – 82 nut (ref.)

11/32 in.

2126782	Valve assem.	1 Req.
2151667	Guard	1 Req.
AAF501-C-10-24	Screw	3 Req.
AAF305-1032	Nut	3 Req.

5194616-23 Line (Ref.)

View **H** 14

Looking forward

FIG. 1-84 (*c*) Typical pencil-drawn production installation drawing, with dimensions.

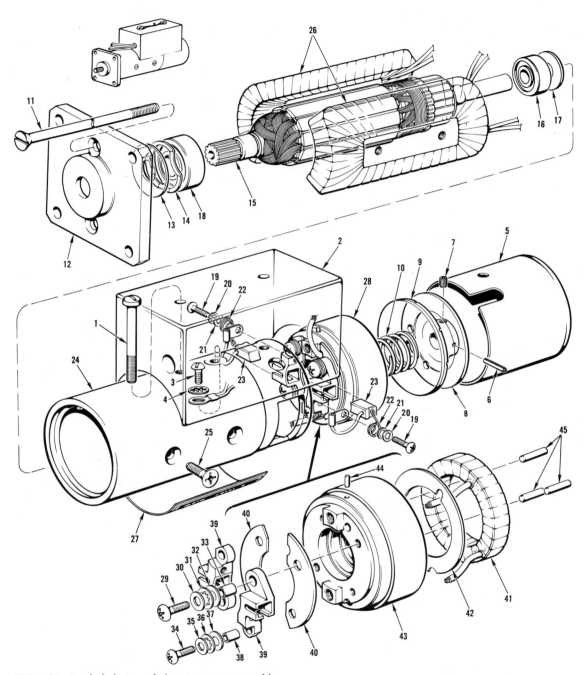

FIG. 1-85 Exploded view of electric motor assembly.

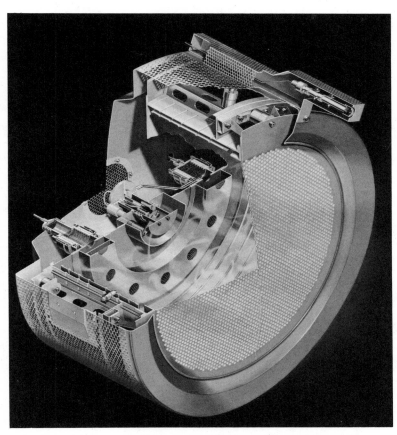

FIG. 1-86 (*a*) Cutaway illustration of NASA Jet Propulsion Laboratory equipment. *(Courtesy Ken Hodges.)*

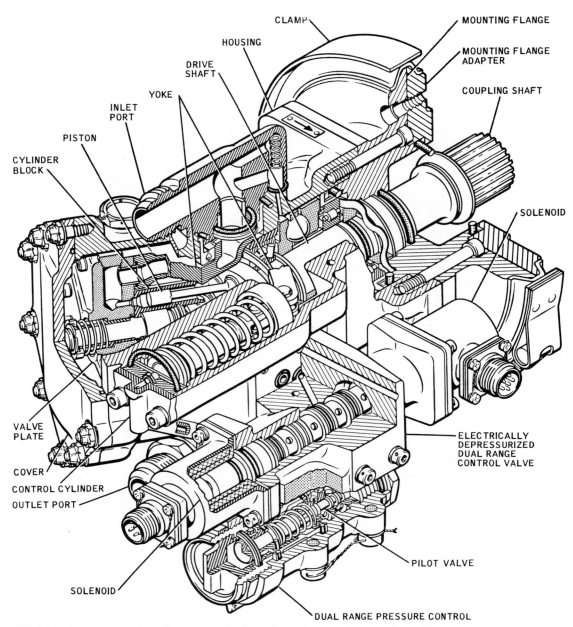

FIG. 1-86 (*b*) Cutaway view of pump. In this line illustration, sections are shown in cutout patterns from shading sheets to code components. (*McDonnell Douglas.*)

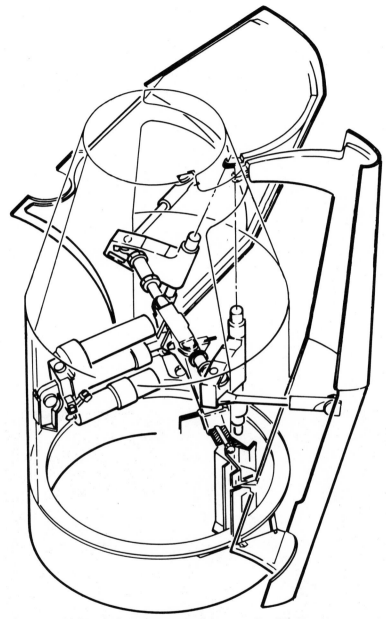

FIG. 1-87 (*a*) Light-line contours in "glassine effect" help relate components of this unit. (*Rockwell International, Space Division.*)

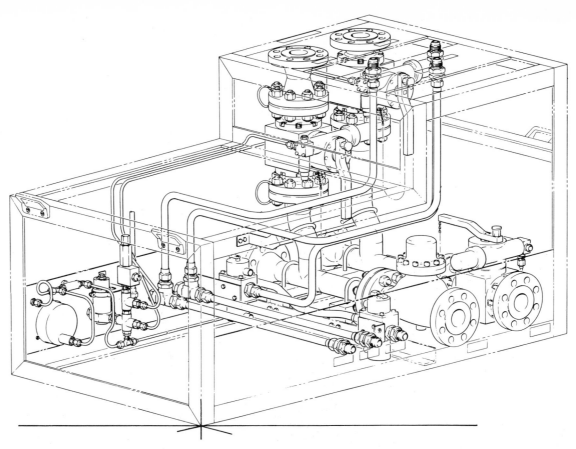

FIG. 1-87 (*b*) Phantom line of one long line and two short dashes is used in this dimetric assembly of piping systems. (*North American Rockwell Corporation, Tulsa Division.*)

FIG. 1-88 Mission study depicts rendezvous of Space Shuttle with satellite station. (*Rockwell International, Space Division.*)

Technical Illustration Methods

In preparing a technical illustration, basic information usually is derived from blueprints and other engineering drawings; from sketches supplied by engineers; from other references such as illustrations of a similar nature reproduced in reports or other publications; or from data sheets depicting components, assemblies, etc. If a scale model or a prototype of the actual object exists, it should be studied, sketched, and photographed. This may be done either by a professional photographer or by the illustrator, who may use a Polaroid camera when fast work is necessary.

After consultation with engineers to clarify the purpose and extent of the illustration and with a technical editor or graphics director to define such factors as drawing size, medium, art requirements, and completion schedule, the illustrator will prepare an accurate preliminary layout to a convenient scale. This layout is usually done in pencil on tracing paper, viewing the object at the angle that mostly clearly depicts its functions and details from the standpoint of the intended user (for example, a mechanic who must look upward in order to remove parts).

Measurements are plotted so as to lay out the contours and proportions of the object accurately. The laying out of these points, involving measurements, calculations, and the conversion of figures from actual scale to drawing scale, is the most time-consuming part of making a technical illustration, but these steps are necessary if the end result is to be an accurate representation for engineering purposes (Fig. 1-89).

Of the various axonometric and perspective systems for showing objects three-dimensionally, isometric is the easiest and most rapid to draw, and therefore is most often employed in art for manuals, handbooks, parts catalogs, and similar publications that require several three-dimensional illustrations that follow a uniform, simple, and standard method of projection. Dimetric projection is second to isometric in terms of speed and convenience. The other systems are listed in the order of their increasing complexity and the time and cost involved in their preparation: trimetric, one-point perspective, two-point perspective, and three-point perspective. Two-point perspective is a highly popular system with most illustrators whenever it is necessary to create the illusion of true three-point perspective without actually using the third, or vertical, vanishing point. One-point and three-point perspectives are mostly used by architectural illustrators to create the effect of scale in structures.

Three-Dimensional Art Aids   There are a number of devices that may be used as aids in laying out three-dimensional illustrations. Some of these aids are:

- Three-dimensional grids (see Figs. 1-58 and 1-65) may be constructed by the illustrator, as described above, or commercially available axonometric or perspective grids may be used. The grid is placed under tracing paper, and points and lines are then traced, using the grid lines as a guide, to create the preliminary layout.

- Isometric and dimetric triangles with foreshortened scales, as well as a variety of special adjustable patented devices, templates, and guides, may be used to save time in producing axonometric projections.

- A special T square with movable arms, known as a *lineaid*, is a device used for laying out true perspective lines (Figs. 1-90 and 1-91a and b). Other patented devices, such as the Klok perspective board, make it possible to shorten time in laying out and scaling perspectives, as well as to accurately draw lines converging to distant vanishing points without projecting these lines outside the limits of the drawing board.

- Plastic ellipse templates are available in every needed size and angular increment, to depict circular objects viewed at various angles without the need for tedious point-by-point construction of ellipses. French curves, ship's curves, and flexible curves are aids used to lay out complex compound curves and similar configurations.

- Small objects, models, and photographs can be placed in a viewer/projector, also known as a visualizer machine, which projects a flat or three-dimensional image onto a translucent working surface for fast, accurate tracing. Such units, which are rather large, are often used in art departments and advertising art studios to save layout time.

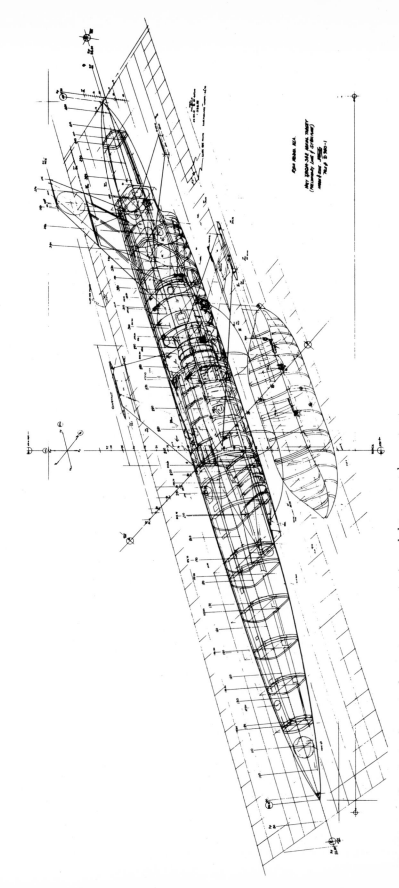

FIG. 1-89 Pencil projection layout in three-point perspective is needed to accurately construct illustration of an aircraft. (*IAM magazine.*)

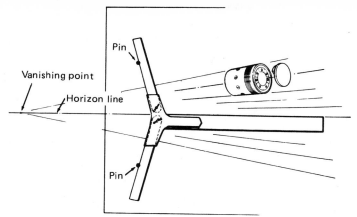

FIG. 1-90 The lineaid.

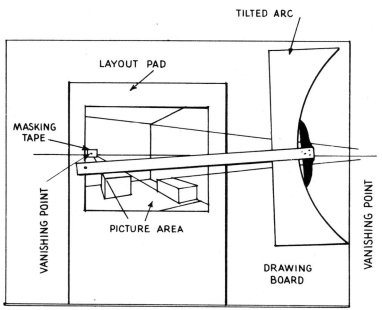

FIG. 1-91 (*a*) Use of a board cut to form a perspective arc and standard T square in projecting lines to a vanishing point.

- A camera lucida is a simple optical device, consisting of a movable arm clamped to the drawing board and equipped with a prism lens through which the artist views a photo tacked to a vertical board. Looking down through the prism, one sees an enlarged image on the drawing paper, an image whose contours are then traced in exact duplication of photographic perspective. Tracing from photographs, however, is not recommended for general practice in perspective drawing, because the camera often distorts distances and receding planes.

In addition, when drawing illustrations for magazines, advertising, and similar purposes, many illustrators use self-contrived shortcuts, or plain "eyeballing"—using one's own judgment as to what looks best—to speed perspective layout. Top illustrators succeed in eliminating projection methods this way, but for engineering work the need is for truthful accuracy rather than a handsome look that might misrepresent the subject.

Once the preliminary layout on tracing paper has been completed in pencil and approved, it must be used to create a finished illustration in one of many techniques or mediums. The following sections will describe the chief methods of preparing a finished illustration.

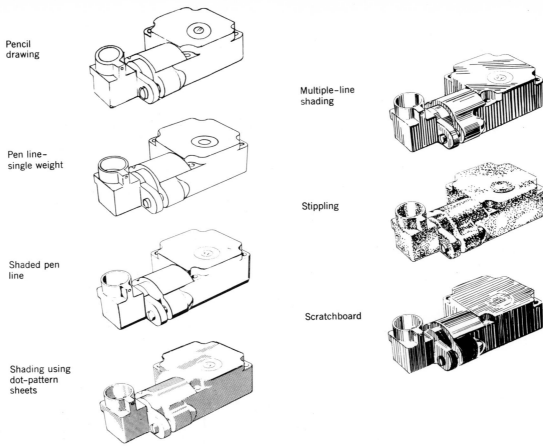

Pencil
drawing

Pen line—
single weight

Shaded pen
line

Shading using
dot–pattern
sheets

Multiple–line
shading

Stippling

Scratchboard

FIG. 1-91   (*b*) Line techniques for illustration.

Preparing Line Illustrations: Methods and Mediums

What we call *line illustrations* are all those illustrations which need not be photographed through a halftone screen in order to register every line and shade in reproduction (Fig. 1-91*a*). Stippling, the technique of placing ink dots on a drawing, and dot-pattern shading screens—printed on clear adhesive film that is cut to shape and applied to a drawing—are devices illustrators use to break up solid areas into dots much as the halftone screen does, gaining the benefits of shading without incurring the higher costs of halftone screen reproduction.

Line technical illustrations are almost always prepared as finished art by tracing over the preliminary layout onto translucent materials such as polyester drafting film, high-grade plastic-coated or transparent tracing vellum, or tracing cloth. The preliminary layout cannot be used for this purpose because it is worn and dirty and has traces of unwanted lines, notes, and corrections. In the final art, it is important to avoid crowding, to allow proper space in views, and in general to follow good layout practice (Fig. 1-92*a* and *b*). Most technical illustrators do a finished drawing in black india ink, using technical pens with their own ink reservoirs and selecting from a set of pens to draw each different line width—a light line for outline, a thick line for the shadow side, etc. (Fig. 1-93).

As a drawing base, polyester Mylar drafting film has become a favorite in recent years. It costs more, but it is tougher, more permanent, and dimensionally stable. It will not shrink or warp with temperature or humidity changes, and the drafting film stands up to abuse better than other surfaces and can be erased any number of times without damaging the surface. For illustration purposes the type of drafting film with a matte surface on one side and a shiny surface on the other is preferred. The velvety matte surface can readily be drawn on in pencil, ink, crayon, markers, and other mediums.

Some illustrations are drawn in pencil, using a medium- to soft-grade lead that is black enough to reproduce clearly in diazo, microfilm, or electrostatic processes. Ink is preferred for microfilm reduction under high-quality military standard legibility specifications or for use in printed manuals for the military services, all of which set very strict standards for sharp linework and clear readability.

Pencil is not recommended for use on drafting film because it smears easily on the hard surface. But many illustrators and drafters use plastic pencils to draw black, non-smearing lines that meet quality standards for publication art. The pencils come in two types: all plastic, or a lead that is half plastic, half graphite, and which slides more easily on the drawing. Plastic lead's black line is much more reproducible for printing purposes than the gray line of graphite pencils, which reproduces poorly and has a shine that reflects light during the copy camera stage of the platemaking process.

Shading in Line Art

The trend in preparing illustrations, especially for government and military manuals, handbooks, and other publications, has been toward simpler delineation, eliminating all but the necessary details and stressing extreme simplicity in shading. Many publications use line drawings done in the same line weight with no shading (Fig. 1-92b). Oversimplification can easily become self-defeating, however, because the nature, forms, and contours of most three-dimensional objects can best be recognized when brought out by shading. Clearly, for most handbook and catalog types of illustrations it would be too time-consuming and costly to shade objects with a painstaking multiplicity of fine, graduated pen lines (the standard shading method used in Europe during the past century has been the costly, graduated pen line). As a result, the most frequently used technique, one which depicts parts clearly without involving undue time and costs in preparing manuals—which can contain thousands of drawings in the case of complex aircraft, weapon systems, and the like—is the "thick-and-thin" method (Fig. 1-94a, b, and c). In this method a uniform medium-weight line is drawn for all contours, with a heavier line two or three times thicker drawn on the "shadow" side of the object. (In illustrations, light comes from a hypothetical, fixed source usually above and to the left of the object, with shadows cast on the right and bottom). Another weight of line, thinner than the contour line, may be used to show the inside contours and details of an object.

For many purposes illustrators prefer to use several lines of different thicknesses to depict shading. Most illustrators use Rapidograph, Castell, Mars, or other makes of technical fountain pens, selecting various pen points from "00" to "4" in line weight as they work (Fig. 1-95).

Advanced Techniques for Shading in Line Art

Other methods of shading may be employed to create subtle tones, textures, and often highly artistic effects while at the same time retaining the cost advantage in reproduction of a line drawing as opposed to halftone screening. Pen-and-ink techniques are popularly used to illustrate informational literature and promotional brochures, as well as technical manuals, parts catalogs, and the like (Fig. 1-96a and b).

Scratchboard Art   Scratchboard is a form of engraving in which the lines are produced by scribing with sharp tools through the black coating of a white board. A layer of special white clay is first applied over a cardboard base. Over the clay a coat of black india ink or tempera is painted across the entire surface, or only in areas defined by a pencil layout. Details of the drawing are drawn in white pencil onto the black, and the white lines are then scribed with a variety of scribing tools to expose the white clay under the ink. Errors are corrected by painting black over the mistake and scribing again. Black lines may be added in white areas using a pen or brush and ink (Figs. 1-97 and 1-98).

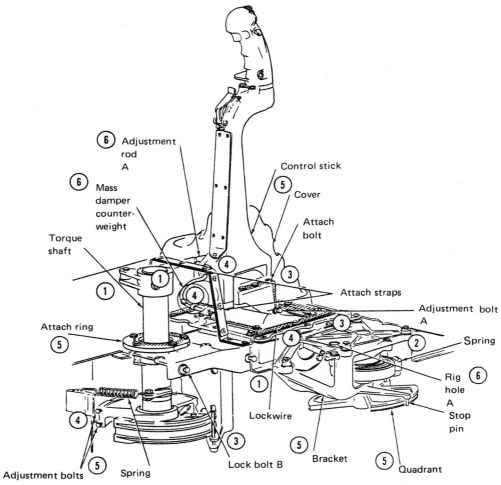

FIG. 1-92 Comparison of good and bad layout practices. Incorrect: (1) Leader lines tangent to and touching edges of copy. (2) Leader lines running parallel to lines of image. (3) Leader lines crossing congested areas or parts called out. (4) Arrow heads poorly placed in congested areas. (5) Poor placement of leader lines in relation to copy. (6) Stacking of callouts (more than one line).

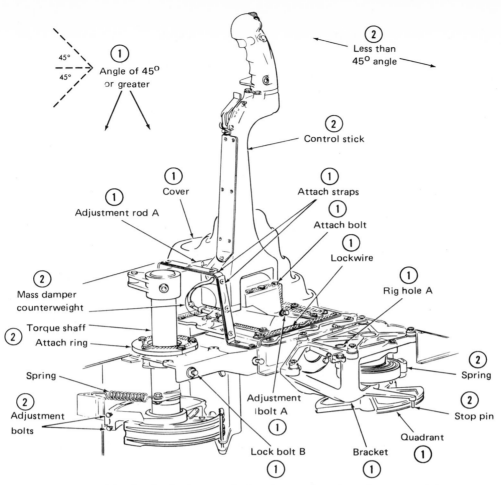

FIG. 1-92 Correct: (1) When leader lines angle sharply, run them from the center of the callout text. (2) More horizontal leaders should come off the end of the text. *(Northrup Norair.)*

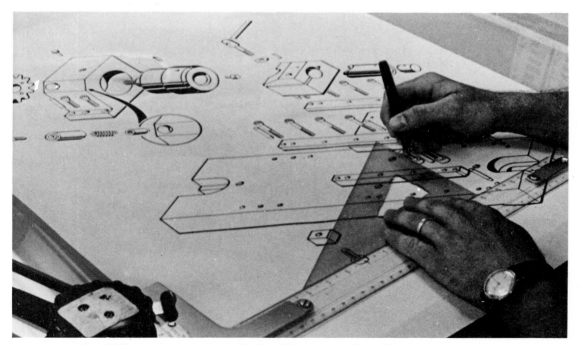

FIG. 1-93 Technical illustration is drawn on drafting film using a technical fountain pen. *(Drafting and Repro Digest.)*

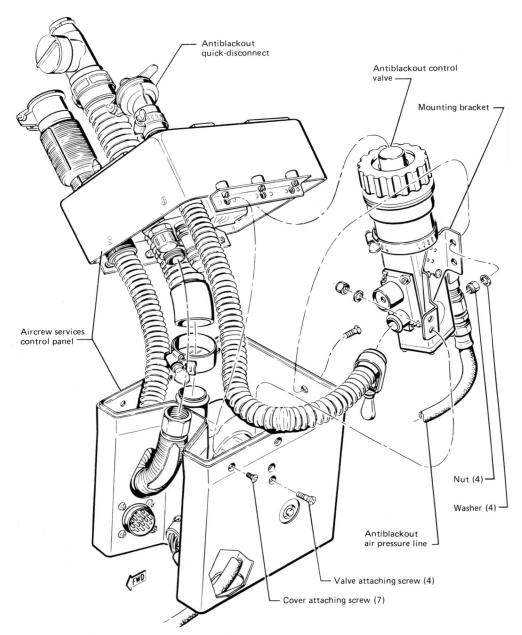

Antiblackout
quick-disconnect

Antiblackout control
valve

Mounting bracket

Aircrew services
control panel

Nut (4)

Washer (4)

Antiblackout
air pressure line

Valve attaching screw (4)

Cover attaching screw (7)

FWD

FIG. 1-94 (*a*) Exploded-view handbook illustration rendered in "thick-and-thin" technique based on light line and heavy line to indicate shading. *(McDonnell Douglas.)*

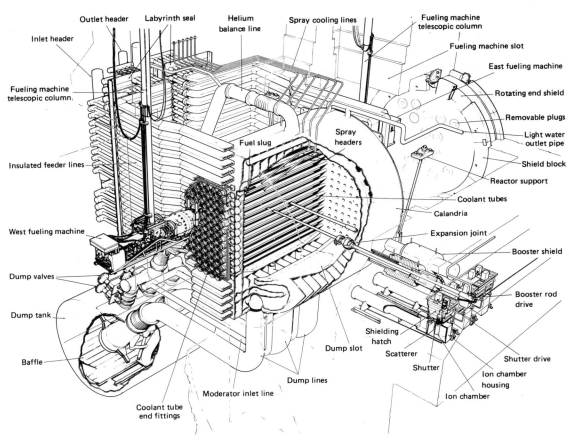

FIG. 1-94 (*b*) Thick-and-thin line technique and three-point perspective add contours and depth to this cutaway view of a nuclear power plant. *(Canadian General Electric Co., Ltd.)*

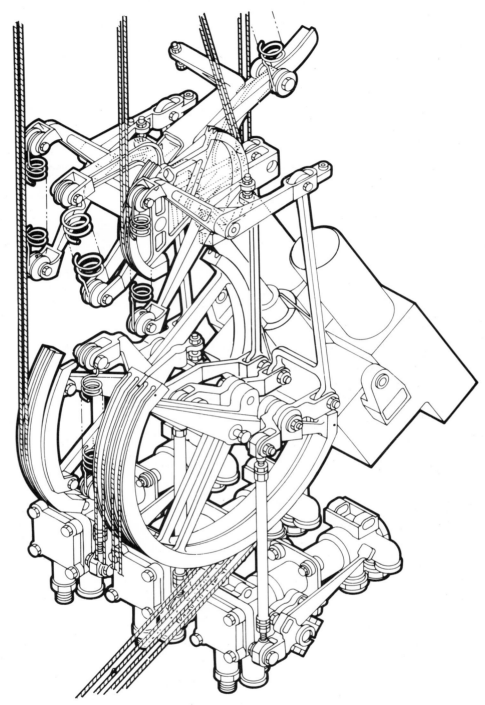

FIG. 1-94 (c) Heavy outline with light interior linework is effective in unifying a unit whose multitude of components might otherwise look confusing. *(McDonnell Douglas, Aircraft Division, artists F. Bohnert and C. Fortmuller.)*

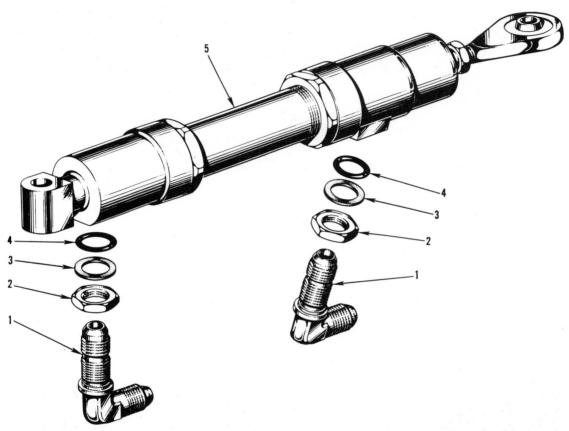

FIG. 1-95 Many line thicknesses are used to depict cylinder parts in this ink illustration. *(Rockwell International, Los Angeles Division.)*

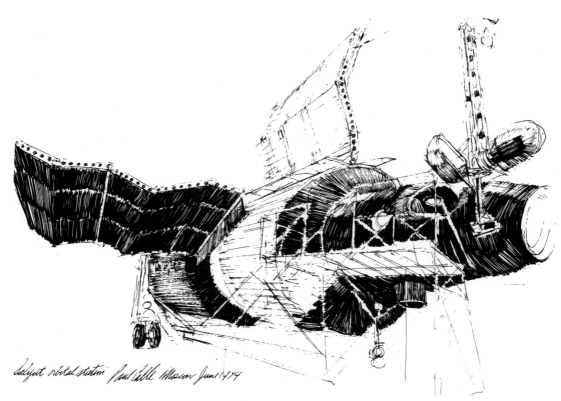

FIG. 1-96 Sketches with india ink fountain pen by artist Paul Calle. (above) U.S. astronauts and (below) Soviet Salyut Orbital Station, in 1975 Apollo-Soyuz Project Flight. *(NASA.)*

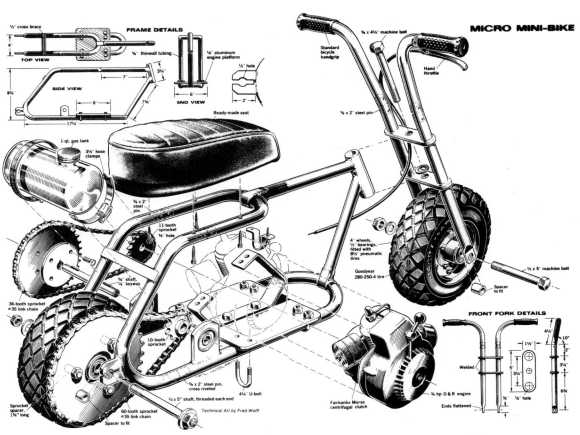

FIG. 1-97 Pen-and-ink illustration of a mini-bike for *Popular Mechanics* combines stippling with fine-line linear shading. (*Popular Mechanics magazine.*)

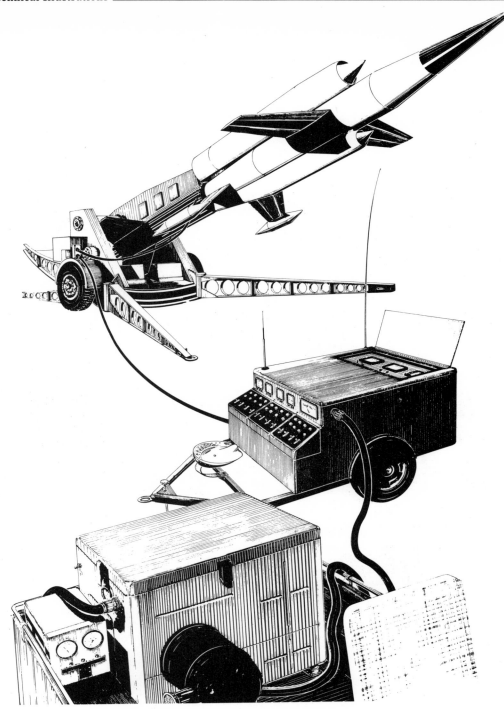

FIG. 1-98 Scratchboard illustration for an engineering firm's brochure.

Stippling In stippling, the illustrator applies ink dots to the paper, spacing these dots widely apart to give the effect of a light tone or placing them closely together to suggest darker values. Stippling requires very patient and exacting work in order to produce smooth effects without clumps appearing here and there. Although stippling and scratchboard are elegant and eye-catching in appearance, both are costly and lengthy processes used mainly in commercial art and advertising but rarely in technical graphics (Fig. 1-99).

FIG. 1-99 Scratchboard illustration with stipple effect for advertisement. *(Courtesy Bill Carr.)*

Special Boards Shading effects that can reproduce well in line may also be created by using a special type of illustration board that has raised dots on its surface, over which the artist can draw with a lithographic pencil or crayon so that the black peaks and white valleys give varied tones and textures. This type of illustration is frequently done for newspaper political and sports cartoons and for similar purposes, and occasionally for technical art presentations that are intended for publication.

Shading Sheets Widely available in art supply stores are a great variety of dot, line, stipple, and other textures printed on adhesive-backed clear film sheets. The shading sheet is laid over the illustration and the area to be shaded is first trimmed out with an art knife and then applied to the surface with a burnishing tool.

Broken Textures Various pebble-grain and other broken textures can be created by use of the dry-brush technique of dragging across the drawing a brush from which most of the ink or paint has been wiped, or by dabbing an inked sponge, foam, or similar material against a drawing to mottle the surface. Such effects are often used in decorative spot illustrations and in technical art to add tones and shadings that add visual interest but are printable as line illustrations (Fig. 1-100). A great variety of effects is possible (Fig. 1-101a to d).

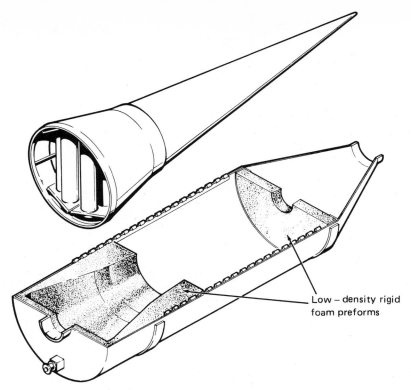

Low – density rigid
foam preforms

FIG. 1-100 Texture-board technique used to show contours in an ink illustration. *(Philco-Ford Corporation, Aeronutronic Division, artist D. Brown.)*

Preparing Tone Illustrations: Methods and Mediums

Continuous-tone illustrations are those which are drawn in a series of blended tones and shadings that require photocopying through a halftone screen in order to make plates for printing reproduction. Tone illustrations may be drawn using pencils, crayons, pastels, or similar tools; painted with watercolors, gouache, tempera, casein, or acrylics; or sprayed with an airbrush.

Pencil and Drawing Mediums The simplest way to produce a full-tone illustration is by rubbing or smudging graphite pencil, or by drawing in broad strokes of pencils whose grades of very hard to very soft leads create tones varying from a light silvery gray to black. These broad strokes are achieved by forming a flat, chiseled edge on a pencil lead by rubbing it on a sandpaper pad. Also, many artists use flat-lead sketch pencils, similar to carpenters' pencils but of artist's grade and available in 2B, 4B, and 6B leads. By drawing with pencils ranging from 4H to 6B, delicate gray tones can be produced that are much more attractive than those made by smudging or bearing down with just one pencil (Figs. 1-102 and 1-103).

Other drawing mediums are charcoal, pastels, pastel pencils, and crayon. Pastels in stick and pencil form are often used to create gray or colored advertising layouts for a client's visualization and approval. The effect of finished art or photographs is sketched onto a page or package. Pastels are also used in drawing many types of industrial illustrations, technical proposals, and the like, either alone or combined with other mediums. Pastels are also used on black or colored paper to prepare design-concept illustrations as aids in developing new products such as toasters, shavers, and automobiles. These illustrations are quickly sketched and are bold, colorful presentations. Pastel color particles easily come off with handling of a drawing, however, and a pastel drawing must be sprayed with a fixative to protect its surface.

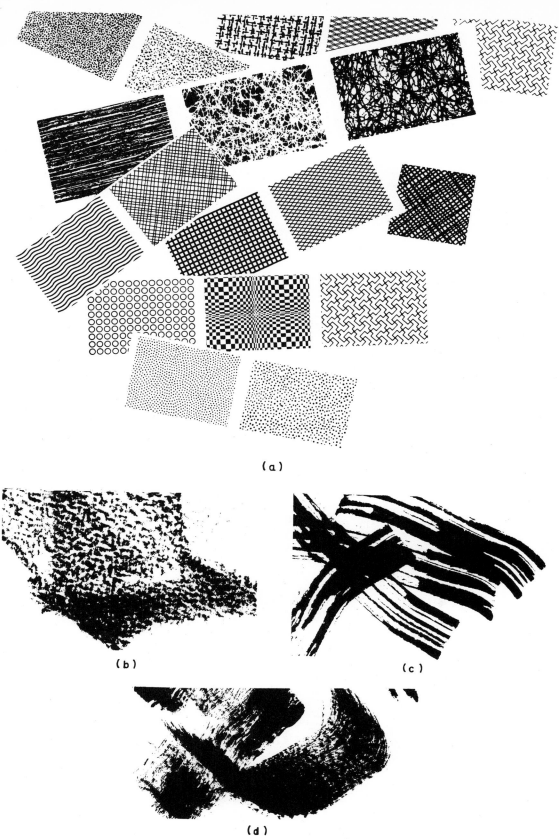

FIG. 1-101 Texture effects for line drawings. (*a*) Self-adhesive pattern sheets. (*b*) Crayon on raised-surface paper. (*c*) Ink with steel calligraphic pen. (*d*) Dry-brush effect.

FIG. 1-102 Pencil illustration.

FIG. 1-103 Pencil layout on tracing paper, for color illustration. *(Courtesy Bill Carr.)*

Markers   Fine-line marker pens and broad-nib felt markers are also effective draw-ing instruments for many purposes. They are essentially most useful as sketching tools and are a favorite with artists who must prepare layouts for packaging, audiovisual presentations, displays, or technical meetings. A marker is a fast, responsive tool in making "instant" charts to be used immediately or to sketch a "talking rough" to depict an industrial product design or the like in a group discussion. The brilliant colors dry instantly and may be used on papers, boards, films, slides, overhead projection trans-parencies, opaque or clear plastic sheets used for mapping, or overlays on technical art. For drawing on paper, a special marker paper is available that will not allow color to bleed through to underlying sheets on the pad (Figs. 1-104 and 1-105).

FIG. 1-104   Art director Jack Reinhardt holds up layout of a folder done with colored pastels and markers, outlining the uses of the Space Shuttle. *(Rockwell International, Space Division.)*

FIG. 1-105   Cartoon poster is done entirely with colored markers. *(IAM magazine.)*

Markers come in a large variety of colors and tones, and by starting with light colors and overlaying them with strokes of deeper hues, additional tones and values are created to enhance the bright range of colors in the picture. Popular uses for marker illustration include layout of labels, package designs, and the like, because the color goes smoothly on slick papers, metallic foil, and formed plastics. In drawing quantities of large display charts, diagrams, and lettering tabulations to be used in technical, financial, and other presentations, a favorite tool of many illustrators is the refillable fountain-pen marker, which can be used continuously for hours before refilling from a marker ink bottle. By using several of these pens, each with its own color, or a supply of varied-color disposable markers with round or chisel-edge nibs, colorful, clean displays can be produced as fast as the hand can draw. In doing large charts and diagrams, very broad strokes can be laid down using a T-shaped felt nib in the pen to create a wide stroke for bar charts, large titles and headings, or blocks of color.

Painted Illustrations

Paintings may be rendered in any of several mediums or in a combination of mediums. All the mediums used in technical illustration are water-based; the long time needed for oil paint to dry rules it out as a practical commercial or industrial art medium compared to the rapid-drying aqueous mediums. (A new oil-based fast-drying medium, called alkyd colors, now promises to erase oil's disadvantages.)

Most tone illustrations are rendered on illustration board, usually of heavy or extra-heavy weight (thickness) to prevent warping of the board under applications of aqueous color. The preliminary tracing-paper layout is traced onto the board by placing between them a sheet of prepared tissue with a rubbed graphite surface and transferring the line image by tracing over it with a hard pencil. Once this is done, the image is further refined on the board by more pencil drawing until the exact delineation in all details appears as a light pencil layout on the board, which is now ready for painting.

Watercolor Watercolor is formally defined as any transparent or opaque water-based paint medium, but to the public as well as to most artists the term connotes, through popular usage, transparent color painting. In this sense and as used here, watercolor is a mode of painting in which a series of color washes, applying diluted watercolor paint or drawing inks, is brushed onto watercolor paper or illustration board, creating a luminous, sparkling effect through the transparent tones reflecting the white paper beneath.

Starting with light tones, successive color washes are painted one over the other once each wash has become dry. Special effects that are caused by running of paint, pools of wet paint, and settling of pigments are created by brushing saturated color into an area of wet paper or still-wet color, in the "wet-on-wet" technique. Because each overpainting of a color wash darkens the picture's values, lightening can be accomplished only be gentle scrubbing with a small, wet sponge to "lift" the color from an area before repainting in lighter tones, or by overpainting with opaque color. Highlights may be added by scratching through colored spots with a knife or a razor blade to expose the white paper underneath, or by adding touches of white paint (Fig. 1-106).

Transparent effects of color are attractive in any type of illustration, and many illustrators add to the eye appeal of their work by deliberately using both transparent and opaque colors in conjunction with each other.

The Opaque Paint Mediums Tempera, gouache, casein, and acrylics are all opaque paints that may be used to render full-tone illustrations. All dry rapidly, helping the illustrator to speed completion of the art. Tempera and gouache are very similar; both are creamier when mixed with water, both go on more smoothly, with better blending of tones, and both brush more evenly for painting fine details, than acrylic or casein paints. However, many illustrators prefer to use casein or acrylic paints; unlike the other colors, which dissolve and produce muddy tones when fresh, wet colors are overpainted, casein and acrylic are insoluble when dry, permitting any number of added coats, changes, or corrections without disturbing the basic paintings (Figs. 1-

FIG. 1-106 Illustration rendered mainly in transparent watercolor, with touches of opaque gouache for highlights. *(Courtesy Ken Hodges and Rockwell International, Los Angeles Division.)*

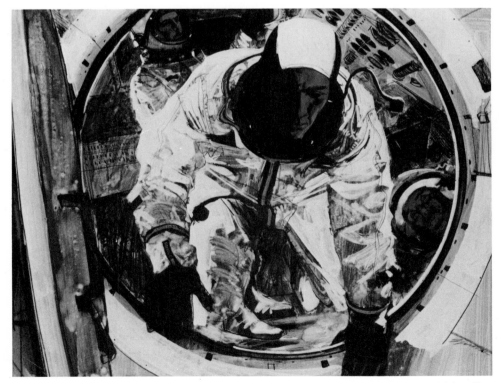

FIG. 1-107 Acrylic painting in watercolor technique. *(General Dynamics, Convair Division, Roy Gjertson, artist.)*

107, 1-108, 1-109, and 1-110). Casein paint (binder made from skim milk curds) has a pastel, velvety look preferred by many artists, but the colors are not widely available in art stores. Acrylics are highly popular, may be purchased anywhere, and come in a wide range of bright colors in both tubes and jars. A range of accessories includes matte and gloss medium, which may be added to the paint water or overpainted as a hard, durable varnish; an acrylic gel for thickening the color; and acrylic gesso for

FIG. 1-108   Architectural rendering done in tempera. *(Thomas Strat & Associates, Inc.)*

FIG. 1-109   This color illustration in watercolor and designer's gouache is a mission study showing the Space Shuttle in a planned landing approach at Cape Kennedy. *(Rockwell International, Space Division.)*

FIG. 1-110   An illustrator paints the final touches of color in gouache, using a photograph as a reference for accurate detail. *(Rockwell International, Space Division.)*

preparing a white base for painting on hardboard or other surfaces. Its great asset is that the acrylic polymer–based plastic paints are permanent, can stand flexing and rough handling, and will not darken or chip with age or misuse. As a result, many illustrators prefer acrylics, either as the sole medium or combined with the other water-based paints for both fine art and commercial and industrial graphics work.

Collage   Collage is an art form in which papers, fabrics, and other materials, including in some instances three-dimensional objects, are glued onto an art surface, usually in a way that they combine with painting or drawing in various mediums to create a picture. Developed by avante-garde nineteenth-century painters, the use of torn newsprint, sticks, strings, and "found objects" created a unique visual effect seized upon by post–World War I graphic designers and used with great effectiveness in magazine, advertising, and industrial illustration and in other forms of contemporary graphics. Collages are made by bonding pieces of material to an art panel with acrylic medium or polyvinyl (white) glue, adding rendered colors, and finishing with a protective coating of acrylic medium or varnish over the entire picture (Fig. 1-111).

Airbrush Art

The airbrush is a miniature spray gun used by artists to produce smooth and even textures, blends, and gradations by applying a fine spray of ink or diluted watercolors. About the size of a marker pen, the airbrush is attached to a hose that supplies pressurized air to a nozzle in which color contained in a small cup is ducted by suction into the airstream, creating a spray. Sometimes an artist will create an illustration entirely through the use of airbrush, but usually the tool is used to add delicate tones and finishing touches to full-color illustrations that have been painted by brush with tempera, gouache, or other mediums. Airbrush methods are also used extensively to retouch photographs, emphasizing details of interest and subduing obtrusive backgrounds or distracting elements.

Use of the airbrush requires special skills and training, but the airbrush artist almost always uses it basically as a tool to supplement and enhance work in other art techniques. Even a picture rendered entirely by airbrush requires conventional artistic

FIG. 1-111 Collage design made with strings and cardboard cutouts cemented to canvas board with acrylic medium, and painted with heavy acrylic paints, symbolizes electronics and physics interactions. *(Courtesy Ken Hodges.)*

skills and knowledge for sharpening details with pen and brush and for painting in certain flat areas.

Although painted technical illustrations are almost always done on kid-finish illustration board, as are some airbrushed works, more often airbrush renderings are done on heavyweight matte-surfaced photoprint paper that has been dry-mounted onto thick illustration board. The photo paper has a velvety surface much more receptive to the smooth effects and transitions of airbrushing than are other paper surfaces. The artist's penciled tracing-paper layout can be photocopied onto the matte photoprint paper, providing a black line image that is an effective guide for cutting masks and for defining color transitions.

Of all art processes, airbrushing takes the most time, mainly because at each step the color spray must be masked off to confine its tones within a sharply defined outline, much as chrome parts are masked off on an automobile for spray-painting. Each separate area of the drawing that has different tones and represents a different visual plane or part must be carefully trimmed out and removed from a sheet of special thin frisket tissue that is pasted over the entire drawing section. A prepared tissue with its own adhesive is commercially available, but many illustrators prefer to use their own formula; the artist first coats the frisket tissue with a thinned-out solution of rubber cement and bonds it to the drawing, then cuts out the "window," the area to be airbrushed, and then removes any dried adhesive with a rubber cement pickup eraser. The artist sprays the tones in this aperture, and the edges of the mask keep the color out of surrounding areas. After the paint has been applied, the remaining frisket is removed, the drawing is cleaned of rubber cement, and a frisket is applied to another area and the process is repeated. Most of the time and cost of an airbrushed illustration depend on how many frisket masks must be stuck on, cut out, and removed.

Beginning with the largest masks, work progresses on to the smaller ones. Small details are painted in using brushwork with tempera color. In the final, finish stage contours are sharpened with fine lines drawn with a ruling pen or a brush, and final touches for highlights and black accents are done with a brush.

A great deal of airbrushed technical illustration consists of cutaway views with sections removed to clearly depict all the inside components in their working relationships to each other (Figs. 1-112 and 1-113). When this cutaway area comprises most of the

FIG. 1-112 This airbrush cutaway illustration of an engine is a combination of airbrushed tones and hand-rendered lines, sections, and details. *(Ford Motor Company, Graphic Arts Department.)*

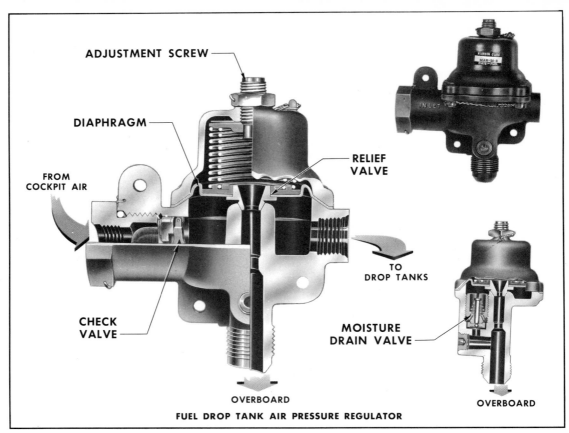

ADJUSTMENT SCREW

DIAPHRAGM

FROM
COCKPIT AIR

RELIEF
VALVE

TO
DROP TANKS

CHECK
VALVE

MOISTURE
DRAIN VALVE

OVERBOARD

OVERBOARD

FUEL DROP TANK AIR PRESSURE REGULATOR

FIG. 1-113   Airbrushed orthographic assembly illustration.
(Rockwell International, Los Angeles Division.)

illustration, the artist prepares the entire art, usually combining painting with airbrush work. However, when the cutaway is only partial, and a sizable portion of the exterior remains exposed, the recommended practice, whenever the machine or prototype is available, is to have a large photograph taken of the unit and do artwork only in the cutaway area. On this photograph, enlarged to the proper scale, the artist overlays a sheet of tracing paper. On this tracing paper the layout is done in pencil, following blueprints and other references, of all interior details as they would appear. A diazo whiteprint is then run off from this layout. The diazo print is then cut to fit the outline of the cutaway section and pasted onto the photo exactly in true position; this paste-up is then sent back to the photo lab to be photocopied, producing a photo of the machine with a white cutaway area inscribed with layout lines and dry-mounted on heavy illustration board, ready for airbrush treatment. Only the interior cutaway area requires full airbrush treatment; the surrounding continuous-tone areas need only retouching.

In another version of this process, the artist takes the initial enlarged photograph of the entire machine, and with graphite or colored pencil designates exactly the areas to be deleted. Photo lab technicians then bleach out those areas on the photoprint, leaving a blank white space for the artist to superimpose the sectional layout.

Airbrush Photo Retouching   Nearly all photographs taken for technical product advertising, and almost all equipment photographs taken to illustrate repair, maintenance, operation sequences, correct assembly, etc. in manuals and handbooks, are inexact and confusing to the eye. To clarify them requires retouching by a professional artist experienced in photo retouching and possessing the art knowledge needed to amplify and isolate the particular areas and operations explained in the text. Also, even on new machines or equipment, there may be nicks, dents, stampings, stains, etc., which tend to confuse the detail and so must be removed by retouching.

When it is determined that a photograph must be retouched in order to clarify detail, eliminate distracting machine markings, or bring out key points of interest, the artist sends the initial 8- × 10-in. glossy print back to the photo lab with instructions for a larger matte print to be made and mounted on artboard. In retouching this photograph, the artist usually airbrushes semitransparent tones onto the photo, rather than an opaque coat, in order to preserve as much as possible of its authentic appearance. In areas of interest, light-and-shadow contrasts are accented, contours are sharpened with pen line and brush, and backgrounds are subtly subdued with a blending gray tone over darks, lights, and hard edges. Where linework, such as call-outs, is required, the artist adds a clear acetate overlay over the retouched photo, keying the overlay to the base photo with register marks. On the acetate, the artist places adhesive-backed arrows and nomenclature, to avoid tampering with the art surface and also to allow for any changes of linework later without risking damage to the art. For printing, the photograph is halftone screened, and the plate with this image is then "double burned" to include the line image of the overlay (Fig. 1-114).

Most skilled airbrush artists use many shortcut methods to do their work faster with fewer steps but without detracting from the end results. These include:

- Artists keep on hand a large set of clear acetate sheets having a wide variety of contours and holes cut in many angles and curvatures, to use as movable, easily positioned masks. An accomplished artist can use such masks in all but the most complex drawing areas as a substitute for cutting and cementing friskets.

- Nonadhering sheets (known as *Solotone*) are placed over photographs as overlays. These continuous-tone gray sheets with a variety of opacities may be used to create phantom effects, to deepen tones, or to add or subdue backgrounds to enhance contrast. These sheets can render many airbrushing operations unnecessary.

- To mask off small areas that are difficult to frisket, a liquid frisket may be applied with brush or pen. After airbrushing over the area, the liquid frisket is peeled off like a rubber sheet to expose the unsprayed zones.

LETTERING

In the applied arts, a picture must have the practical goal of communicating ideas to others, and this is done through graphic design, in which the artist arranges illustrations, white space, color areas, blocks of text, and headings to most effectively express a given message. As a result, the artist must know lettering and type intimately. The artist must use knowledge and imagination to select, manipulate, and in some instances create, letterforms in such a way that the ideas expressed in the words are somehow reinforced by the style of lettering or type used and by the way these fit into the overall design so as to add meaning, mood, and impact to the ideas. Lettering is used in every aspect of commercial and technical art: books, films, promotional and corporate literature, audiovisual presentations, displays, and advertising. Lettering can take many forms, and although today's designer can use preprinted letterforms such as typographic and dry-transfer alphabets, hand lettering is still a necessary discipline in many applications.

Freehand Lettering

In lettering by hand, there are two types, the freehand letter and the constructed letter. The chief method of freehand lettering used by the technical artist is showcard lettering, sometimes called "showcard writing" because the letters are drawn as fast as a person would write. A favorite tool for this is the red sable rigger brush, a round brush varying in size from No. 1, the smallest, to No. 12, the largest. Another favorite is the single-stroke red sable with a flat ferrule.

For freehand pen lettering, lettering pen tips come in four basic styles: square tips

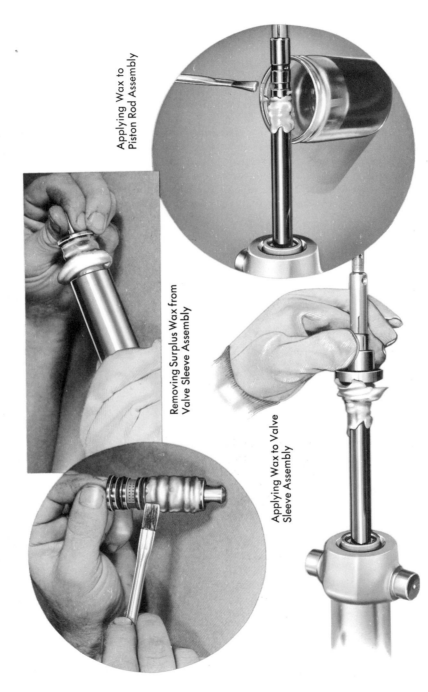

Applying Wax to
Piston Rod Assembly

Removing Surplus Wax from
Valve Sleeve Assembly

Applying Wax to Valve
Sleeve Assembly

FIG. 1-114 Photographs of parts-assembly procedures are airbrushed, with captions pasted on acetate overlay, for handbook reproduction. *(Rockwell International, Los Angeles Division.)*

for poster gothic and block letters; round for gothic and outline letters; flat for roman and italic shaded letters; and oval for bold roman and italic letters (Fig. 1-115). Also available are "steel brush" pens with split nibs up to ¾ in. wide for large display lettering. For lettering small letters in quantity, lettering fountain pen sets with interchangeable nibs can letter for long periods without frequent dipping into an ink bottle. Marking pens that come in a variety of tips are also an excellent tool for instant lettering.

Paper for showcard lettering comes in many surfaces and colors. Also available are showcard boards, which are thinner than illustration board and have a shiny surface for attractive presentation placards and displays.

Showcard paints, also known as poster paints, are tempera colors that come in jars. The temperas are stirred and diluted with water to a creamy consistency and then applied with a brush. When thinned more, tempera colors also will work in lettering pens, although pens are most commonly used with inks.

For freehand lettering, the letters should be sketched out lightly in pencil along guidelines on the surface so as to assure good spacing between letters and words, as well as proper size and layout, before starting. Spacing is very important; letters should look as if equal space exists between them, even though truly even spacing would create visual gaps or close-ups because of the different forms of the various letters. Letter spacing requires adjustments of all letters, and the eye is the only true guide.

Whereas gothic and roman letters such as A, E, and M are drawn with several strokes, the freehand script letters, similar to those used in handwriting, are often drawn with a single stroke. Scripts are usually best drawn with a flexible pen that makes hairline strokes with light pressure or wide strokes if the pen is pressed harder. For the smaller types of freehand lettering, which are often decorative in nature, with thick-and-thin letters that are often based on Old English or uncial-type construction

FIG. 1-115 Italic lettering with Speedball chisel-nib pen. *(Courtesy Prof. Samuel S. Anslyn, Glendale College, Glendale, Calif.)*

FIG. 1-116 A student learns lettering by practicing constructing Roman letters following a basic alphabet such as that found on the column of Hadrian, at left. Letters are outlined on tracing paper, traced, and inked. *(Courtesy Prof. Samuel S. Anslyn, Glendale College, Glendale, Calif.)*

FIG. 1-117 Constructed lettering, in the modern style of the Roman alphabet, is inked for use in advertising design. *(Courtesy Prof. Samuel S. Anslyn, Glendale College, Glendale, Calif.)*

(known as calligraphy), the artist can choose from a variety of wide-nib pen points to produce work for certificates, diplomas, posters, and charts.

Constructed Lettering

In recent years, phototypographic alphabets and other technology have been able to offer designers many alternatives to hand-rendered letter designs. Constructed lettering is now used mainly for unique typographic headings on book jackets, for corporate identity programs requiring special alphabets used to individualize advertising, and for new alphabets commissioned by phototypographic firms. Creative constructed lettering, in which the artist originates or varies letterforms, is usually designed from a detailed, precise pencil layout, followed by many fine strokes of a pen. Once the letter has been constructed in outline form, its hollows are inked in with a larger pen or a sable brush. Many alphabets that appear to be a spontaneously brushed script are constructed in this manner. Alphabets vary from those which closely follow classic letters (old style; Fig. 1-116) to modern adaptations (Fig. 1-117).

Mechanical Lettering

To create uniform appearance, size standards, and better readability, mechanical lettering, which is done using templates, may be used. There are two basic types of templates. One is a plastic guide with holes punched through the plastic for letters and numerals. With the guide aligned along a straightedge to a penciled guideline on the drawing, a technical fountain pen is inserted into the hole for each letter, tracing its contour, and then the guide is slid over to the next letter (Fig. 1-118).

The other type of template has letters indented on it instead of holes. The templates

FIG. 1-118 Typical hole lettering templates show a wide range of alphabet styles that can be used.

are made of laminated plastic with characters engraved onto the face to form guide grooves for a scribing tool. The scriber holds a reservoir-type pen; this remotely inks the letter on the paper as the tracing pin at the far end of the scriber is guided along the letter grooves in the template. Templates come in many typefaces, but the most commonly used face is simple gothic (Figs. 1-119 and 1-120).

,ABCDEFGHIJKL·MNO,QR

(%)'ABC 456789· abcdefghijklm

DEFGHIJKLMN

abcdefghijklmn 123456789

ABCDEFGHIJKLM

YZ&:;!?()''012

HIJKLMNOPQRSTUVWXYZ

&0123456789$

ABCDEFGHIJKLMNOPQRSTU

§§$¢0123456789

FIG. 1-119 Lettering using an indented alphabet template with a scriber (top), with some of the available lettering styles shown below.

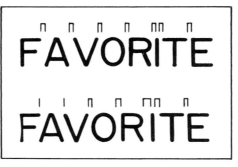

FIG. 1-120 Spacing between letters must be judged by eye—the top line is measured to be evenly spaced but looks badly spaced, while the lower line, using visual judgment, appears well spaced.

Dry-Transfer and Cutout Lettering

For preparing pages with blocks of body text, the artist can call for the copy to be set to specifications and furnished as proofs by linotype, phototypesetting, or, as in many in-house operations, reproduction typewriters. For headings and other large lettering, more creative aids than those set in the rigid format of cast metal type are available. The handiest and foremost of these aids are dry-transfer and cut-out lettering sheets imprinted on clear film. They are available in a wide selection of fine, reproduction-quality type styles in sizes from 6 point to 214 point and can be manipulated into endless combinations and variations to fit design requirements. The letters adhere to any dry, smooth surface. Dry-transfer letters may be superimposed on an illustration, a photograph, or a transparency to be used in overhead projection presentation.

To use dry-transfer letters, peel off the backing sheet. Draw guidelines on the art surface, and then position the letter sheet in relation to it to line up the letters. With a letter exactly overlaying the guideline, rub over the letter with a ballpoint pen or a burnisher. This bonds the letter; to make it stick permanently, put the backing sheet over the letter and burnish firmly. Letters may be lifted off with a piece of transparent tape or erased with a soft eraser.

Cutout letters have an adhesive backing; after peeling off the backing sheet, position the sheet with a letter where it should be placed, press down lightly to tack it in place, and trim the letter out with an art knife. The letter is then burnished to bond it firmly (Fig. 1-121). Letters, symbols, colors, and the like can be assembled to create a composite mechanical (Fig. 1-122a). Often a complex layout can be composed entirely of pressure-sensitive sheet material, with no inking required (Fig. 1-122b). This is an inexpensive, rapid art method for use with the most common industrial printing processes (Fig. 1-122c).

Photolettering

Another valuable aid to the designer is photolettering. An artist can now have on hand one of the many portable tabletop phototypesetters on the market, to speedily produce all kinds and sizes of headline copy. The machines work by casting a light beam, much like a photo enlarger does, through a negative image onto a strip of photo paper or film, which is then automatically processed to emerge as a sharp, clean letter proof. In many machines the letter images are on wheel-shaped masters rotated by finger to each letter, like a dial phone. The wheels are easily interchangeable, so an artist can keep a whole library of typefaces and type sizes at hand (Figs. 1-123 and 1-124).

For artists who must handle considerable text in the layout and production of camera-ready paste-up pages (known as *mechanicals*), having both lettering systems han-

FIG. 1-121   The letter combinations at the top illustrate the overlaps, off-line positioning, and mixtures of contrasting type styles that can occur using cutout and dry-transfer letters, mistakes that are impossible with other forms of typesetting. The words below show a few of the hundreds of type styles available in these sheets.

dily available is considered the best solution. Often, designs for advertising, technical, and institutional literature end up as finished mechanicals in which hand-lettering, dry-transfer or cutout letter headings, photolettering, and blocks of body text are all combined. Each type of lettering has its advantages for certain graphic situations, and each has its place in contemporary visual communication.

FIG. 1-122 (*a*) An artist applies dry-transfer lettering to artwork. *(Courtesy Prof. Samuel S. Anslyn, Glendale College, Glendale, Calif.)*

COLOR PLANNING AND USE

The choice between color and black and white, and the extent to which color is used in an art project, can have a great effect on the impact of the communication message and, depending on the medium in which the art is reproduced, on costs as well. Unlike fine arts painting, in which the artist uses color for its own sake, the graphic artist in commerce or industry employs color selectively, basing the applications of color on practical criteria. As used in technical or promotional graphics, color serves these purposes:

- **Catches the eye** Any art that promotes or advertises, from a sales folder to an industrial product display, uses color as a device to capture attention, to focus the eye on important items, and to hold viewer interest through the pleasing, attractive use of color.

- **Provides information** Modern catalogs must depict, as accurately as possible, color photographs of products so that customers can see exactly what colors, shades, and tones are used.

- **Serves as identification code** Color may be used in technical illustrations and diagrams to identify and differentiate components, systems, flow, and other elements. For example, a flow diagram of a jet engine might show red areas indicating the passage of hot gases, blue denoting cool airflow, green for oil, and so on.

The costliest form of color, for printing, is continuous-tone reproduction of a full-color painting. The printing is done by successively overprinting color separations, made by filming the illustration through color filters to produce separate screened plates of the primary colors. In the printing process, first a plate containing only the screened yellow portions of an illustration is printed. Over this the red plate is printed so that where the red and yellow halftone screen dots are juxtaposed the effect to the eye is orange. The blue plate follows; blue and yellow dots side by side produce a green optical effect, and blue next to red dots makes a purple color. Finally a gray plate, containing all the gray tones in the picture, is overprinted. In printing, microscopically accurate registry of the plates is necessary so that the tiny halftone-screen dots will fall into an exactly even pattern instead of overprinting each other to create a mottled look called *moiré*. The plates must also print at a precise angle to each other.

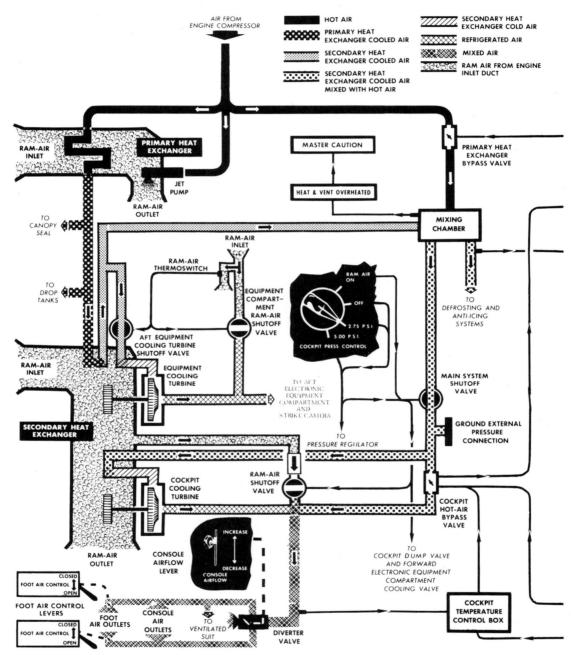

FIG. 1-122 (*b*) Flow diagram for black-and-white handbook uses a legend of patterns and screens applied from tapes and shading sheets to denote hot and cold air and other flow situations. (*Rockwell International, Los Angeles Division.*)

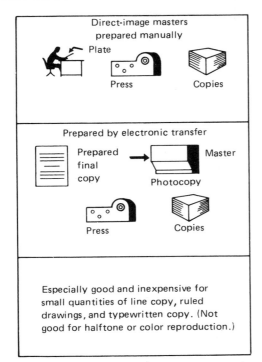

Direct-image masters
prepared manually

Plate

Press

Copies

Prepared by electronic transfer

Prepared
final
copy

Master

Photocopy

Press

Copies

Especially good and inexpensive for
small quantities of line copy, ruled
drawings, and typewritten copy. (Not
good for halftone or color reproduction.)

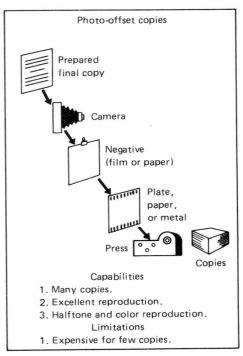

Photo-offset copies

Prepared
final copy

Camera

Negative
(film or paper)

Plate,
paper,
or metal

Press

Copies

Capabilities
1. Many copies.
2. Excellent reproduction.
3. Halftone and color reproduction.
Limitations
1. Expensive for few copies.

FIG. 1-122 (*c*) The most inexpensive reproduction processes, requiring the simplest art preparation, are direct-image copying and photo-offset lithography.

FIG. 1-123 An operator dials a circular font of a photolettering machine to produce strips of words on paper or film. The cabinet at right holds a selection of interchangeable fonts in many styles and type sizes. (*Drafting and Repro Digest.*)

The use of preseparated color in artwork is less expensive than producing full-color work, but more costly than black-and-white art because of the need for extra art, plates, and press operations. Each portion of a drawing that will appear in a color is drawn by the artist on a separate overlay. The overlay is made directly into a printing plate without need for the color-separation process (Fig. 1-125).

Both commercial and military producers of technical handbooks, manuals, and other nonpromotional literature do not use color because of the extra time and cost involved, with one exception—providing an effective code of visual identification. Even for this purpose the use of color, with its extra art, overlays, plates, and presswork, is

FIG. 1-124 Large block letters in photopaper strip, set up by the operator on this photolettering device, will be used to make a poster. *(Courtesy Prof. Samuel S. Anslyn, Glendale College, Glendale, Calif.; photo credit, Scott Erickson.)*

FIG. 1-125 Artists confer on placement of registered blockout film overlays to designate preseparated color areas on art pasteup. *(Courtesy Prof. Samuel S. Anslyn, Glendale College, Glendale, Calif.; photo credit, Scott Erickson.)*

frowned on except in unusual cases. In most instances black-and-white patterns, such as a variety of patterns of diagonal lines, dots, cross-hatched lines, and the like, are visually as distinct from one another as colors would be, so these textures are often used on drawings as a code in place of color. Hundreds of black-and-white texture and pattern adhesive-backed sheets and tapes are available and can be stored within the artist's reach, to be readily retrieved and used on drawings without the need for painstaking manual drafting (Fig. 1-122b).

Also, because these pattern sheets may be purchased in colors in addition to black and white, by careful planning they may be selected to serve a double purpose. An illustration may be rendered using color patterns which reproduce in black and white for printing purposes, but may also be photographed to produce a color slide for use in audiovisual presentation.

An added color tool for artists is 3M Color-Key, a graphic arts imaging material that enables the user to create numerous color variations on film and in opaque or transparent hues from a piece of black-and-white art. The film is exposed and processed in a portable contact printing frame, and the color films may then be used for packaging design mock-ups, film and video gels, 35-mm slide art, transparencies for overhead projection, overlays used to make color charts, and poster art. One of the films produces opaque white lettering for overlaying color or black-and-white photos.

As in the case of color slides, any use of color that does not involve printing costs little or nothing while adding immeasurably to graphic impact and communication value. All artwork prepared for direct use as display charts, posters, and other graphics for viewing at meetings, lectures, training and engineering discussions, and the like should contain as much color as possible. And in practically every audiovisual presentation color is mandatory; a projection screen, unlike the page of a book, reflects white light which is painful to the eye, so that color tones convey a much more acceptable message to audiences than do black-and-white images.

If there is no color in the photographs or basic artwork to be used in slides, color may be added through colored titles, colored-paper tint blocks, and other paste-on devices. In the case of black-and-white original art, slides may be assembled with colored gelatin films (gels) sandwiched into the slide frame or reproduced onto colored, rather than black-and-white, diazo film. There are many ways to introduce color into projected film presentations, all of which can help make the message a successful one at low cost. These will be discussed in more detail in Section 2.

PREPARING ART FOR PRINTING

Promotional and technical literature such as sales brochures, advertisements, booklets, and the like call for three stages in assembling camera-ready art, or mechanicals, that can be photographed, made into printing plates, and printed by one of the printing processes. These stages are:

- The preliminary layout (rough).
- The finished layout (comprehensive layout, or "comp").
- The paste-up (or final) dummy; or camera-ready art.

The artist takes the written copy and, following instructions, successively creates the roughs, a final layout, and finally the camera-ready copy, which includes every element of final art and text for the printer. Editorial and management approval is, of course, required at each step.

Simple Technical Publications

Many publications, such as internally distributed reports, studies, and standards books, as well as parts catalogs and repair and instruction books used by mechanics, shop people, and the like, are prepared as simply and inexpensively as possible (Fig. 1-122c). Often such literature is created using typewritten text as original reproduction copy rather than typeset material, often with special reproduction typewriters using typefaces closely simulating standard typeset alphabets. However, many in-house publications facilities produce work of high typographic quality on automated phototypesetting systems.

Whatever repro typing system is used, text pages lacking artwork are often typed directly onto special plastic or paper offset plates, while pages that include artwork, photos, tabular matter, or special headings, as well as those which may be subject to future alterations, are pasted up on preprinted forms. Because typewriter text is slightly larger than most typeset text, paste-up pages are often made to 1¼ to 1½ times the size of the standard 8½- × 11-in. page used in publications. Reduction of the over-size paste-up page to standard size improves the appearance of artwork and text, reduces the book's page content, and often erases small blemishes that might appear if same-size original copy were used.

In the simpler publications, the typist follows instructions from the writer on how much space to leave for an illustration and types a figure caption below the space. On the page, the illustration is then pasted in place. If the art copy is oversized, or if it is continuous-tone art or a photograph, a black-line box (keyline) is inked in the space, or a piece of black paper or masking film is cut to size and affixed in the space. The art is then marked with proportionate crop marks and dimensioned in inches to the proper scale to fit the paste-up. The artist keys the art to the page with a letter or page number designation. The art and the paste-up page become one in the production-negative stage of printing, when the stripper at the printer's assembles both negatives together for platemaking.

The Preliminary Layout

The first rough layouts are prepared as small sketches in pencil, showing possible arrangements of text, pictures, and headings. There is no attempt at detail—picture elements are shown as simple masses, and text is indicated by blocks or a few parallel lines with no attempt at straightness or accuracy—and thus the whole aim is to try out every possible placement of copy to find the best one for the given purpose. The rough layout selected from these experiments should meet these aims:

- Include all copy elements.
- Show the graphics to the scale that will best reproduce and will depict with strongest visual impact the ideas most important in the publication.
- Meet standards of good composition, spacing, and readability.
- Allow enough "air," or white space, around copy elements to provide room for changes, additions, and deletions in the final art stage.
- Characterize the general concept or character of the publication.
- Read from the same direction as the text without turning the book (Fig. 1-126).

The Finished Layout

The final layout, or comprehensive, prepared for approval by management, the editor, or client, should fairly closely resemble in scale and appearance what the final printed result will be. The layout is usually drawn on layout bond paper with ink, layout chalks and pastels, colored pencils, and marker pens. Headlines and lettering closely simulate the intended printed art and may be sketched in using pen and ink or created from dry-transfer or cut-out pressure-sensitive lettering sheets applied to the layout. Illustrations may be sketched in, or if the photos and art renderings to be used in the work are already available, photostats or other reduced-scale copies may be pasted on the layout for greater authenticity. Text blacks are indicated with ruled lines, or if the client seeks a really exact facsimile of the finished product, as is often the case for advertising, package designs, etc., fake text called "greeking," available in clear pressure-sensitive sheets, is trimmed and bonded to the layout.

The Paste-Up Dummy

For many publications such as technical manuals, magazines, handbooks, and the like, it may be advantageous to have the printer assemble pages from negatives by piecing these negatives together, or "stripping," and to print from those, rather than to prepare camera-ready art for each page. If the printer assembles the elements for page make-up, the paste-up dummy serves as the guide. The paste-up dummy is a double-page

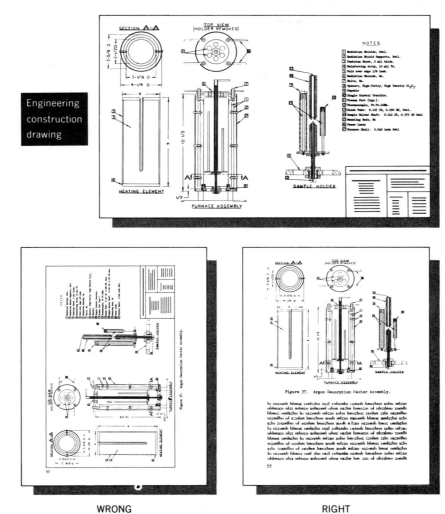

WRONG   RIGHT

FIG. 1-126 Engineering drawings, schematics, and other art should be laid out to read without turning the book sideways, even if it requires layout across extra pages. *(NASA SP-7008.)*

form, usually the same size as the page to be printed, and preprinted with blue lines defining borders, margins, text column areas, register marks, allowance for illustrations to "bleed" off the page, and other repetitive elements. Around the page borders, 2 in. or more of blank paper allows room for instructions and notes to the printer.

On the dummy form, the editor or production artist assembles text, photos, headings, etc. by cutting these from page proofs, photostats, and other finished copy and pasting them in position. Photos and artwork are indicated by pasted-on pieces of paper or pencil outlines with letters and numbers keyed to original art, which is scaled and marked to correspond.

The paste-up dummy of the publication, along with the separate pieces of artwork, is not itself used in platemaking, but serves as an exact guide to the printer, who follows dummy instructions in assembling negatives of all final approved copy elements into a complete unit for printing.

Camera-Ready Art

The other way of preparing final page art, one that ensures closer control and allows for subtle adjustments and improvements in the paste-up, is to assemble art copy that can be directly shot by the copy camera and printed with a minimum of makeup by

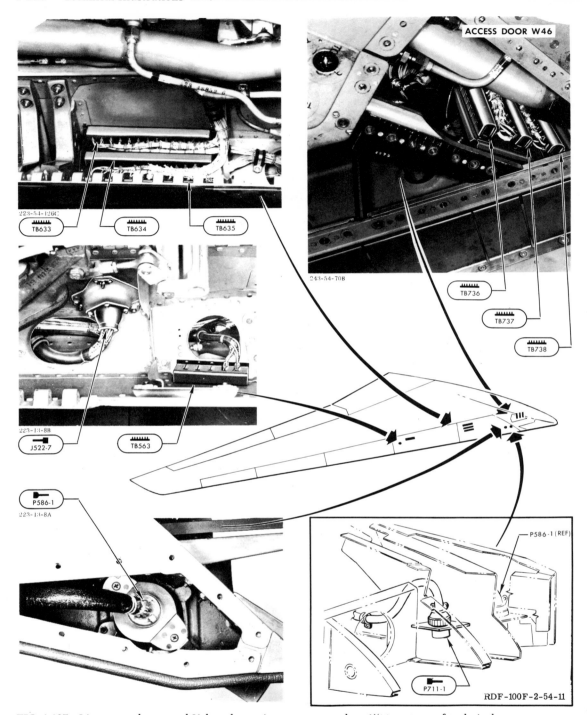

FIG. 1-127   Line art and screened Velox photoprints are mounted on 1½× pasteup of technical manual page. *(Rockwell International, Los Angeles Division.)*

the printer. This is called *camera-ready art* or *mechanicals*. The headings, text matter, and simpler small line illustrations are pasted, usually same-size, onto an illustration board with rubber cement, having been cut and trimmed to fit with an art knife from text proofs, photolettering, etc. Linework and areas to be imprinted in a flat color are drawn or laid out in red translucent blockout film on a separate overlay sheet of tracing paper or acetate registered to the base art (Fig. 1-125).

Most original art and photographs are not included on mechanicals. Instead, they are indicated as ink outlines, or keylines, or as blocks of black paper, red masking, or blockout self-adhesive film trimmed to proper scale and keyed by page numbers and letters to the originals. Line and halftone-screened negatives photocopied from original art or photographs are stripped by the printer into the page negatives prior to plate-making. In many catalogs and manuals, however, it is common practice to paste both reduced copies of line art and halftone-screened Velox positive photocopies directly onto the mechanicals (Fig. 1-127).

Publication Art Guidelines

Some recommended practices in laying out and pasting up publication art are as follows:

1. Whenever possible, try to lay out publication pages for repair manuals and other technical books so that each illustration is as closely adjacent to the text describing it as possible so that it can be studied without turning pages. This is a great help to mechanics, assemblers, and others whose work involves simultaneous reference to both text and visual aids.

2. Avoid placing pictures sideways on a page. They should always read the same way as the text, even if it becomes necessary to add a page to accommodate an overlong drawing or to otherwise complicate the layout.

3. There are several art aids, all more accurate and much faster than an artist's own mathematical calculations, for figuring the proportional sizes in reducing or enlarging art copy for paste-ups (Fig. 1-128). Several scaling devices, such as the Scaleograph, have grids or increments that give exact dimensions. Sizing art may be done very simply by ruling a corner-to-corner diagonal on a tracing paper overlaying an

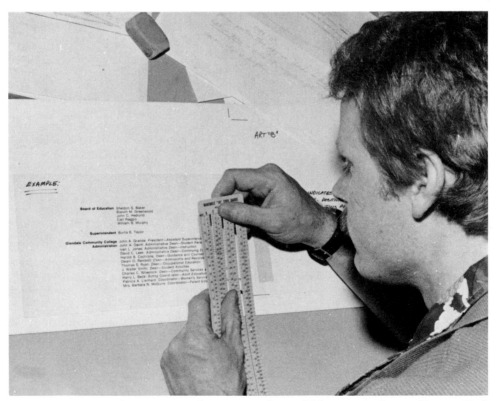

FIG. 1-128 A type scale saves time in fitting type to the available space in a layout. *(Courtesy Prof. Samuel S. Anslyn, Glendale College, Glendale, Calif.; photo credit, Scott Baird.)*

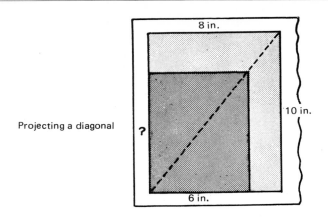

Projecting a diagonal

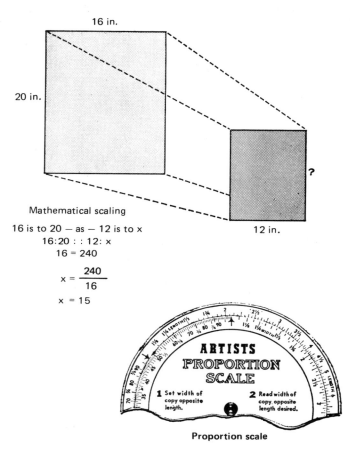

16 in.

20 in.

Mathematical scaling

16 is to 20 — as — 12 is to x

16:20 : : 12: x

16 = 240

$$x = \frac{240}{16}$$

x = 15

12 in.

ARTISTS PROPORTION SCALE

1 Set width of copy opposite length.

2 Read width of copy opposite length desired.

Proportion scale

FIG. 1-129 Projecting a diagonal, mathematical scaling, and a proportion wheel scale are three shortcuts to proportioning art copy. *(IAM magazine.)*

original, then transferring the tracing to the dummy and pin-pricking along the diagonal to the required width to fit the paste-up or dummy (Fig. 1-129).

An additional tool for mathematically minded users is the electronic calculator. Irregular shapes and vignettes that do not lend themselves to conventional proportioning methods may be exactly delineated to a desired scale in the paste-up by using an optical mirror device, the camera lucida, or by pasting a vandyke print or photostat in the proper place and marking the print *for position only* as a precaution.

4. Use a grease crayon or a litho pencil to indicate sizes and other instructions on all photographs, artwork photocopied onto glossy print paper, plastic overlays, or films. The crayon comes off easily if changes must be made.

5. In giving a photo retoucher, an artist, or an engraver instructions for altering original art (silhouetting, portions dropped out, lettering surprints, etc.), draw outlines and instructions in light blue pencil on a tissue flap over the drawing.

6. In artwork or photos showing people, have them facing into the page, not looking out of it. If necessary, flop the art, unless it includes text matter or other items that would read wrong in reverse.

7. Finished publication art represents a costly investment, and must be secured against becoming damaged or dirty in storage, retrieval, or transit. Artwork and paste-ups must be mounted on stiff illustration board, and then covered with a tissue paper (for corrections) flap directly over the base art and an outer flap of brown kraft paper. Each drawing, overlay, and outer flap must be clearly, uniformly marked with titles, control numbers, etc. (Figs. 1-130 and 1-131).

8. Leave ample white space around headings, special notes and warnings, and all other elements that may have to be corrected or added to in future issues.

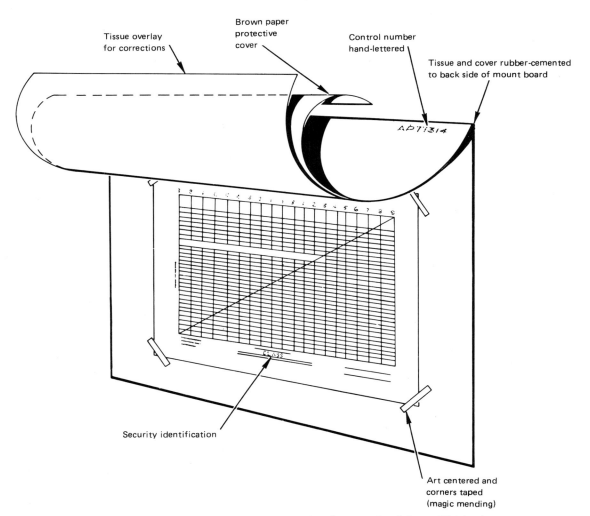

FIG. 1-130 Mounting and flapping camera-ready art. *(Rockwell International, Space Division.)*

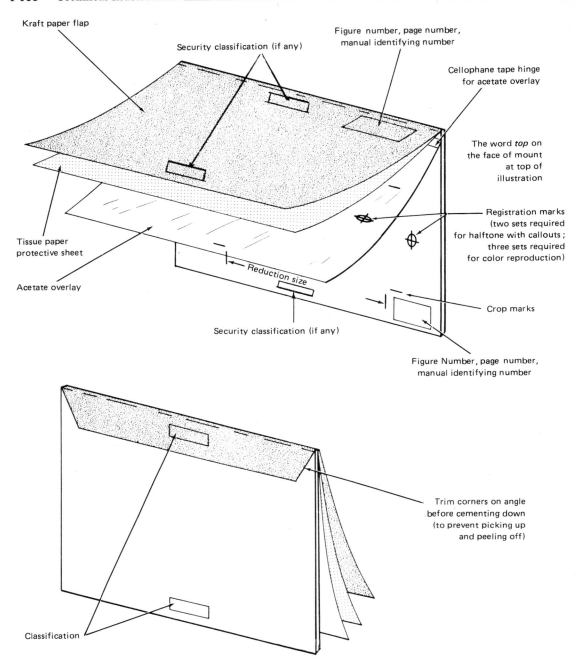

FIG. 1-131 Identifying, marking, and covering artwork including overlays to military specifications. *(MIL-M-38784.)*

9. Be sure to use only clean, sharp original art, copies, or photographs; avoid using:

   a. Small-sized photos. Order a 5- × 7-in. or 8- × 10-in. photo, and if the detail is fuzzy, retouch the photo.

   b. Second- or third-generation diazo prints or photocopies of photocopies.

   c. Preprinted orange-and-green–line engineering graphs. Retrace the main grid increments in ink, eliminating the fine lines in between major divisions as well as all but the essential plots and curves.

   d. Pencil-on-vellum tracings, which are poor original copy and are likely to have broken lines or missing detail in the print. Have a pencil tracing reproduced on wash-off film or photo paper that can be retouched to restore missing lines, and use this print as original copy for publication.

e. Text for paste-up copy that has been typed with a cloth typewriter ribbon. Repro typing should be done only with special plastic or carbon ribbons made for that purpose.

f. Full-color charts or illustrations for originals in black-and-white reproduction. Blues may drop out, yellows and reds will register as solid black, and other distortions may be produced by the camera.

Photo Handling and Scaling for Printing

It is easier to handle and mark photographs when they have been mounted on heavy white mounting board or illustration board, allowing a margin of at least one inch of white board outside all edges of the photo for crop, bleed, and size markings and other instructions. Rubber cement or paste may be used, with paste often preferred for heavy photos with a pronounced curl that rubber cement will not hold. The best procedure is dry-mounting, in which a resinous dry-mount tissue is sandwiched between the back of the photo and the mounting board. This sandwich is then inserted in a dry-mounting press for a few seconds, where clamped tightly under heat and pressure, the tissue adhesive melts and bonds the photo to the board.

Caution: Before mounting, make sure to look on the back side of the photo and copy off print or negative numbers, credits, identification, and other data onto the back of the mounting board.

The following precautions should be followed carefully in handling photographs:

- Do not write or sketch with pencil, or make scaling layouts, on tracing paper directly over the photograph, unless using a very soft pencil, such as a 4B, which will not indent the emulsion surface of the photo beneath. The photocopy camera easily picks up indentations; for the same reason, never use paper clips on photos.

- Cover all mounted photos or art with a paper-flap overlay. Retouched photos should be protected additionally by an inside flap of glazed paper.

- Photos should not be handled or mailed in a face-to-face position, or rolled unless the print surface is on the outside.

- Keep waxy or oily substances away from photos, and avoid leaving fingerprints on the emulsion surface. On an unretouched photo, finger touch produces oily spots that repel retouching solutions, inks, and paints. On a retouched print, fingerprints can scratch and spot retouched areas—especially airbrushed areas consisting of a delicate film of paint.

- Transport or mail photos, even if they are just going across the street, sandwiched between heavyweight cardboard.

When scaling and marking a number of photos for the lithographic camera operator or the engraver, look the photos over with a view toward standardizing as much as possible on size reductions. By enabling the copy camera operator to position *groups* of photos on the copyboard for a specified reduction, rather than taking a succession of individual shots, much time and money can be saved. Figure photo reductions in percentages (easily done using a proportional circular scale). Thus, for example, instead of marking photos, say, 72, 78, 84, 87, and 90 percent, mark them all a flat 80 percent, and make whatever slight adjustments are necessary to the working layout.

When designing technical literature using many photos from diverse sources, the best recommended practice is to group the pictures by types that will best suit a given copy camera exposure—for example, all sharp, glossy prints in one group, all soft, matte prints in another group, and all high-contrast prints in yet another. Better reproductions will result, and the photo technician will not have to experiment many times with different camera settings to achieve copy results that represent the optimum compromise. In the case of groupings, it is also advisable to trim mounted photos and art as closely as possible so as to make room for a maximum number on the camera copyboard for each exposure.

Section 2

Technical and Audiovisual Presentations

George A. Magnan

Audiovisual presentations of every kind are the single most-used tool for commercial and industrial communications today. They take such forms as engineering briefings, interplant and interdepartmental information sessions, financial meetings of directors and top executives, orientation lectures, technical training, and new-product development proposals. Audiovisual presentations are used in every aspect of advertising and sales promotion, including filmstrip shows at trade expositions, sales demonstrations, presentations of proposed advertising campaigns by ad agencies to clients, and educational slide shows on rear-view projectors by doctors and other specialists explaining the nature of their services.

An audiovisual presentation may range from a dentist's slide show to a patient showing dental techniques, to an intimate round-table discussion among a few managers, to an elaborate color slide or motion picture presentation to many hundreds of people in an auditorium. A great many methods of visual presentation including every art technique, plus special ones devised for film art, as well as a broad variety of projection, screen, and control equipment, are available to match the diversity of audiovisual applications. A top-quality slide presentation is probably the most sophisticated of these and may involve advanced graphic arts film processing, animation, or special cameras to prepare the slides, and can create excitement and the illusion of motion through complex multiscreen imagery with computerized sequencing.

PLANNING TEXT AND FORMAT

The proper audiovisual medium must be chosen from among the many types and variations available, and this choice is largely dictated by such factors as: the number of people who will be present; the distance between the speaker, the visuals, and the farthest viewer; the nature and size of the conference room or hall to be used; and the facilities, such as controlled lighting, lectern, speaker system, and availability of electrical receptacles.

In most large companies, because the preparation and delivery of audiovisual presentations is a highly important and extensive activity, presentation coordinators and specialists are available to help the presenter. Sometimes this aid can take the form of assisting, through professional writing, in verbal presentation. The presentations specialist can also help analyze the best way to present the program and will advise on how to prepare the visuals and set up a schedule. Such a specialist is familiar with the equipment available, the time it will take to prepare visual material, and the most effective presentation that will fit the allotted budget. The specialist also attends to such practical details as arranging for the hall or conference room, obtaining and transporting equipment, controlling the lighting during the presentation, and notifying participants.

If such assistance is not available, the audiovisual coordinator will need to make arrangements; plan and schedule the presentation; and contact the artist, photographer, and others involved in preparing the visuals well in advance. This is necessary in order to allow time for art preparation and cycles of processing time in photo or reproduction facilities. Adequate time must be taken to plan the nature and sequence of visuals to most effectively express the key ideas in the presentation. These key ideas, the salient points of the verbal presentation, should be reduced to brief phrases; as appearing on visuals, the message is more dynamic when each line of lettering begins with a verb, as in "take immediate action," (*active* voice) rather than a noun, as in "action should be taken immediately" (*passive* voice).

These ideas, roughed out in outline form on paper to plan the visuals, should pace the message and condense the story to be told. Try to confine each visual to a simple message with as few words as possible and a single diagram or illustration; if lengthy sentences and subcategories must be included, it is better to break them into two or more sequential visuals. Similarly, illustration matter should be confined to expressing no more than one easily grasped idea, so that a complicated chart or illustration, for example, must be broken down into a series of thoughts and sectional views in order to avoid presenting one visual for an exceptionally long time. The average chart should be read and absorbed in 30 to 60 seconds; any longer than that will lose the attention of the audience and break the flow of the speech.

The best presentation gets its point across quickly and graphically. A few charts that emphasize the message are more effective than many charts. Resist the temptation to reinforce the main message with an abundance of statistical proof presented in graphic chart or tabular form; for every viewer who becomes more convinced, ten may be confused. Effort should instead be directed toward screening from a mass of statistical and factual material those facts which are most significant and need to be brought out to give a clear and realistic picture. To translate this picture into relationships that viewers can readily grasp and that managers can use in discussion and as a basis for decision making is the whole purpose of visual communication. In any formal visual presentation this responsibility is one that should be shared jointly between the presenter and the illustrator. The illustrator can create effective visuals only from a well-written script that organizes the talk so that facts, ideas, and concepts will be presented in an orderly sequence. Artistic visual aids will not substitute for a planned presentation; one cannot take a group of visual aids and write a script around them. The charts, and the decision whether many, a few, or none at all should be used, should be made only after study of a well-prepared script and preparation of an outline that determines what needs to be illustrated.

For the presenter planning visual aids, here are the steps to be followed:

1. Locate the story in the documentation.

2. Make the story as simple, understandable, interesting, and convincing as possible.

3. Visualize the high spots so that trends and relationships can be grasped quickly and accurately.

4. Discard all nonessential details—trim out lists of topics, subtitles, nonfunctioning cartoons and diagrams, and decorative illustration.

5. Plan to explain, add to, or build around visual aids—never read them to the audience.

Scheduling

Scheduling all elements of a presentation along a time span to a firm completion date is the best way to ensure that the presentation will be properly prepared to suit the need. Through the schedule, each person or group involved will be able to do a good job without extra overtime costs or the costly revisions that always result from a hurried operation. Also, there will be sufficient time to review the finished material before the presentation date.

A checklist of the main steps in scheduling should include:

1. A rough draft of the proposed presentation is the basis for noting the key points; sketch the key points on standard 5- × 8-in. index cards or on preprinted planning cards in rough pencil form, and note by number the place of each in the whole presentation.

2. A list of sources of services and data, such as art, photo records, library work, etc., with an estimate of the time each will require.

3. A work schedule breakdown of the time needed for each operation such as art, photography, and reproduction.

4. Notices to individuals and groups as soon as possible, enabling them to plan ahead and readjust their work loads.

5. Follow up on each phase as work progresses, since one delay can back up other operations and endanger the deadline.

6. Reviewing the finished presentation, allowing sufficient time for changes and improvements.

7. Rechecking the time and location of the presentation to make sure no changes have occurred.

Instructing the Artist

About the same time the draft of text for the presentation has been done by the presenter, but before polishing it in final form, the presentation artist should be consulted and the project explained in full detail, leaving the artist time to plan ahead. The presenter should work with the artist as follows:

1. Provide the illustrator with all the reference visuals that are available—blueprints, maps, drawings, sketches, clippings, photos, etc.—and with information as to where to locate missing elements.

2. From the rough layouts, check the labeling that is to appear on charts or slides, bearing in mind the need for message clarity, the benefits of limiting words to as few as possible, and the need for large lettering so the audience can read visual text. Captions should be as short as possible, and each visual should express only one major idea or point.

Although all presentation visuals, to be effective, must be simplified and exaggerated, this holds especially true for lettering and colors of projectior materials. Everything shown must be readable to the person in the last row. The message must be brief, the facts condensed. Because a projection screen cannot be turned around, as one would a book, no titles, labels, or captions can be angled or placed in a vertical position. Every slide or overhead transparency must have a horizontal image in a rectangular shape.

Projection films must be designed for almost instant impression, because they flash on the screen in fairly rapid sequence. For better readability, titles and captions should be prepared in lettering of a maximum size in the available space. "White space," which may be a valuable design element on a publication page, is glaring and undesirable in a projected visual.

3. Review the layouts as they are completed, making sure that the ideas are presented in the correct order by reviewing numbered layouts if dealing with a short presentation, or on a storyboard (which will be explained later) if a slide sequence is the medium.

4. Once all the details of the layouts have been approved, use the time allotted for producing finished art to polish the script, and try not to interrupt the artist until all the final art is complete.

The Dry Run

Whatever visual aids are to be used, when all are finished, the presenter "dry runs" the material at least once (Fig. 2-1). At this time visual aids are keyed to proper placement in the script and the speaking time is clocked. If flip charts or display boards are used, make sure that a proper easel is available and in place, adjusted to the right height for good audience visibility, and sufficiently illuminated. A pointer should be placed on or near the easel.

When projected visuals are used, the projectionist is provided with a marked copy of the briefing and rehearses the cues and signals from the speaker.

At the dry run, the illustrator should be invited to take a critical view of the material in use for lettering size, color effect, and readability of illustration details at a distance before making any further adjustments to improve visual impact.

The lighting of the room should be adequate for good visibility, but not overbright or distracting. One person should be given responsibility for controlling the lighting, from full illumination for talks and floor discussions to subdued lighting during film projection. That person should become well acquainted with the locations and functions of switches. Other practical and important considerations are controlling other distractions, such as people moving around, microphones, cords, and equipment. Noise from projection equipment must be kept to a minimum.

Positioning a Projection Screen

One aspect of previewing a presentation is often neglected: determining the proper placement for the projection screen during the dry run, and ensuring that the screen is located there during the final presentation. Too often a screen may be set up anywhere at the last minute without regard to audience needs. Make sure the screen is high enough so that the audience has an unobstructed view. Where the floor is without a slope toward the lectern and no platform is available, the bottom of the screen should be about 4½ feet from the floor. Do not raise the screen too high, however, since this creates audience discomfort. The screen should be tilted inward slightly toward the audience at the top, at right angles to the ray of light cast by the projector, and generally perpendicular to the viewer's line of vision. This adjustment prevents distortion of the image, or "keystoning."

Accepted standards of distance between screen and audience related to screen size are:

1. Distance to the closest viewer: Two times the width of the screen.
2. Distance to the farthest viewer: Six times the width of the screen.

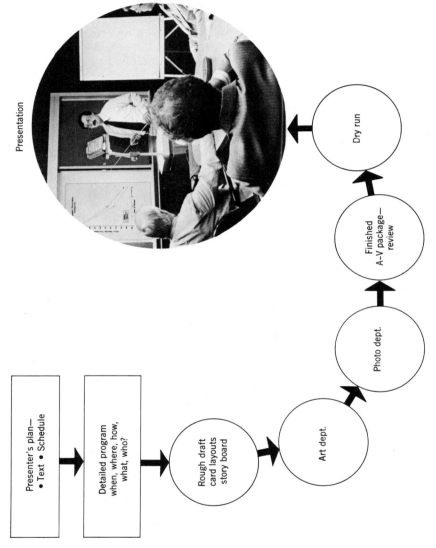

Presentation

Presenter's plan—
• Text • Schedule

Detailed program
when, where, how,
what, who?

Rough draft
card layouts
story board

Art dept.

Photo dept.

Finished
A–V package—
review

Dry run

FIG. 2-1 Steps in a presentation.

3. Widest angle of view:

  a. Beaded screen: 22° from the centerline of projection.

  b. Matte-finish screen: 30° from the centerline of projection.

  c. Lenticular screen: 40° to 50° from the centerline of projection (Fig. 2-2).

There are two general types of projection screens available: front projection screens, with opaque surfaces, and rear projection screens, which have translucent surfaces. Front projection screens come in beaded, matte, and lenticular surfaces. (The lenticular screen is made up of tiny lens elements that confine the reflection or transmission of light and permit wide-angle viewing.)

If a screen is matte on one side and polished on the other, turning the polished surface to face the viewer will provide good contrast but will reflect room light. This will require complete darkness in the room. If the matte side is used, facing the audience, there will be slightly less contrast in the picture but viewing can be done in subdued room light, which is generally preferable to darkness.

Location of the Projector

The aim of the presentation planner should be to so arrange the projection system in relation to the screen and audience that the presence of the operator, as well as the projector with its fan noise and light flickers, do not distract viewers (Fig. 2-3). In large

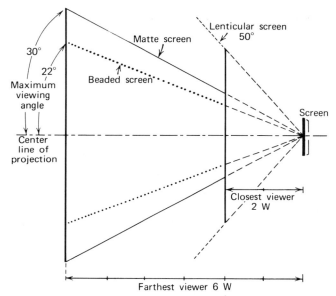

FIG. 2-2 Audience distance from screen.

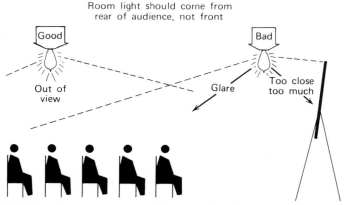

FIG. 2-3 Placement of room lighting—a key factor.

halls accommodating hundreds of people this may be done with a very large screen and a projector located at the rear, behind the audience, in a booth; most presentations, however, are to relatively smaller groups, and the projection system is in the worst location—in the center aisle, toward the middle of the seated audience.

There is one alternative, however, if the room layout permits it: the best way to keep the mechanics of a presentation out of sight of the audience is to put the projector *behind* the screen. This hides the machine and operator and reduces projector noise and other distractions.

For rear projection, a translucent screen must be used so that the image will go through the screen, registering on the side facing the audience. Slides must be reverse-mounted in the projector to enable right-reading from the front. Portable, adjustable translucent screens are available in a number of sizes.

For guidance in locating the screen and projector in relation to the audience, the following table is offered.

Distances (from Projector to Screen), Based on Width (W) of Projected Image in Feet

16-mm movie, 2-in. focal length, 0.380-in. aperture	$5.25 \times W$
35-mm single-frame film strip, 5-in. focal length, 0.906-in. aperture	$5.5 \times W$
35-mm double-frame film strip, 5-in. focal length, 1.34-in. aperture	$3.75 \times W$
2- $\times$ 2-in. slide, 5-in. focal length, 1.5-in. aperture	$3.35 \times W$
2¼- $\times$ 2¼-in. slide, 7-in. focal length, 1.75-in. aperture	$4 \times W$
2¾- $\times$ 2¾-in. slide, 7-in. focal length, 2.25-in. aperture	$2.25 \times W$
3¼- $\times$ 4-in. slide, 12-in. focal length, 3-in. aperture	$4.15 \times W$
5- $\times$ 5-in. overhead projector, 12.5-in. focal length, 4½-in. aperture	$1.45 \times W$
7- $\times$ 7-in. overhead projector, 12.5-in. focal length, 6½-in. aperture	$1.925 \times W$
10- $\times$ 10-in. overhead projector, 14-in. focal length	
10-in. aperture	$1.4 \times W$
9½-in. aperture	1.425
9-in. aperture	$1.555 \times W$
Optimum distance to farthest viewer	$6 \times W$
Optimum distance to nearest viewer	$2 \times W$

DIRECT PRESENTATIONS

In choosing the means to be used for tying in the verbal address with a visual message, the presenter's first decision must be whether to use a direct presentation in which the graphic displays themselves tell the story—such as boards, flip charts, or posters—or a projected presentation in which artwork is reproduced onto films or transparencies that project onto a screen. For many purposes presenters prefer the first because it is simple. The presenter can usually handle charts personally during the course of an address, or erase and redraw visual material if using a chalkboard, and is thus not dependent on processing of films, the availability of projectors and operators, or other factors that could go wrong. Artwork can be changed and used immediately without the added cycle of reshooting and slide making. Also, for one-shot, low-budget presentations, direct chart making is much faster and more economical than projected visuals.

Boards

The simplest device for combining visual aids with a talk is the classroom-type chalkboard or blackboard (Fig. 2-4). The chalkboard is limited to uncomplicated drawings and diagrams which do not require precise delineation; colored and fluorescent chalks can be used to add to the effect. The drawings may be quickly erased and new ones easily substituted; trial-and-error diagrams and ideas from the classroom or audience can be jotted down swiftly; and chalkboard's greatest asset is offering a two-way vehicle for feedback and group participation without the need for special equipment. Although primarily a classroom tool, the chalkboard is often preferred by engineers in industry for their technical presentations. One uncommon use is in the styling studios

FIG. 2-4   The simplest form of presentation is the chalkboard.

of major car manufacturers, where full-size renderings of cars are prepared by illustrators using pastels and other erasable materials in proposing new designs during management sessions reviewing new models for acceptance.

Two visual aids often used for trade-show displays and for technical training programs are flock boards (flannel boards) and magnetic boards. The flock board has a coating of fuzzy felt or flannel, or of a special plastic fiber (Velcro), which holds placards and lightweight three-dimensional items, models, etc., on the board. Key words, names, small drawings, etc., are drawn on a cardboard strip with the appropriate backing, which sticks to the flock board as the lecturer puts each strip on in succession. The speaker may go back to previous cards when the need arises (Fig. 2-5a and b). Similarly, magnetic strips may be stuck onto the surface of a metallic board.

Flip Charts

Flip charts, often called flip-over charts, are in widespread use for almost every type of presentation. Flip charts are prepared on flexible papers that can be turned just as one would flip the sheets on a large pad of paper. The charts are prepared in many sizes ranging from 11 × 17 to 30 × 40 in., according to the audience size and viewing distance involved. For flip-chart presentations that will be shown only once or twice, the charts are usually prepared on ordinary sheets of drawing paper, but for purposes where they will be used many times over a long period of time, a more durable material is called for, such as map cloth or one of the several synthetic papers now available. Both materials resist tearing, are easily cleaned, and will last indefinitely.

For small roundtable discussions, the charts are made from 11 × 17 to 18 × 24 in. in size and are usually bound in a hardcover ring-type binder supported by a tabletop easel. The speaker flips each chart vertically from the front of the book, up over the ring binder, and to the rear while discussing each successive idea depicted.

Larger flip charts are usually attached by means of a clamp to a special floor easel, and as with the smaller units, the speaker flips each chart over while proceeding through the presentation (Fig. 2-6). Easels, paper pads, clamps, and so forth are commercially available, but many large companies prefer to have portable kits of flip-chart components that are designed and made to their own special requirements for use by their traveling engineers, executives, salespeople, etc.

Flip charts are simple to use and more quickly produced than are other visual aids. Lettering may be applied with crayons, felt pens, a Speedball pen, or one-stroke brush methods (Figs. 2-7 and 2-8). Illustrations can be either drawn directly on the paper, or added by splicing art, photographs, printed matter, etc., into a window cut in the chart and taped in place from the back.

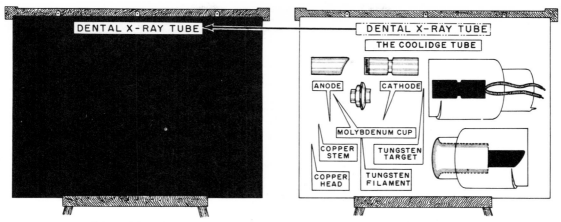

FIG. 2-5a   Parts of flannel-board graphics.

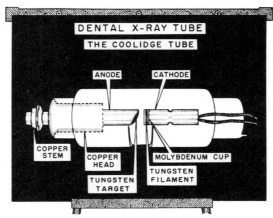

FIG. 2-5b   Parts in place to make complete flannel-board chart.

FIG. 2-6 Flip charts mounted in a three-ring binder are placed on an easel for floor demonstration. *(Northrop Norair.)*

OVERSEAS TEAM

CONTRACTOR WOULD PROVIDE TEAM
- SUFFICIENT IN NUMBER
- APPROPRIATELY SKILLED

TO ACCOMPLISH:
- GSU LEVEL REPAIR
- MODIFICATION
- SUPPLEMENTAL TECHNICAL ASSISTANCE

FIG. 2-7   A quick, simple flip chart done with one-stroke brush lettering.

FIG. 2-8   An artist letters a flip chart with brush and poster paint.

Flip charts offer a ready solution to demands posed by the so-called one-shot presentation, which consists of inexpensive, quickly-prepared visuals needed in a hurry and to be used for only one occasion. When needed immediately, flip charts can be done on an assembly-line basis, with one artist producing lettering on colored paper strips; another making illustrations; and another cutting, combining, and taping elements together to provide complete charts according to a master layout.

Drawings of machines or other objects that are available as tracings on paper or drafting film are handy aids; blue- or black-line diazo prints can be run off, and eradicators can be used to remove unwanted linework from the print. The artist then directly applies color with crayons, brush and ink, or felt pens to the print and adds headings and other text matter with Speedball or other wide-nib pens or with one-stroke brushes. Heavy outlines and details may be traced with an india ink technical pen to bring them out. Mistakes and changes on a flip chart are handled by drawing a new version on a paper, placing it over the flip chart exactly in position, and then cutting a block through both papers with a razor blade or art knife. The new item is then inserted and taped into position from the back of the chart. Photoenlarged copies also are useful time-savers (Fig. 2-9).

For flip charts that will be handled repeatedly over a long time, the tougher map cloths and plastic-coated papers are recommended. To show a sequence, such as assembly of a unit, on a single chart, a progressive series of transparencies, each showing a new step in transparent color, may be bonded to the larger chart, also in flip-over format.

For transporting, most flip charts are rolled up in a heavy mailing tube or some other cylinder; small flip-chart sets are kept in loose-leaf presentation binders (Fig. 2-10).

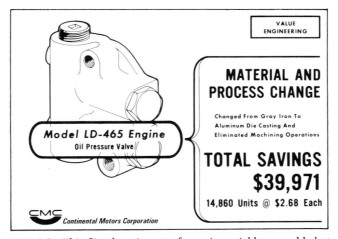

FIG. 2-9 This flip chart is part of a series quickly assembled at 8½- × 11-in. size from pasteup of typeset copy and handbook art, and then photographically enlarged to 24- × 30-in. size using the finished photoprint as the flip sheet.

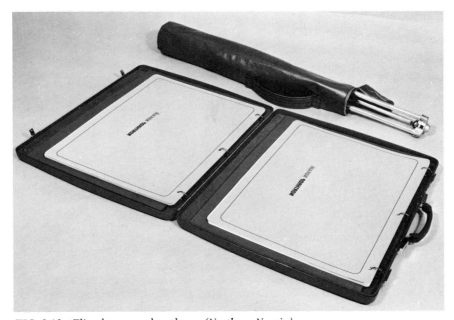

FIG. 2-10 Flip chart travel package. *(Northrop Norair.)*

Poster-Type Display Charts

Another type of chart for presentation is the rigid poster or placard. These may be variously known as briefing charts, management control charts, or wall charts. Display charts may vary from 18 × 24 to 30 × 40 in. in size, according to the size of the audience. As another directly used visual, display charts offer the presenter the advantage of simplicity as well as providing the chart artist with more options in art treatment than flip charts allow.

Display charts are usually prepared on illustration board, a chipboard surfaced with durable bristol paper, which can be worked on in any art medium and can withstand heavy erasures. Photos, prints, and even lightweight three-dimensional objects can be bonded or wired onto the board. Being rigid, the display chart can be overlaid with various stick-on materials and self-adhesive sheets of preprinted color, pattern, and lettering film that would crease or peel on a flexible chart (Fig. 2-11a to e).

Any lettering method or style can be used, from Speedball pens and brush lettering to headings made by photolettering machines. Adhesive-backed tapes can be used to provide lines in charts, instead of drawing the lines by hand. The chart is laid out using a nonreproducing blue pencil, in case it should be photocopied later, and then tape is run along the chart lines where desired. Black tapes as narrow as 1/16 in., in a dispenser, are rolled onto the board to form the desired linework. Heavy lines in a choice of colors and patterns, such as solid-line, short-dash, long-dash, and many other combinations may be put on a chart. The use of these stick-on materials is a great aid for making charts that may be subject to revision, since they can be lifted and repositioned readily as compared to reworking ink or paint lines. Changes in headings or blocks of text are made by pasting a new strip of lettered paper over the old one on the chart. Photographs are dry-mounted to the board.

For charts that are updated periodically, such as progress charts, a clear acetate overlay is laid over a blank base map containing only elements that are not subject to change. Tapes, color sheets, etc., are put on the overlay and changed at will without messing up the whole chart. For frequently handled display charts used for such purposes as technical training, new employee orientation, and the like, the acetate overlay or a laminated plastic over the chart provides protection against grime and scuffing of the art surface.

A relatively new type of rigid display board now becoming popular with presenters is a white board of porcelain-enameled steel. This board has a soft-sheen surface that accepts special water-soluble markers readily, and markings are easily wiped off with a damp cloth. Also, magnetic visual aids cling to the steel surface.

A great advantage of rigid art boards is that they make it possible to use overlays as step-sequence visuals, a useful aid in training and similar purposes (Fig. 2-11d). It is often advantageous for the speaker to show the basic chart, map, machine, or other subject by flipping over a series of hinged clear film or acetate sheets, beginning with a basic unit and showing on each added overlay, on which further art and copy has been rendered, the additions as they relate to the underlying design. As many as six or eight of these flaps, hinged at the top with tape and aligned by register marks to the art beneath, may be attached to depict a logical sequence.

Many presenters prefer, for certain occasions or types of presentation goals, to use display charts in traveling lectures and demonstrations. For this purpose 20- × 25-in. charts are found to be a highly convenient size—not too large for travel packaging, and proportionate to the standard 8- × 10-in. photo print size or a standard 4- × 5-in. negative so that charts can be handily photocopied for printing or other reproduction.

Traveling Presentations

For a traveling presentation, thin, lightweight charts are most convenient to package, carry, and handle while demonstrating. Three-ply bristol board or an equivalent material in stiff card stock is recommended for this use. Another fine material for this purpose is Fome-Cor board. This is a polystyrene foam sandwich laminate between white clay-coated smooth board facings. Lightweight, strong, and rigid, the surface of Fome-Cor is excellent for poster colors, markers, inks, and other charting mediums, and color

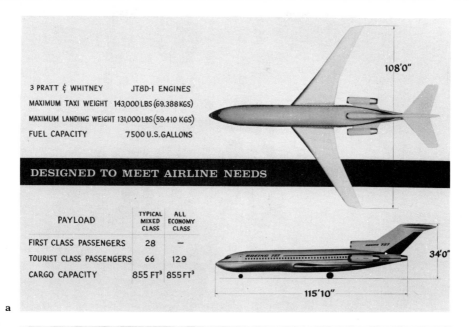

a

b

FIG. 2-11a, b   These display charts were made by pasting strips of colored paper and cutout aircraft illustrations on illustration board, and then adding titles with dry-transfer lettering and text with freehand Speedball pen lettering. *(Boeing Company.)*

sheets, photographs, letters on film, and other paste-on elements bond well to it. Traveling cases in smaller chart sizes are available from art supply stores, and in companies which routinely put on road shows or travel presentations for technical or sales purposes, special travel cases or disposable corrugated cardboard chart holders are made available to users.

Despite the greater bulk of display charts on travel tours, many presenters prefer them over projected displays because of their basic simplicity. The presenter does not have to learn how to set up and use a complicated projection device, screens, etc., nor is it necessary to arrange for an operator to do so. The charts can be used anywhere,

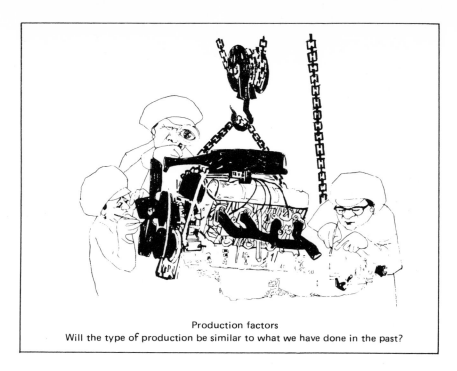

Production factors
Will the type of production be similar to what we have done in the past?

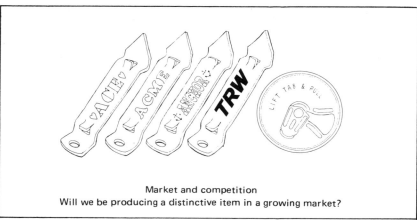

Market and competition
Will we be producing a distinctive item in a growing market?

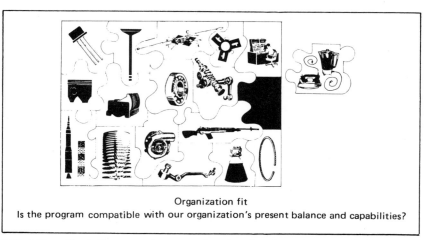

Organization fit
Is the program compatible with our organization's present balance and capabilities?

FIG. 2-11c Charts from a series on market planning show that a simple black-and-white illustration treatment can create an effective, sophisticated presentation. *(TRW Systems, artist Alan Daugherty.)*

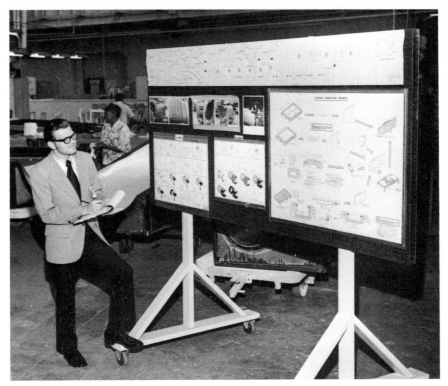

FIG. 2-11d Display charts illustrating important aircraft assembly and installation procedures are used in production shop areas. *(Rockwell International, Los Angeles Division.)*

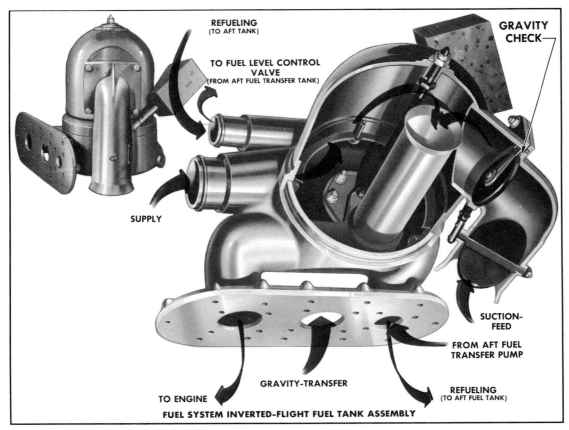

REFUELING
(TO AFT TANK)

TO FUEL LEVEL CONTROL VALVE
(FROM AFT FUEL TRANSFER TANK)

GRAVITY CHECK

SUPPLY

SUCTION-FEED

FROM AFT FUEL TRANSFER PUMP

TO ENGINE

GRAVITY-TRANSFER

REFUELING
(TO AFT FUEL TANK)

FUEL SYSTEM INVERTED-FLIGHT FUEL TANK ASSEMBLY

FIG. 2-11e This technical training display chart is used to instruct technicians in maintenance procedures. *(Rockwell International, Los Angeles Division.)*

including places where projectors, or electrical sockets are not available. Many users favor rigid charts over flip charts because, not being bound into a fixed-sequence book, the charts may be used in any order of progression.

PROJECTED PRESENTATIONS

Each type of presentation has its applications, and for most purposes the growing use of many types of audiovisual projection reflects its value to presenters as the most compact and convenient visual aid. There are two main types of projections; the still presentation of a succession of fixed images, and motion pictures. Use of stills, as slides, filmstrips, and overhead projections, far outstrips that of movies, which are employed for many commercial sales and institutional messages but are relatively complex, costly, and time-consuming to produce. However, closed-circuit television (CCTV) is a form of the motion picture that is coming into widespread use in many organizations for various types of presentations.

Developing a Storyboard

In developing a slide or filmstrip series, the artist must study the first available draft of the text in order to plan a pictorial sequence, in collaboration with the writer of this first draft, in order to portray the ideas in their proper sequence and emphasis.

This pictorial plan for an integrated series of visuals coordinated with text takes the form of a *storyboard*. Just as in the making of sound movies, animated cartoons, and television commercials, the storyboard is the only logical device for framing an outline and planning the words, pictures, and succession of ideas so that all develop simultaneously. To begin with, the speaker, writers, and other concerned parties prepare a series of storyboard cards, or planning cards. These planning cards are 3- × 5- or 5- × 8-in. cards, and one card is filled out for each visual step. The card is numbered; a rough pencil sketch shows the action; and a description, including text to be narrated by the speaker, is written beneath (Figs. 2-12 and 2-13).

From this group of planning cards, inserted in proper sequence from top to bottom in rows of grooves on a 3- × 4-ft or larger storyboard, the continuity of the presentation as a whole can be visualized. The storyboard, then, is the key to the presentation, to be used for reference by everyone involved in it (Fig. 2-14).

With the storyboard plan approved, the technical artist takes the planning cards to the drawing board and prepares final layout cards showing pictorial details, colors, and lettering laid out with markers, layout pastels, gouache, or some other medium as the art will appear in slide form. Typed final text is attached to each card. With the agreement of the presenter and other parties to the presentation, the artist then uses

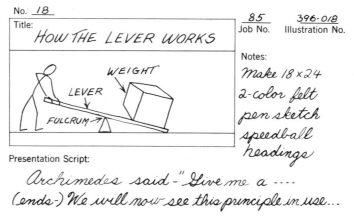

FIG. 2-12 A planning card for a slide presentation, preprinted to include a rough sketch and information, is 4- × 5-in. or 5- × 8-in. size.

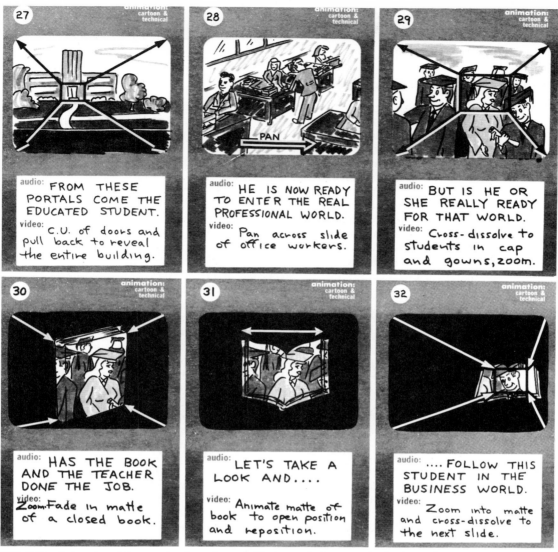

FIG. 2-13 Animating slides can be done through filming techniques, as is shown in this storyboard of slide sequences, with audio and video notes to indicate camera tricks that will be used to create motion effects. *(Robert P. Heath Productions, Inc.)*

FIG. 2-14 A storyboard. Planning cards are fitted into numbered slots in sequence to provide a single focal point for presenters, writers, artists, etc.

the final layout as a guide to create the finish art to be photographed for projection (Fig. 2-15).

Artwork for Slides and Filmstrips

As in all types of presentation visuals, art for slides or filmstrips must be simplified in content, leaving the message brief and the facts condensed. Pictorials and lettering should be large and unmistakably legible to the last row of the audience. Illustrations must be functional in design, each stressing only one central idea and using only familiar symbolism. Lettering on slide art should be kept to a minimum; if lengthy text is necessary, it is better to divide the message onto two slides, repeating the subject title, rather than cram it all onto a single congested format. Text should be composed chiefly of abbreviated phrases or sentences.

Artwork prepared for slide or filmstrip projection may be relatively small, in sizes that may vary from 6 × 9 to 16 × 20 in. according to the artist's requirements. Artists who are assigned to make art for projection on a production basis use a uniform, standard-sized art format with borders, positions for titles, and other guidelines and information printed in nonreproducible blue line on bristol board or light illustration board. Such a format allows uniformity in lettering size, design, and use of photographs.

Using a smaller format than that used in most display charts, artists are able to use cut-and-paste techniques at will and to include such fragile mediums as pastel, non-durable tissue, and similar materials, as the art will in most cases be handled only once as camera copy. Instead of painting areas of color with tempera paints, the artist may cut out shapes of the desired tints or color tones from Color-Aid, Pantone, or other special art color papers designed for such purposes and affix them to the art board with rubber cement. For illustrations, the artist may cement or dry-mount photostats, photographs, copy prints of illustrations, diazo prints, or, for many presentations, pictures clipped from magazines or other printed publications. The artist can often directly use 8- × 10-in. glossy photographs, cropping and mounting them to fit the need.

Headings can be hand-lettered, using a Speedball pen or one-stroke lettering brush methods, or lettered mechanically with Leroy or other template aids. The format size is convenient for professional-looking typeset letters, which are also faster where a quantity of artwork for projection is involved. Lettering may be typeset words imprinted on clear overlay films using in-house photolettering machines such as the StripPrinter, Photo-Typositor, or A-M Headliner, or foundry type may be ordered

FIG. 2-15 Artwork is placed on the copy board of the camera for photography, and will be made into a 35-mm color slide.

through a typographic service. If only a few words are needed on the art, lettering can be transparent lettering sheets from which individual letters are dry-transferred, or adhesive-backed sheets with letters cut out and burnished in place.

Another method enables the artist to create opaque white or colored lettering, as well as colored pictorials or design areas, using a graphic arts process called 3M Color Key. Using a simple exposure and developing method that can be performed in the art room, the artist takes any black-and-white art copy and makes it into a colored image on clear plastic, which is then mounted over the basic slide art as an overlay. Typeset copy using any of these methods has a much superior appearance and is more legible and uniform than hand lettering.

When time is of the essence and a series of slides requiring a large quantity of text is called for, the fastest and most economical solution is to first type the copy in strips or blocks of text on an IBM or other reproduction typewriter using a paper ribbon, then paste or tape the blocks on a small format, about 7 × 9 in. A same-size copy negative is produced from this format, and by photographing this negative using combinations of back and front lighting, with colored gelatine films laid over the clear spaces of lettering or design areas, bright color effects in lettering and background appear in the finished slide film (Fig. 2-16a to e).

Using new developments in camera technology, an advanced method for creating 35-mm slides and filmstrips has been developed that creates multiple colors and many special effects within an optical color printing camera, instead of pasting all art onto a board and photographing that. Artwork is prepared as a series of registered overlays, one for each color; using black-and-white pin-registered art and type successively placed in the special-effects camera, a wide selection of colors and dozens of effects such as posterization, neon glows, rainbows, motion effects of zooming and twisting, and many others may be incorporated into the final processed 35-mm film. The process creates more brilliant effects than ordinary slides because the optical printer process,

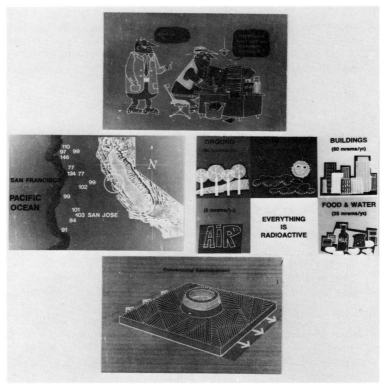

FIG. 2-16a Examples of 35-mm color slide winners of slide art contests in *Industrial Art Methods* magazine illustrate the virtually limitless variety of graphic techniques used for statistical, scientific, sales, and other purposes.

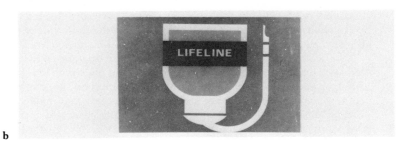

b

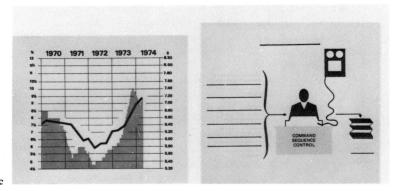

c

FIG. 2-16b Very simply designed, and effective in symbolizing industry employee services, are the slides used in an orientation presentation for new employees. *(Boeing Company.)*

FIG. 2-16c 35-mm color Super-Slides shown here are a larger type of 35-mm slide that is often used.

using transmitted light through color filters, produces a much more color-saturated look than does the reflected light of conventional art copy under the camera. Used by major corporations to produce top-quality projected presentations, the process is presently available only at a few service firms in big cities (Fig. 2-16d).

Slides

Slide presentations, far from being a "poor man's movie," have many intrinsic advantages over motion pictures for technical training, sales promotion, briefings, and many instructional purposes. Viewers are better able to absorb information presented visually if the image remains motionless until they have studied every detail, and the presenter can control the duration of time a slide will show on the screen to coincide with the spoken message. An added advantage is that the presentation may be stopped at will at any given point to allow for audience feedback, the exchange of questions and answers with the slide showing on the screen as a point of reference.

Of the still projection methods, 35-mm (2 × 2 in.) slide presentations are by far the most popular; 35-mm cameras, projectors, film, and slides are widely used by the public. The equipment is highly portable and is found almost everywhere, and standard 35-mm slides can easily be inserted into presentations of art copy such as charts, titles, and illustrations that have been copied to 35-mm slide format. As many as 150 or more slides in numbered sequence may be contained in a standard or circular slide tray. The ability to change the sequence easily is of great importance in the case of technical slide shows or other industrial presentations, because a slide show for a technical proposal may very well involve preparing slide artwork as the product's design is being created on the drawing board by engineers. Because finished slides are easily inserted in any desired order in the slide tray, the order of their appearance can be switched right up to the moment of presentation—a vital asset for many occasions.

The presenter can advance slides at will, hold a slide while group discussion takes

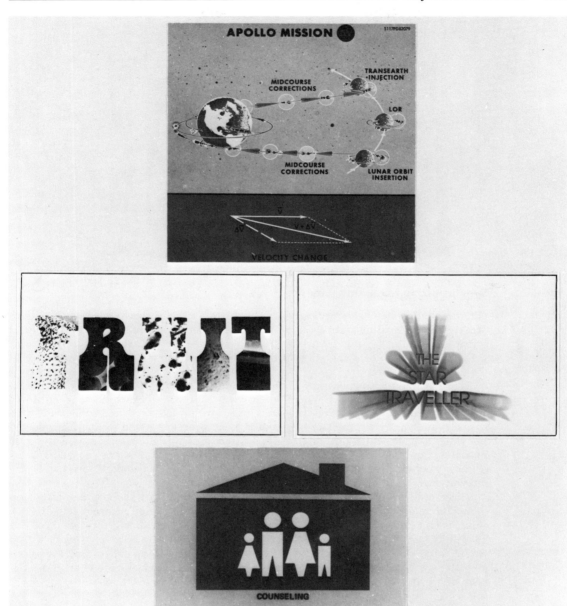

FIG. 2-16d New techniques using optical color printing methods can impart many special effects of motion and multiple colors to slide art, similar to effects used in motion pictures and television. *(Pix Productions, Inc., Santa Ana, Calif.)*

place, and backtrack to a previous slide if necessary. Increasing the flexibility of slide presentations for group interchanges of information are "random-access" projectors, which permit the presenter to project any slide in the tray, by number, onto the screen by pushing the desired button. Slide projectors come in a great many types and sizes, ranging from very simple units in which one slide is inserted at a time to ultrasophisticated machines with many automatic and special features. Perhaps the most compact projector is the portable desk-top unit, which unfolds from a carrying case to project a foot-wide image on a rear projection screen for viewing by a few people; the largest machines, on the other hand, can project movie-screen sized images for viewing by hundreds in a large auditorium.

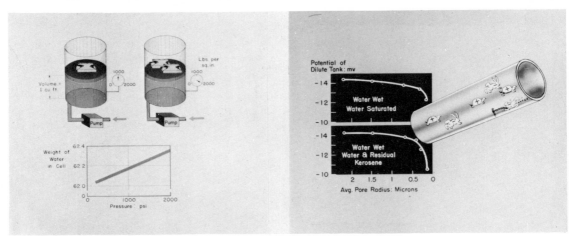

FIG. 2-16e   Lantern slides are often used for scientific presentations. *(Esso Production Research Co., artist Virginia Ware).*

Synchronizing Slides and Sound

For most industrial, sales, technical, and similar uses, the speaker presents a "live" oral discourse, reading from a script or spontaneously following an outline. A formal script is keyed to the appearance of slides, which can be changed at will by a hand-held control switch or by an operator, who gets cues from a script copy numbered at the desired intervals.

There are numerous occasions, however, such as product displays at trade shows, or technical training courses, where one-way repetition of an exact message will serve a better purpose. A tape recording of a talk can be delivered in the absence of a speaker, and the recording can be planned to be precisely timed to the projection of the slides. Music and sound effects can be added at this time, either with in-house recording equipment or with the aid of commercial services such as recording firms or radio and television stations. Any music and sound effects should be used with permission only.

The simplest way to key a taped narrative to the slides is to cue the slide changes on the script, and then switch slides by cue according to the script. Alternatively, the cue changes can be audible, with a little click or tap in the tape being the signal for the next slide. A smoother presentation with no audible signals can be done by using a tape whose sonic cues automatically trigger the slide change. Units called tape recorder–projector synchronizers, available commercially, can actuate the slide-change mechanism of a projector to accord with the cues on the tape.

Slide Sizes

The most commonly used slide is the 2- × 2-in. mount with 35-mm film inserted (22.9-mm inside frame). Another 2- × 2-in. mount is used to contain the popular 126-size film in cartridge form, for use in the Kodak Instamatic type camera, which shows as a 26.5-mm square frame. Still another 2- × 2-in. mount format is made for holding 127-size film, called the super-slide, with a larger, 38-mm square inside image.

Less often seen is the 2¾-in. square slide containing a 2¼-in. square film in the 120 or 620 size. This size, as well as the so-called lantern slide, which is 3¼ × 4 in. and within which any film size up to 2½ × 3¼ in. can be inserted by using a mask, is commonly used for scientific and highly technical presentations (Fig. 2-16e). The large film image can project finer details onto a larger screen than can the 35-mm slides.

A special slide transparency especially useful to teachers in holding class discussions immediately after a photograph has been taken is the 2⅛-in.-square Polaroid film. After exposure in the camera, the film tab is pulled to start development as with Polaroid films; the film is removed, dipped in a quick-dry hardener, and snapped into a plastic mount, all within a few minutes.

Mounting

The most common form of slide mounting is the inexpensive cardboard mount for 35-mm films, which is the type of mounting normally used by photography stores and film processors. It is also possible to order slides that come in more durable mounts of plastic or metal. Valuable slides requiring additional protection in handling, especially those used for oft-repeated presentations, may have the film sandwiched between thin glass sheets. Metal frames and glass are especially needed for lantern slides, as the large-sized film would quickly warp from the heat of the projector unless held flat by the glass sandwich.

Slides may also be mounted at home or in the office, using cardboard mounts into which the film is inserted and then sealed; or they may be mounted in plastic or metal frames, which encase the film by snapping into place.

For visual variety and emphasis, particularly in projecting a series of slides showing color photographs, masks may be used as an attention-getting device. Masks may be bought in kits that include a variety of cutout inside forms, such as a circle, star, triangle, and the like, which are laid over the film in assembling the slide so that the projected image on the screen will be a geometric shape instead of the usual box. In addition, custom masks may be cut if an unusual shape will enhance the slide.

Techniques and Accessories

In a booklet called "Slides with a Purpose," issued by the Eastman Kodak Company, several suggestions are made that can aid the slide presenter to put on an easier, more professional-looking show. Use of the "dissolve control technique" is recommended. This involves use of a device which, when used with two projectors, controls the intensity of light from each, fading the image from one projector while the other projector comes on. This makes a smooth visual transition from one slide to the next, with no abrupt changes from the pictures to a black or white screen in between.

Another suggestion is the use of black slides. These can be made either by inserting an opaque material in or taping over a slide, or by cutting out 2- $\times$ 2-in. cardboard squares. The black slides are inserted in the slide tray as spacers, in places in the talk during which no visuals are shown, and as beginning and ending slides. The black slides are a nondistracting substitute for flashing a bright white light on the screen at these intervals, and much easier on eyes that are accustomed to darkness.

For evaluating slides, comparing them with others, slide editing, keying them to planning cards, and similar purposes, a slide illuminator is recommended. This interior-lighted translucent plastic box can hold dozens or hundreds of slides at an angle in recessed rows for viewing. A desk-top slide viewer with an interior light source is the best way to preview a slide for possible flaws before it is projected on a large screen.

Filmstrips

The filmstrip is a series of full-frame or half-frame images prepared by copying presentation material onto a continuous strip of 35-mm film. As many separate stills as one can fit into a standard 80-slide projector tray can thus be contained on a tiny roll of film that takes up much less space. Although slides and filmstrips project the same kind of picture on the screen, the difference between separate units and a continuous strip call for weighing the advantages and disadvantages of each for the particular graphic display the presenter has in mind. The filmstrip's tiny size is a great asset when it comes to presenting a lengthy subject that might run to hundreds of visuals, as well as in packaging and mailing when many films and repeated mailings are necessary. Also, the pictures on a filmstrip are in a fixed sequence, so that individual visuals cannot be misplaced or lost or presented out of order or upside down. The presenter is not distracted just before the presentation by having to make sure that all the slides are inserted in the correct manner in a tray and can thus concentrate on the address. The filmstrips are cheaper to produce because they involve no time or cost of inserting and securing films into individual mounts.

Since filmstrips advance by means of cogged wheels as in a movie projector, sequences of visuals can be shown more rapidly than slides, which involve movement of an entire tray or carousel of slides to show each one. Filmstrips have many advantages over slides for presentations that are lengthy, with a great many visuals; which must be shown simultaneously in many places in exact duplicate form; and which are to be repeated many times as sales, educational, and industrial messages.

On the other hand, filmstrips are not recommended for presentations calling for updating changes, corrections, and additions. While graphic alterations to a slide involve only that slide, any change in a single frame of a filmstrip means processing a whole new strip, or cutting and splicing to remove and replace the frame. Also, the picture area in most filmstrips is smaller than that of slides. This is because most filmstrips are shot to half-frame size (18 × 24 mm), as compared to the full-frame size (24 × 36 mm) of the average slide. This size is sometimes called double-frame because it is double the size of motion picture film on 35-mm film. The smaller picture on a filmstrip means that the picture quality, when projected at the same size as a slide, is slightly less; and the picture has less light.

Filmstrips, being in a fixed position, cannot be rearranged at will in different sequences in order to adapt a presentation to the needs of a variety of audiences. And all frames must be horizontal on the screen, whereas in a slide presentation vertical-format slides can be used whenever it is an advantage for a visual to appear in that mode. This gives the artist more creative freedom in preparing slides.

Sound

Slides and filmstrips can be accompanied by sound. The narration can be read aloud or recorded, and an audible or subsonic cue can be used to switch the next visual on screen. In schools, silent filmstrips are more commonly presented because they offer the greatest flexibility to the teacher in giving lessons. The silent filmstrip does not interrupt, and it can be stopped at any point for class discussion or the teacher's comments. Projected visuals can be repeated, bypassed, or reviewed. However, filmstrips delivered with sound are the accepted mode for most business, sales, and industrial presentations, especially when the audio message reinforces the words shown on the filmstrip. When recorded, filmstrip sound is provided by a phonograph record or a magnetic tape cartridge. Cues signal the operator when to advance the film, or the film may advance automatically.

The most common type of sound filmstrip presentation involves a small audience and a portable tabletop unit. The filmstrip is projected onto a small screen—which comes in the carrying case—usually by rear-view projection, or onto a separate larger, screen. Pictures change automatically, as the sound on a recording contains inaudible cues along the audio track that activate the frame-changing mechanism.

Preparing Filmstrips

The art process of creating visuals for a filmstrip is identical to that of making art for slides, including the same story outline, sketches, storyboard, and final artwork assembled onto art boards and photographed on 35-mm film. In fact, most filmstrips begin as a set of slides, which permits experimentation to determine the best sequence, eliminate unwanted visuals, and perform other editing. When editorial adjustments result in the approved optimum set of visuals in the best sequence, the slides are duplicated onto a single 35-mm film. The alternative method for producing the filmstrip is to photocopy the art flats and other visuals in the correct order onto the film roll in a 35-mm camera, but this is time-consuming and poses difficulties—particularly the inability to shoot several frames of the same art in order to gain duplicates or to find the best results among several different exposures.

Overhead Projection

Of all projection systems, overhead projection is the most effective for two-way communication. Unlike all other projection methods, the overhead projector reflects an image indirectly by means of a mirror, a condensing lens, and a light source in the base of the projector that reflect up through a horizontal glass table. The light passes through the glass and through the large visual transparency placed on it; the projection beam is bent 90° to project a picture on a screen. The chief advantages of this projector are that the presenter faces the audience at all times during the presentation and personally uses the projector to change the visuals during the narration; and that the projected image is so strongly lighted that the room need not be dark in order for it to be seen, as with other projection systems. Most important for good rapport and communication, the speaker and the audience are face to face, clearly visible to each other as they exchange information, questions, and answers (Fig. 2-17). And the process of preparing and reproducing the visuals is fast and simple.

Overhead projection is used mostly in teaching. Transparency slides covering courses in every subject are available commercially, and since transparencies come in sizes up to 10 × 10 in., they can also be prepared directly by the teacher using ink, markers, crayon, and other mediums. Because the slides are in the open, where the presenter can easily reach and manipulate them, it is possible to "animate" the screen images by moving cutout arrows on the surface of the slide or by using a pointer to indicate key points of interest. The speaker can work with crayons on the surface of the slide, and later the markings can be rubbed off with a soft cloth to prepare for subsequent presentations.

Simple Line Transparencies

Prepared easily and inexpensively, line transparencies are recommended for charts employing only words or simple drawings lacking detail or precise delineation. The presenter can often do the work personally without calling on an artist. General guidelines are as follows:

1. Draw the original on 8½- × 11-in. material. If the surface is vellum, polyester drafting film, or some other translucent base, the drawing may be reproduced as a transparency in any diazo whiteprinter, using the appropriate foil. If drawn on a white, hard-surfaced paper, transparencies may be run off in a transparency maker or an office copier.
2. Lay out the chart to a 7- or 7½- × 9-in. format indicating the border with crop marks or an outline in blue pencil.
3. Lines, graphs, and illustrations may be drawn with a soft graphite pencil, a black plastic pencil, india ink, or a litho pencil. All typing, inking, shading, etc. should be directly applied to the form; paste-ups should not be added because their cut lines are picked up by copying machines. Color can be manually added to the transparency (Fig. 2-18).

FIG. 2-17 By using overhead projection, a presenter has the advantage of facing the audience. *(Rockwell International, Space Division.)*

FIG. 2-18 An instructor prepares a simple diagram with scale and litho pencil on a mounted overhead-projection transparency.

4. Headings and text may be mechanically lettered with a Leroy or similar scriber and guides. Body copy may be typed onto the paper original using a large sans serif face, such as the IBM Orator, or it may be typed onto pressure-sensitized translucent film for transfer if a film base is being used. Letters should be at least ¼ in. high. (If the original can be read from 10 ft away, it should project with good legibility.)

5. Headings and copy can be produced using a tabletop photolettering transparency composing machine to put crisp, printed-look words on adhesive-backed clear phototape. The photolettering machines are available from 3M, Kroy Industries, A-M Headliner Division, and other manufacturers.

6. Transfer letters, symbols, and tapes are easily added to an original or transparency by carefully positioning each unit and then rubbing it over the art with a burnisher or other tool to transfer the printed form to the base surface.

Simple transparencies as described above are often preferred by presenters because corrections or changes can be made to the film while it is being projected.

Film-Positive Transparencies

Film-positive transparencies should be used when the art is complex or detailed, when it must be drawn oversized, or when it includes full-tone photographs. In developing art for film-positive transparencies, unrestricted use can be made of paste-ups and all kinds of art aids, both translucent and opaque. However, since these transparencies can be produced photographically, continuous-tone black-and-white or color photographs can be used in the artwork; color or complex details may be added by overlay.

Standards   Because visuals on the screen are only as good as the originals they are prepared from, the following standards are recommended to help ensure that they communicate clearly:

1. For projecting extra-large originals, such as computer printouts or special charts, the originals may be reproduced onto an oversize transparency film (8½ × 11¹¹⁄₁₆ in.), which is then projected without a frame. Art copy too large to fit on this film must be photocopied to standard art format size.

2. Everything on the visual should be instantly recognizable; if not, it should be labeled, especially in charts and diagrams.

3. Keep the message simple, limiting each original to one point or comparison. Condense and abbreviate text, breaking paragraphs into sentences and sentences into key terms and phrases. Avoid using more than six or seven lines of copy and no more than that number of words per line.

4. Make sure that all linework is dense and black (india ink, carbon typewriter ribbon, printed cutouts, clip art, etc.) to ensure good reproduction.

Although either horizontal or vertical format may be used, the horizontal format is generally considered best for audience viewing. In either case, titles should be located at the top and information in the upper two-thirds of the format for best retention.

Frames

Although an overhead transparency can be projected unframed—as in the case of the oversized films cited previously—frames should be used for all but the most spontaneous one-shot presentations. Frames block light around the edge of the visual, make the unit rigid for handling and storage, and provide a handy border for identifying indexes and notes. The frames come in standard sizes, of stiff cardboard, and the visual is easily attached. First the frame is placed upside-down on a flat surface; then the transparency is centered face-down on the frame; and it is attached securely by taping all four corners.

Of the several sizes of slide mounts, the type most often used is a hinged cardboard frame with its outer margin measuring 10×12 in. and the inner aperture measuring 8×10 in. The two ways to prepare both slides and overlays, as described previously, are to draw each by hand on transparent Mylar or acetate, or to photocopy opaque art onto a reproducible positive film which is then used as the slide (Fig. 2-19a).

Overlays

Overlays help simplify difficult concepts and also enable the presenter to build the story told by the visual into a logical progression. Overlays involve two or more imaged transparencies used in a sequence over each other.

Transparent overlays may be hinged in place over the base slide transparency; in the first, or open, position only the basic art is shown, but as the instruction progresses each overlay is laid down in proper position to show an additional phase or more components of a visual situation in a step-by-step technique. Step-by-step visuals of this nature are ideally suited to instruct such subjects as technical training of industrial workers, assembly and disassembly of equipment, evolution of an organization, or proposed additions to a building complex. An overlay is also an excellent device for

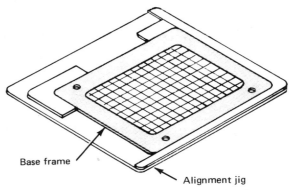

Base frame

Alignment jig

FIG. 2-19a For smooth sequencing of overhead-projection presentation, an alignment jig is used for sandwiching a single transparency into the light frame.

making a new-to-old or present-versus-future comparison on the same transparency, or for simultaneously showing several solutions to a problem (Fig. 2-19*b*).

One may add as many as six or seven overlays to the original slide in a successive sequence, using different colors, tones, and patterns to differentiate each from the others (Figs. 2-20 and 2-21).

Adding Color to Overhead Transparencies

There are three ways to add color to transparency film: color foils, color adhesive film, and manual use of liquid colors and pencils (Fig. 2-22).

1. The simplest, quickest way to add color appeal to a clear transparency with black linework, without using special processes in making the original, is to reproduce the transparency onto color transparency film, either as a color positive or a color negative. Placing a blue foil over black-on-clear film adds color while subduing white glare in the projected image. Or, the original may be reproduced as a color-on-clear film. Combining the two adds to the color effect; thus, a transparent red heading sandwiched with an unimaged blue foil produces a purple color.

   Color foil can, using a technique called "billboarding," be used to highlight a specific area on a visual. A sheet of imaged or unimaged color film is taped to the mounting frame over the imaged transparency. By cutting a section out of the color foil (being sure not to cut through the base film beneath), attention is focused on a desired area. A variety of color transparency films are available commercially from 3M Company, Keuffel & Esser, Scott Graphics, and other manufacturers. Imaged color films in up to 50 colors may also be made by the artist using the 3M Color-Key Contact Imaging process, in which the Color-Key films are developed on the spot, using the firm's portable tabletop exposure unit (Fig. 2-23).

2. Because large color areas cannot be applied evenly to film transparencies using transparent inks or crayons, they can best be added as cutouts of pressure-sensitive

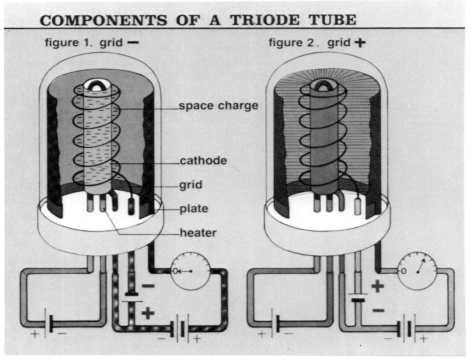

FIG. 2-19b Typical colored overhead-projection transparency used in school courses. *(Instructional Dynamics Corporation, Inc.)*

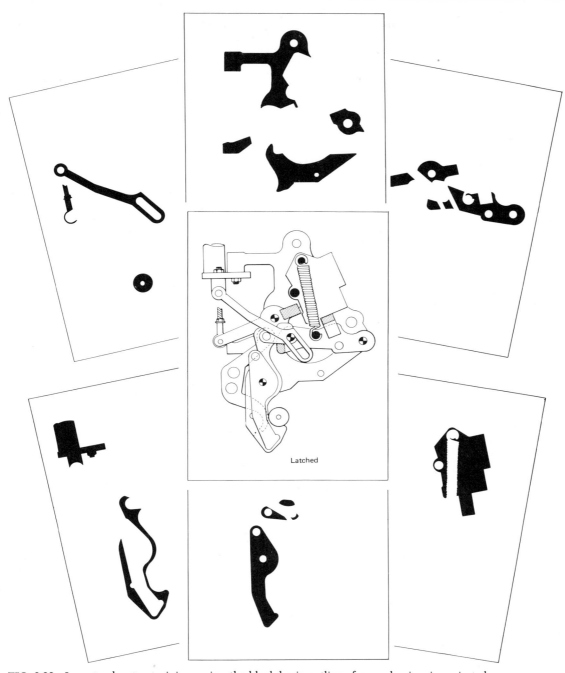

Latched

FIG. 2-20 In a step-by-step training series, the black basic outline of a mechanism is projected on the screen; then, one by one, the instructor adds each successive different-colored overlay, explaining each addition as it appears. *(McDonnell Douglas, Aircraft Division.)*

color transparency sheets such as Bourges "Cutocolor," Zip-A-Tone, and Meca-norma. Color is added to prepared transparencies by placing the framed visual face down, then trimming a piece of color adhesive film to roughly cover the desired area. After aligning an edge of the color film to an edge on the image, the backing is peeled off and the film applied to the visual. Carefully cut around the image while being careful not to cut through the visual, remove excess color film, and the exact shape will show up in color.

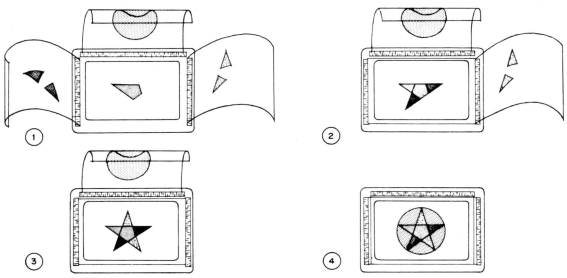

FIG. 2-21 Unlike bound-together static transparencies, overlays can be used to show a sequence of parts, functions, additions, etc., by successively overlapping the base transparency with overlay flaps in any desired order.

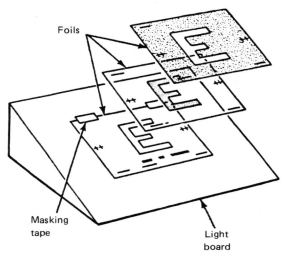

FIG. 2-22 Laminated static transparencies are put together in exact registry on a light board and taped together before assembly as a unit in the mount.

3. For adding small areas of color to a transparency, a variety of markers, pens, and pencils specially made for working on films is made by several manufacturers, including Koh-I-Noor, 3M, Schwan-Stabilo, and others. Both fiber-tip pens and broad-nib felts can put bright, luminous colors on film in both permanent and water-soluble form. Colored inks may be brushed on film in small portions, but will show uneven streaks in large areas when projected on the screen. A no-crawl solution, available for work on plastics, should be added to inks or dyes not specially made for this application.

Transparency Materials and Techniques There are many materials that are additional aids to presenters and artists in creating overhead projection visuals. These are:

- **Plastic sheets** Many manufacturers supply transparent plastic sheets whose surfaces have a special tooth for drawing upon with pencils, crayons, and other mediums. A translucent-color pencil known as Lumochrom draws an image that shows on the projection screen as a bright color.

- **Tracing paper** A good grade of this translucent, inexpensive paper can be drawn upon with pencil, grease crayon, ink, and pastels, and most other mediums as well as tapes, adhesive letters, etc. bond easily to it.

- **Diazo reversal foil or photographic negatives** A negative image can be produced from a positive foil using reversal diazo film, and a photo negative can be used as a master to produce a right-reading positive foil without extra photo processing.

- **Transparentizing** If a printed page of photos or art is blank on one side but not transparent enough for good diazo reproduction, a transparentizing solution can be applied to the blank side. This may make the copy transparent enough to reproduce directly onto a diazo foil.

- **Transferring pictures** A specially coated plastic may be placed face to face with a magazine picture (printed on a clay-coated paper) and then put under heat and pressure in a dry-mounting press to laminate the plastic and picture together. The paper is then soaked off with warm water, leaving the picture impressed on a clear transparency suitable for overhead projection. Another method, using rubber cement, is effective in lifting pictures onto transparencies to create excellent diazo overhead projection visuals.

Animation in Overhead Projection

As mentioned previously, by moving arrows, other shapes, or pointers along the surface of a transparency on the lighted stage of the overhead projector, the effect of

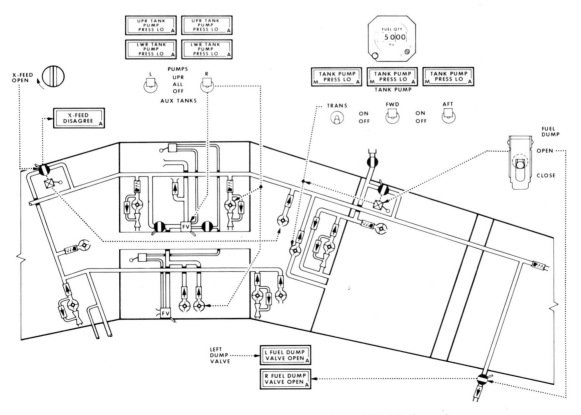

AIRCRAFT FUEL SYSTEM FUNCTIONAL DIAGRAM (DUMP)

FIG. 2-23 This diazo training aid is in black line on blue film, with a yellow screen overlay defining the wing area showing as green on the screen. *(McDonnell Douglas, Aircraft Division.)*

motion is projected on the screen. For mechanical instruction, animation showing stages of movement in mechanisms, such as in a piston engine, a hydraulic system, etc., can be created by registering an overlay so it fits between slots added to the base slide and moving the overlay by push-pull motions. A more versatile form of animation of this type may be used by making models of mechanisms with movable components cut from colored, translucent flat plastic, which can be laid on the projector stage and manipulated.

More sophisticated effects can be achieved with light polarization techniques. By using two sheets of polarized film and rotating the upper one, visuals appearing on screen can fade in and out. In another process, a motor-actuated polarized plastic unit on the projector admits polarized light onto specially treated plastic strips on the face of the slide, creating various effects of motion in the screen image (Figs. 2-24 and 2-25). Using this method, effects such as swirls of changing color, radiation, and linear movement can be helpful on technical diagrams and other art to depict combustion, sliding action, gas flow, fluid movements, and the like.

Opaque Projection

Opaque projection is the only projection method that allows opaque copy such as papers, books, forms, and three-dimensional objects to be displayed directly instead of

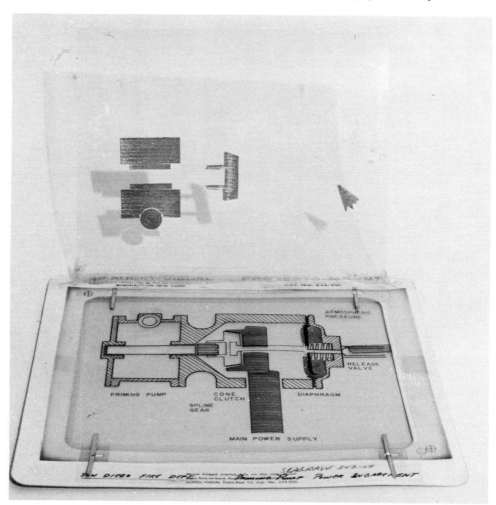

FIG. 2-24   By pasting strips of specially treated plastic material on a transparency overlay, as shown here, effects of motion are created when a light-polarizing device on the projector is used during projection.

FIG. 2-25 An instructor adjusts a circular light-polarization accessory on an overhead projector.

from transparent film versions. In industry, schools. research and development institutions, and the like, opaque projection enables presenters who are not artists to project visuals onto a screen without chart-making effort and advance preparations, except for gathering the material to be projected. As such, opaque projection is commonly used for all kinds of instruction, indoctrination, and training purposes.

The presenter does not need to have copies or films made to set sizes; an open book, a machine part, a photo, a magazine clipping, or anything else that will fit may be placed in the stand of the opaque projector and projected in full color and detail on the screen.

Because the screen image is from reflected rather than from direct light, the room must be almost dark. Another disadvantage is the heat given off by the light bulb, which can damage the opaque paper copy if it is left for very long in the machine, unless the unit has a good cooling fan.

One important use for the opaque projector is for tracing. For preparing large charts, displays, or poster-type boards, which require copying small illustrations, diagrams, etc. from reports and other printed publications on a greatly enlarged scale, the small original can be projected onto a larger board and traced in pencil, for later rendering in exact proportion and scale in ink. Commercial artists frequently use this device to save time in laying out large advertisements, illustrations, and designs from approved small sketches and layouts.

One popular unit, the Seerite, projects a picture image up to 6 × 6 in. and comes with a motor, a fan, and a slide tray that lets the user change copy without moving the projector or refocusing. The projector will project its 6- × 6-in. image as a 44-in.-square picture at a distance of 72 in., and at 144 in. the visual becomes 88 in. square. Another machine, the Vu-Lyte II, has an endless-belt conveyor that crank-feeds continuous copy and incorporates an arrow that can be moved anywhere on the screen as the speaker accentuates key points.

Multimedia Presentations

Multimedia presentations are the ultimate in projection sophistication, because they call for several projection systems to display screen images in combinations that require substantial planning in order to coordinate visual sequences in a smooth continuity. Using this method, still pictures and motion pictures may alternate on the same screen; several images may appear on different parts of the screen simultaneously; or a series of changing images may appear at random on several screens in order to create an overall impression, mood, or feeling of action and change in a presentation rather than a straightforward display of a factual nature.

A fairly recent technique, multiscreen projection may involve projecting images onto any number of screens at the same time. According to experts, the simultaneous recording of many visuals on the mind makes a more powerful impression than does a single screen, and when properly integrated, more than 700 images can be shown during a 10-minute period without the presentation seeming to be chaotic or rushed. For example, in a presentation about strawberries, eight pictures of strawberries might be shown simultaneously, ranging from a macro-image of a single berry to a long shot of a field, the picking, and the packing—all creating a stronger impression than one would get from eight sequential views of strawberries.

Developed and designed by audiovisual specialists, many large multimedia presentations used in government, industry, and education are presented in specially constructed audiovisual rooms. Many new auditoriums built for museums and other institutions presenting films to the general public incorporate multimedia arrangements of a number of screens and several types of projectors that may be operated through a control center. Such elaboration is not a necessity for successful multimedia, however, especially in education and training, where each method can still be used to best advantage but slower, simpler sequences may make for more effective instruction. For example, at a major aerospace firm that trains pilots, mechanics, and other specialists in proper flying, repair, and other operations on its aircraft for international airlines, multimedia training is done in a special room where slides, overhead projections, wall charts, and motion pictures build one on top of another to provide complete courses of instruction. And teachers have found that by adapting a classroom to include filmstrips, motion pictures, slides, and other media, using two or more together or in a succession keyed to the nature of the course can be a superior, attention-getting device for teaching students.

Both front projection screens and rear projection screens may be used in the multimedia approach. With rear projection, the equipment may be placed behind the screen and more lighting can therefore be used in the classroom for the benefit of students in reading or taking notes. Mirrors must be placed between a rear projection screen and the projector in order to reflect a right-reading image that can be read correctly by the audience.

A great deal of preparation and many practice runs are needed to ensure a smooth flow of audio and projected images in a multimedia presentation. This often calls for teamwork between the graphic artist and the audiovisual specialist, who may operate multiple-projector setups in various configurations while coordinating all tape recorders, projectors, and other audiovisual equipment with electronic monitors.

CLOSED-CIRCUIT TELEVISION

Closed-circuit television (CCTV) has become a widespread communication technique in many large organizations, including school systems that use CCTV to provide not only formal curriculum instruction but also subsidiary news, sports, and social events reporting to college groups and the like. Local news of immediate and topical interest, recorded on video tape, can augment telecast course material, making CCTV a flexible, up-to-date teaching medium of the highest quality.

The software, or educational material, for CCTV courses may be converted to television by means of a filmstrip, a movie, or a slide projector, plus a television camera. Reception from one source is called *uniplex;* multimedia is called *multiplex.* One way to televise 16-mm movies is to use a self-contained rear-screen projector, such as one made by the Kalart Victor Company, that projects a picture from behind a screen for photography by an external television camera. The most effective method for telecasting films is to project the film directly into the lens of a television camera equipped especially for this.

For use in schools, lecture halls, and auditoriums, CCTV may be projected onto large front- or rear-view screens in images up to 9 × 12 ft in size. The most common application for large organizations in business or industry is to place several television sets throughout the audience for convenient viewing by all.

Although visuals for CCTV embrace many creative arts, including writing, directing, acting, and live photography, there are also many applications for illustrating and special graphics, such as diagrams, technical art, charts, and title stills.

Artwork for CCTV

The nature of the screen and type of transmission offers many possibilities for creative art expression, but also several limitations (Fig. 2-26). In preparing art for television, the artist should keep the graphics simple and strong. Lines should be heavier than those used for other mediums, with more contrast in backgrounds and areas of tone. The cardinal rule for all television lettering, illustration, and backgrounds is: Keep it simple and bold.

A gray scale depicting tones from white to black in 10 percent increments, as used also in graphic arts, is needed to plan contrasts between backgrounds, lettering, and illustrations. Photographs and tonal illustrations used on television must provide more contrast than conventional visuals, because much of the contrast is lost in television transmission, which tends to blend the tones.

Because the television screen image has the horizontal proportion of 3:4, the art format must have this proportion. The art may be any size, but it must allow for a larger picture than the art rectangle. This extra margin, or bleed, is to ensure that the camera will not pick up a format edge, and also to prevent lettering, etc., from being too close to the inner frame of the television's picture tube, which will create the illusion of the lettering rolling off the screen due to electronic distortion. A convenient size both for preparing art flats for television and for their handling and storage is 15- × 20-in. boards, on which the art is mounted.

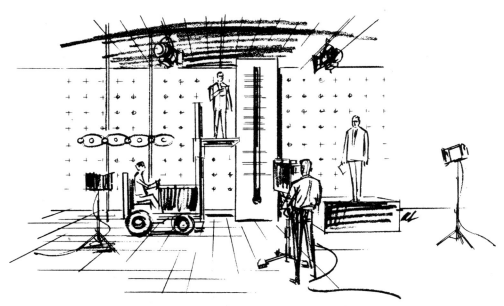

FIG. 2-26 In a typical stage setting for a CCTV production, narrators stand beside a large thermometer chart. *(Rockwell International, Space Division.)*

There are many CCTV techniques for creating visual effects. For example, 35-mm super slides may be used in a special projector to superimpose white letters over static or live action. By assembling dry-transfer, cutout, typeset, or other lettering on pressure-sensitive film on a special dropout-blue background, a slide may be shot that enables the white title to appear over the action as a television image. Some special effects may be created electronically in the television system, and others must be designed by the illustrator (Fig. 2-27).

Take Cards

Art cards, which in television work are called *take cards*, are artwork used in front of the camera. They include titles, lettered signs, photographs, and special effects cards called zoom, tilt, and pull cards. Ranging from 20 × 30 in. to 30 × 40 in. in size, these cards are usually reproduced on photosensitized blue paper. Flip charts used as take cards are often made 36 × 36 in. in size and are drawn on tinted stock to reduce the glare for better transmission (Fig. 2-28).

Zoom and tilt cards are special, large cards used to superimpose copy over a live-action background, using arrangements of white lettering on black cardboard in formats keyed to camera techniques that achieve the desired effect. Another type of card, the pull card, is made from art board with windows that expose lettering or art copy when the masks are pulled, in order to emphasize a message or to add animation.

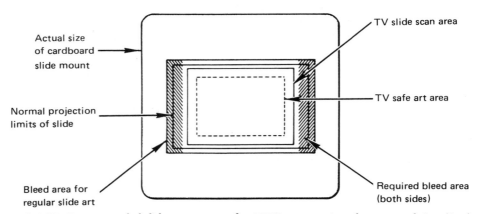

FIG. 2-27   Recommended slide mount areas for CCTV presentation, shown actual size. *(Rockwell International, Space Division.)*

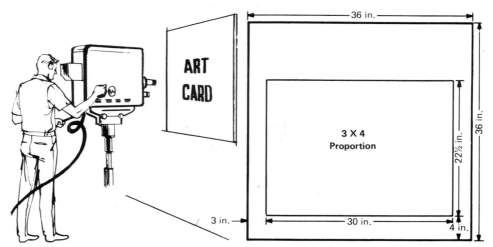

FIG. 2-28   Recommended practice calls for 20- × 30-in. or 30- × 40-in. "take cards" (artwork used in front of the television camera); and 36- × 36-in. flip charts, rendered on tinted stock to reduce brightness for television transmission. *(Rockwell International, Space Division.)*

Continuous-tone black-and-white slides present a complete image for television transmission, and color slides and existing color art may also be used. For black-and-white CCTV, the latter should be reviewed by consulting a special color chart that converts all colors of papers that may be used in color-slide or presentation art to television gray-scale equivalents. This is a precaution because some colors that look light in tone may reproduce very dark on television, and dark colors may look light gray.

MOTION PICTURES

Industrial firms occasionally produce their own motion pictures in-house for such purposes as new employee indoctrination, reporting on contractual progress for government agencies or other contractors, and the like. Most large organizations employ professional industrial film producers to make motion pictures for public relations, public information, and sales purposes. These films are usually of high quality for general showing to large groups. For the technical types of in-house motion pictures mentioned, the same types of artwork made for display card, color slide, or filmstrip presentation are also required for titles, verbal summaries, table charts, diagrams, etc., which are shot with a motion picture camera on 16-mm film and spliced wherever appropriate into the reel of live action. Motion pictures that are made by professional filmmakers may also require art and design by graphic artists, but this type of artwork, which includes animated cartoons, live action with cartoons superimposed, lettering that changes form and color, and other special effects, is a field all its own and an integral part of the industry's unique, complex camera and processing technology. As such, the art methods used in this field cannot be covered in this text.

TECHNICAL AND PRODUCT EXHIBITS

Designing a technical or product exhibit is a special branch of graphic design calling for the ability to visualize three-dimensional forms in much the same way a sculptor does—in real space, instead of in terms of relating these objects to a two-dimensional paper plane, as is the case with technical illustration. It also requires, in addition to design and art experience, considerable knowledge of construction techniques, display materials, lighting, and audiovisual equipment, because many exhibits may incorporate slide, filmstrip, or motion picture projections augmented with sound.

A successfully designed exhibit brings pictures, headings, text, and objects together within the framework of a three-dimensional construction in a way that attracts and informs viewers. An exhibit is always promotional in nature, whether it is a museum exhibit or a sales presentation, in that its main goal is to interest an audience in the subject on display. A museum display is intended to inform viewers; for example, a mineral display may present colorful, entertaining facts about mining history and about the special properties and uses of each mineral. Window displays in stores are another kind of exhibit. Technical exhibits are usually sponsored by manufacturers of products or services and are prepared for use in trade shows; in expositions held in conjunction with conventions of major professional, management, and technical associations; at scientific conferences held at universities and colleges; and on similar occasions. Many industry displays are designed for community centers, airport terminals, etc., to foster a favorable public image as well as to stimulate sales. Most companies have lobby exhibits and, very often, special in-plant exhibits designed to educate employees as to good safety practices and so forth (Fig. 2-29a, b, and c).

Companies that specialize in designing and building any kind of display may be called on for exhibits requiring special quality and such complexities as multimedia shows and unusual construction. For many purposes, however, companies use their own in-house art staff to design and sometimes also to build product and other types of displays in conjunction with plant engineers, carpenters, and other professionals.

In such instances the corporate graphics and marketing staff must be involved. The advantages of creating exhibit designs in house include close contact with the company's production methods, product lines, and model changes; familiarity with man-

FIG. 2-29a, b These United Way displays were basically designed from panels with mounted photographs, put together in easily assembled structures using interlocking devices and an aluminum-tube framework. *(Pitney-Bowes Corp., Corporate Industrial Design Department, designers Sandor F. Weisz and Carl Russo.)*

FIG. 2-29c Trade show display. *(McDonnell Douglas, Aircraft Division.)*

agement thinking on corporate visual identification programs and the "look" which the displays should have as part of such overall programs; the graphical continuity resulting from having all displays designed by the same source; and the ability to produce displays on short notice.

Designing an Exhibit

Before the design of an exhibit is undertaken, all information about its specifications, according to the following questions, should be at hand:

- What dimensions—width, depth, and height—will the exhibit be, and within what area of floor space?
- What is the message of the exhibit?
- What traffic, in terms of attendance, is expected to visit the exhibit?
- What budget is available?
- Will the exhibit be reused? (Traveling shows should be specially designed and made of lightweight materials, for compact packaging, fast setup, and ease of transport.)
- What visual elements, in what combinations (photos, artwork, real objects, audiovisual displays), will best present the concept?

The exhibit design, like all designs, starts with small pencil layouts on tracing paper, so as to arrange the design elements in experimental placements and evolve an attractive, suitable composition. Pastels, colored pencils, and markers are used to shade the drawing and to try out color schemes. Just as in all graphic design, a clean look incorporating simplicity and boldness may involve selecting a few elements from a wide variety of art, photos, and objects; it is important not to try to demonstrate too many things at once. Unless hard choices are made, an exhibit design can assult the eye with a confusing mélange of diversified shapes, colors, placards, and objects.

On the other hand, it is possible to draw a beautiful layout of an exhibit that will not look at all like the original sketch when constructed. Unless the artist has learned to add the language of multidimensional structural design to his or her own vocabulary of graphic design and aesthetics, it pays to consult an architect or package designer if possible before going ahead with the design. Such practical advice on best uses, characteristics, methods of forming, bending, etc. of just about every type of display material is invaluable. Package designers and industrial designers are often excellent exhibit designers in their own right because of their knowledge of the contemporary modular approach to three-dimensional design based on the subdivision of space into volumetric units.

Models

Before embodying the layout concept in a detailed, comprehensive layout for final approval before construction, a scale model can be an invaluable visual aid, a double-check to verify whether the attractive design on paper can or should be built with that geometry (Fig. 2-30).

A small scale model can be constructed right on a drawing table or a desk out of materials close at hand. Geometric shapes are easily constructed by cutting and folding various white and colored papers and boards into blocks, polyhedrons, and other configurations. Polystryene foam plastic, a common packing material, can be cut and trimmed into any form with a sharp knife or a razor blade. Pebble-grained matte board may be cut to simulate the fiberboard, Masonite, or other background boards used in the actual display. Wood veneer–surfaced matte boards are excellent for imitating plywood panels, wooden shelves, etc. Clothes-hanger wire and strips of balsa wood are ideal for forming structural members.

Two- or three-ply bristol board is a stiff, thin material that can be used where it is necessary to show thinner surfaces. Irregular shapes can be made from balsa wood or by molding a bit of modeling clay. Metal panels, ribs, etc. can be shown by sticking pieces of high-sheen metallized papers, available in art supply stores, onto model sur-

FIG. 2-30   This award-winning table-model display of a nuclear power plant features interior cutaway views and color-coded components to show flow of water, air, etc. through the system. *(Battelle–Pacific Northwest Laboratories, art director J. L. Doyle, designer Harold Carlile.)*

faces; this usually looks more realistic than painting those same surfaces with the fuller-looking metallic paints.

Poster paints, stock-on color film, transfer letters, tapes, and so forth can add to the illusion. Pieces of thick clear acetate make good windows, and with imagination other model parts may be created from that same material. Portions of models may be quickly bonded together with white glue.

The purpose is not always to make a professional model, accurate in detail; the model can be quite rough as long as it enables the artist to fully visualize the finished display from all angles and reveals any design discrepancies that might otherwise appear only when the final units have been constructed.

Final Layout

One or more detailed color layouts showing all facets of the exhibit as it will appear, with descriptive notes and details concerning the colors and lighting, must be approved by the exhibit sponsors before construction. Once approved, the sketch and available scale models then serve as references in making construction drawings for use by carpenters and others.

Exhibit Panels

There are too many types and materials available for making exhibit panels to name here, but the basic traditional panel unit is the standard 4- × 8-ft Masonite panel on a 1- × 3-in. white pine frame. Plywood is often used also. The panels are sprayed with colored lacquers. Other popular panel materials are plastic-coated plywoods, vinyl-coated steel and aluminum sheets, and sheet plastic, which may be used for bright effects when translucent color lighting is desired. Acrylic sheets are available in every color and in opaque, transparent, and translucent form.

Exhibit Animation

Moving displays are often used merely as an eye-catcher, as in the case of cartoon figures going through various motions in food displays in supermarkets and the like.

On the other hand, animated displays may have great educational value when used for such technical purposes as showing sequences of mechanical action, the flow of fluids or gases, and similar applications. In exhibits in museums of science and industry in most major cities, animation is used for such purposes as: rotating an arrow around a model of the Mobius strip, showing that the strip has only one surface; turntable displays that highlight the beauties of fine mineral specimens; and showing bubbles moving through colored liquids and lights tracking the interactions of chemical elements and the like.

Animation is not always a solution, and it should not be used as a substitute for effective, attractive design in bringing people to an otherwise dull display. Its applications are limited. But if the display subject is one that can be animated, and when animated presents a real message about the product or process being shown, several appropriate methods of adding motion may apply:

- **Mechanical animation** Usually shown in separate parts overlaying each other in related motions, these effects call for design by display engineers or other specialists, as well as specially fabricated arms, cams, etc. that are actuated by small electric motors.

- **Electric lights** Flow, dash-and-break motion, and other effects can be shown using lights on timed cycles with low or normal voltage, and can be incorporated in an exhibit by an electrical specialist.

- **Polarized animation** This can give the appearance of mechanical animation in all effects, using a rotating Polaroid filter and a simple light source.

- **Dioramas** These are three-dimensional exhibits with motion, changing colors, etc. produced by mechanical and electrical mechanisms.

- **Animated films** Movies and animated short films may be projected as color pictures onto rear-view projection screens, even under normal room lighting.

Art and Lettering

Artwork for direct use in exhibits can be produced as opaques, translucents, or transparencies, as well as in three-dimensional or flat form. A common method is to prepare typeset designs and then have them photographically enlarged to the desired scale. The prints are pasted or dry-mounted to panels for opaque display, or they may be used in negative or positive film form, sometimes combined with color transparency sheets, in light boxes or black walls.

By using the screen process, materials can be photographed or trimmed by hand to make the originals. Artwork can be done in four colors, using screen process work to print from screens of photographically reproduced black-ink proofs run from four-color process separations.

Hand lettering and art can still be highly effective for exhibit work, even in this age of advanced reproduction techniques; for many purposes, hand-lettered headings and small spot-renderings can add a unique, distinctive touch to a display at a lower cost than obtaining and mounting reproductions.

Available from manufacturers of transfer letters are display-sized die-cut lettering in a variety of colors. These may be used to set headings or body text on panels. The lettering may be of the pressure-sensitive type, or it may be applied from sheets from which each letter is impressed on the panel by burnishing. Another type of letter is the three-dimensional character, which may be cut out from wood, metal, or plastic, or cast from plaster.

For exhibit art there are obviously as many possibilities as there are materials and combinations, and there is no limit to the the effects that can be achieved through the creative imagination and ingenuity of the artist.

The Economic Dimension of Layout Drawings[1]

George E. Rowbotham

Layout drawings are road maps to optimum design and economy—without them emerges the agonizing substitution of haphazard drifting for quality design and drafting. Layout drawings are well-conceived precision graphics plans developed in the process of establishing a new or changed design; they are drawn to an accurate scale with only the critical tolerances and basic dimensions. They are prepared as a means of showing the configuration, contours, shapes, sizes, clearances, relationships, materials, and manufacturing and assembly of a group of parts or the assembly of subassemblies.

A layout drawing is not a manufacturing-released drawing; rarely is it released for production. However, it may be used to convey assembly information to the experimental shop. It may be issued as advance information to production for planning, costing, and educational purposes. It is a graphics document; it is the basis from which engineering detail and assembly drawings are produced by manual effort or by interactive graphics (IAG) or computer-aided design (CAD).

Money wasted is a frequent result of the preparation of detail drawings without adequate layouts. When the necessity for drawing quality and for conformance to contracts is acknowledged, getting the desired end results boils down to factors of time and cost. Experience has proved again and again that the preparation of detail drawings without adequate layouts is synonymous with shoddy practice.

AVOID FAULTY DRAWINGS

The result of faulty drawings is always financial loss—sometimes great loss. Such a menace, indeed, makes it difficult to be competitive. Faulty drawings usher in grief and

[1]Appreciation is extended to *Graphic Science* magazine, for portions of this chapter, and to General Motors Corporation for permission to reprint Figs. 3-2, 3-3, and 3-4, which appear in the *General Motors Drafting Standards Manual*.
Note: Figures 3-3 and 3-4 utilize the SI metric system in accordance with current standard practice at General Motors.

failure—scrapped parts, waste, changes, delays, troublesome departures from contractual documentation requirements which eventually must be corrected at the great expense of frayed tempers, wrecked schedules, overruns, high costs, confusion, and eventually poor customer relations. Management must therefore provide the same backing to the layout effort—the key to quality control in drawings—that it provides to the shop.

Although efficient checking will attain quality drafting, it is not the complete solution. Obviously, the answer in the interest of drafting economy and high output is to reduce the number of potential drawing errors before they reach the checker. This can be accomplished by getting the job started right, with layouts, and on time.

At the start, always, the principal design requirements should be known, understood, and completely defined on graphic layout drawings. Meetings should be held with the engineers and others to make certain that the layout accurately and completely defines the required design concept. Meetings should be held with sales, manufacturing, purchasing, quality control, and marketing personnel to make certain that the design is acceptable from the standpoint of customer requirements, ease of manufacture and assembly, availability of commercial parts, documentation requirements, and the like. Detail drawings should then, and only then, be prepared with adequate supervision and constant consultation. Finally, the drawings should be fully checked, not once, but at all stages.

With this approach, it may take a little longer to get the design and drafting effort started, but the time needed to complete the entire project will be substantially reduced. Consequently, there will be fewer errors, less confusion, fewer changes, less costly, wasted effort, less scrap, less misinterpretation, fewer overruns, and little or no need for the dreaded "panic button." For certain, the dollars-and-cents return will be substantially improved.

NO MORE CRYSTAL BALL GAZING

The detail drawing preparation effort can be reduced in costs, and output increased by IAG and CAD with adequate graphical layouts. This is the current trend.

The endless progress in design and drafting technology has had a mighty impact on the industry. Designers used to think in terms of thousands of drawings, or fewer, involved in a particular design; now they think in terms of millions. Consequently, management increasingly focuses critical attention on design and drafting competence as time becomes a more valuable commodity than ever. Managerial competence will more often be measured by an ability to translate continually increasing amounts of scientific breakthroughs and technological achievements into practical designs. Recent notable advances in management aids include the program evaluation and review technique (PERT) and computerized data processing, to name just two. Improvements are essential because as the economic forces settle into a new equilibrium, the competition will not permit delinquent methodology. Crystal ball gazing, it is said, has at last gone out of fashion. With all this technical brilliance, the question is, will quality design and drafting finally be recognized? Has the day arrived when design and drafting managers will no longer be required to bump their heads against the impractical? Will design and drafting people now make drafting decisions? Despite the crucial importance of quality drawings, the answer is, unfortunately, no. Obviously, management should invoke corrective measures.

Strangely, many "shortcut" addicts survive who advocate the release of unchecked drawings, simplified drafting, the elimination of layout drawings, and other panacean myths. Nothing may seem simpler or more obvious to answer than why quality drafting is needed, even to the point of finding it difficult to exaggerate.

LAYOUT SQUEEZE DILEMMA

Every experienced design-drafting manager knows that "muddling through" and aimless floundering are unavoidable alternates to layout drawings. Put in those terms, it sounds rather grim, but there is no other outcome. It doesn't take too much of a mental shake to envision the harassing problems and costly errors that arise when layouts are neglected. Let us begin by elaborating on the dilemma that can be avoided by producing layouts. Without layouts, drafters and checkers function by verbal orders, which may be conflicting; drafters remain idle while waiting for answers; engineers define and define again engineering parameters to designers, checkers, and others; there is constant duplication of effort in answering such questions as: What material? What fit? What are the goals? What hardness? Should it be a source control or specification control drawing? Where is it all heading? What finish? Should it be done over again? What customer specifications are involved? What type of weld?—and so on. Engineers may become more concerned with instruction than with engineering problems. Equally indicative of the neglected layout is the distressing customer relations that result from malfunctions, overextension of envelope parameters, interferences, adverse assembly conditions, and overruns in schedules and costs. The root of the confusion is the mistaken belief that it is less costly to correct errors than to avoid them. In fact, reams of painful drawing revisions may move in when there is a layout squeeze.

ESTIMATING

Another problem that arises when layouts are eliminated is shortsightedness in estimating. Admittedly, it is difficult to estimate manufacturing costs before detail drawings become available. Yet few companies can affort the luxury of carrying the development of a new product up to the point at which this costly detail is avoidable only to have manufacturing costs negate the product's profitability. Even worse is to lose money or even a contract because of under- or overestimating. No two people will come up with the same estimates—simply because there is no exacting formula to follow. What we try to do is evaluate something that doesn't exist, and where often there is little experience to serve as a guide.

It is known what will happen when water is heated to 212°F with surface pressure of 14.7 pounds per square inch. It will boil. This is because nature gives the same result every time under a given set of conditions. But in estimating, there are no such precise conditions, nor can anyone expect dependable results. There is no way of knowing, for instance, exactly what human performance can be expected, because of its unpredictability. What we can do is provide sufficient understanding of the design by a layout. Then it is possible at least to determine the number and complexity of detail drawings needed. With a layout it is possible to do something better than crystal ball gazing.

Another pitfall to be avoided is to carry the preparation of detail drawings up to a point when this costly detail is available, only to have checking reject the drawings or require extensive redraws or revisions. Estimates will be more accurate if the job is done right and done only once. Estimates are bound to be overrun when redraws and revisions pile up. Such waste and overruns can and must be avoided by competent drafting supervision and by the orderly planning afforded by layouts.

PROBLEM DEFINITION

Much difficulty caused by hurry can be avoided by the simple expedient of getting the job started on time, and started properly. The first step in the evolution of a design is to define the problem. If there is one thing designers and drafters have learned, it is the truth of the old injunction of the eighth-grade mathematics teacher: Don't worry about getting the right answer; what matters is setting up the right problem.

In design and methodology, a problem begins to be solved the moment it can be defined, the moment the right questions are asked, the moment the specifications are known and the solutions to the problem are determined. Problem statement is extremely important.

When the designer faces the problem, the number of facts and requirements is almost overwhelming. Failure to face, and define, the problem will cost valuable time, delay a solution, or prevent workable ideas from emerging. In creative problem solving, it is vital that the heart of the problem be clearly understood. For then it is known what the goals are, what fits, and what is relevant. Often the mechanical action, circuit scheme, fluid flow, gaseous flow, etc., can be portrayed by a schematic diagram. After the problem is defined and the requirements known, layout preparation can begin.

Layouts should be planned with order and should be completed step by step with adequate direction and consultation. Freehand sketches prepared in advance of layouts help to organize thinking, record known factors, and retain the concept in graphic form. Sketches are effective and economical in presenting various solutions to the problem. Very often, time will be wasted in solving a design problem if a scaled layout is prematurely started without adequate exploratory study. All this will not by any means slow progress to a snail's pace. If anything, work moves ahead faster when there is an absence of confusion and panic. The crash program therefore need not be ill-advised. Psychologists tell us that it can have beneficial results as long as it provides motivation rather than frustration. With this orderly and motivating approach, design and drafting may take a little longer to get started, but the elapsed time of the entire effort will be shortened. Certainly it will be accomplished more economically, with notable success.

DESIGN REVIEW

Whereas the designer is expected to cope personally with design problems, he or she must also consult others for specialized help. Probably the prime function of the supervisor is to see that this is done. In layout preparation the designer plays a central role—acting in effect as a sort of clearinghouse. The designer must understand the meaning of the old definition of rhetoric as "the art which draws men's hearts to the love of true knowledge." The designer concentrates on shape, relative positions, size, the interplay of moving parts and circuitry, and interprets requirements by consulting with engineers and others so that they will fit into an integrated whole. The hallmark of the designer's reputation is reliability. A designer has innumerable chances for making errors—in the interpretation of goals, in evaluation, in analysis, in integration, in computation, in circuitry, and in scale. Hence, inaccuracy must be guarded against at every point. The designer must move rapidly and yet cautiously. One effective way to benefit from the combined talents of specialists is by scheduling design review meetings. Such purposeful meetings are generally held prior to each of three important phases, namely: (1) the concept study; (2) the detail layout; and (3) the detail and assembly drawings. The purpose of the committee is to review and to improve, not to redesign or to appraise the designer's ability.

Attention must be focused on *what* is wrong, not *who* is wrong. The review committee contributes knowledge and experience on design, reliability, manufacturing, circuitry, repairability, utilization of existing knowledge, customer requirements, etc., that no one person could possess. It enhances the interchange of ideas. The result of this cooperative plan is a team effort involving many minds and many activites. From the start, the committee steers the designer in the right direction. Value engineering is injected into the first design. Such togetherness results in optimum design at minimum cost, consistent with customer needs.

To benefit from the experience of others does not mean that designers should rely less on their own initiative, resources, and creativity. In the design process, the designer must bear in mind that the more original an idea, the more vulnerable it is to criticism. This is probably because creativity is destructive of the accepted. Perhaps it is because there are more people who have built ramparts of security against the threat of change than there are those who welcome change and innovation. Perhaps too many are inclined to put emphasis on safe mediocrity. In any event, conservative complacency should not be carried to the point at which it will dilute the creative vision. Under no circumstances should progress be stifled. The designer not only must have the courage to stand up for his or her convictions and the wisdom to accept order but also must be sensitive to change and swift to modify. Alfred North Whitehead said it well: "The art of progress is to preserve order amid change, and to preserve change amid order."

ADVANCE PLANNING

Another fruitful virtue of layouts is their use in effectively communicating advance information to manufacturing and purchasing for their planning purposes. This permits a great deal of preliminary spadework. Manufacturing then has a longer "get-ready" period, and purchasing gets an earlier start in source searching, inviting competitive bidding, and, in some cases, advance procurement. Such forward planning is particularly important in the disposition of long-term items. Such cooperation among engineering, manufacturing, and purchasing is essential if tight schedules are to be met at a minimum cost. Here again, layouts are important; they help achieve this important goal.

THE OPTIMUM LAYOUT

Layouts must be complete and yet not overdone. This dictum may at first seem paradoxical. Without proper planning and control, layouts can be either inadequate or unnecessarily complex. Just as it is not economical to prepare detail drawings prematurely, work must be narrowed during the exploratory layout stage. Layout preparation must not be carried to the point at which costly comprehensive design detail is presented only to be rejected by a customer or at the design review meeting. The adequate layout is a balance between requirements, the known costs, resources, and time. At the beginning, at the point of genetic origin of the design, rough study layouts should be prepared in advance of detail layouts. The effort saved in the preparation of study layouts can be put to more productive use in the preparation of the approved layout, from which detail drawings will be generated.

The optimum graphics layout is an entirely adequate layout that is prepared within a minimum amount of time. This can be accomplished by establishing requirements on layout preparation, control, and responsibilities as outlined in this section. There is, of course, a great danger in overemphasizing an adherence to strict procedure. We must not let our thinking become distorted or curtailed by the magical effect of printed forms or restrictive legislation. Procedures are no substitutes for judgment and are useful only in the repetitive situation where good judgment has already been tested and proved. Certainly, engineering progress must not be narrowed, but the planning, discipline, and documentation of designs can and must essentially be reduced to maximum simplicity.

LAYOUT DRAWING EVOLUTION

For effective control, the development of the design phase should be accomplished in three stages. The first stage is the design concept layout (DCL). This is primarily a study layout that graphically presents a scheme for meeting the required basic design parameters. It presents various solutions to the same problem. This permits, at minimum cost, the evaluation of the feasibility or merit of an idea and aids in the selection of the optimum design approach. The second stage is the design approval layout (DAL), a further development of a design concept that has been selected through the evaluation of a number of DCLs. The purpose of the DAL is to present graphically to the customer and others for approval the design selected for development into a detail design. The DAL enables all concerned to evaluate the design from the standpoint of function, feasibility, and size without unnecessary detail. It should be not only to scale but also concise and accurate, showing the general design configuration, envelope requirements, and other elements of the basic design concept.

This advance layout leads to the optimum design approach and more accurate cost estimating. The design detail layout (DDL) is the final graphical design of a component or a group of components complete to the point at which detail and assembly drawings can be effectively generated from it. It should be to scale, accurate, and complete, showing the detail design configuration, and other essential engineering requirements, including envelope parameters, strength and hardness requirements, finishes, materials, and such. Careful consideration should be given to conformance with customer requirements and pertinent specifications, operation, reliability, manufacturability, adjustment, weight, assembly, installation, clearances, maintainability, adherence to standards, and such (Fig. 3-1).

INTEGRATED LAYOUT WORK AND CHECKING

An excellent illustration of effective layout planning is the cavity layout used, for example, in the design of jet aircraft engine fuel controls. These layouts speed drafting and avoid duplication of effort when a group of subunits, such as servomechanisms or relief valves, are enclosed in an extremely complex casting. With this plan, a separate cavity layout is prepared for each mechanism, showing the O rings, valves, pistons, sleeves, springs, spacers, seals, travels, passage locations, and surface texture, as well as dimensions and tolerances for diameters and depths of mating holes in the casting, etc. The checker who checks the mechanism parts also checks and is responsible for the passage intersecting points and for mating internal surface dimensions and tolerances of the casting in which the mechanism functions. The checker who checks the casting drawing does not duplicate this check, that is, does not recheck any internal detail that directly houses the mechanism. This person checks only to make certain that no interferences or breakthroughs exist between the previously checked critical

Layout item no.	New part	No. reqd	Part name	Ref part no.	Material	Heat treat	Remarks	Materials approved	Detailer	Checker
1		1	Valve	26475	AMS 5632 cres	R_C 55-60	————		D.D.	P.Y.
2	Change 26476	1	Spacer	——	——	——	Change as shown		D.D.	P.Y.
3		1	Piston	26471	AMS 5610 cres	R_C 26-34			D.D.	P.Y.
4		1	Seal	26475	Leather krome K-210	——	Source: Chicago Rawhide Co.		D.D.	P.Y.
5		1	Spring	26271	AMS 4725 berylium copper	——	Per shunt memo marked print dated 6/4/—		D.D.	P.Y.
6	Change 26083		Valve assembly complete							

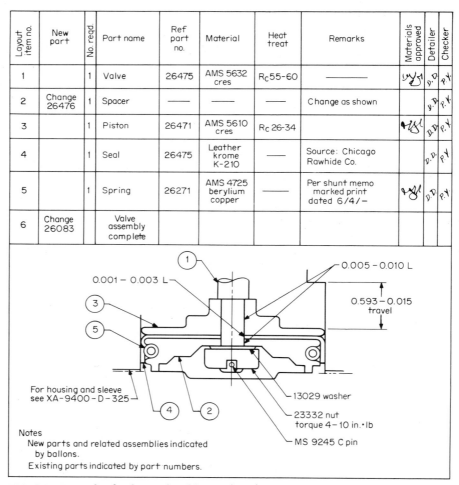

FIG. 3-1   Example of a design detail layout (DDL).

internal casting surfaces that enclosed the various mechanisms involved, that is, checks the configuration. This procedure permits a substantial reduction in the time required to check the casting. For example, on a crash program, when seven different mechanisms and a complex casting are involved, eight checkers can check the same job simultaneously and efficiently. A checking job that might normally take months of work can be completed in weeks.

AUDITING LAYOUTS

Many companies find it worthwhile to audit layouts on a random basis. Usually two or three layouts for each designer are audited yearly. Such an audit is not intended to evaluate design but is concerned primarily with quality layout drawing practice. Layout audit check sheets are often used that list the items which must be considered in the audit, i.e., accuracy, legibility, conformance with standards, completeness, manufacturability, assembly, and originality. Each item is rated by a value such as 0 = Excellent, 1 = Good, 2 = Acceptable, and 3 = Not Acceptable.

The defect average per layout is computed by adding the sum of the defect values and dividing by the applicable number of items on the checklist. Generally, a defect acceptance average should not exceed a value of 2.

OTHER FREQUENTLY USED LAYOUTS

1. *System control layout*   The system control layout (SCL) controls and documents the interfaces and parameters for a unit or system. It includes system configuration, load data, envelope sizes, locations of primary components, etc.

2. *Specialized electronics layouts*   Specialized electronics layouts are prepared for terminal boards and for circuitry, such as printed circuits, welded circuits, and thin-film microelectronic circuits.

3. *Geometric layout*   A geometric layout is a motion analysis of moving parts. It develops paths of motion and determines clearances of parts in the proximity of bodies that are in motion (Fig. 3-2).

4. *Dummy layout*   The dummy layout is prepared independently of the main layout, on a separate sheet. The dummy is generally prepared without regard to any specific position, as it will be oriented on the main layout to simulate the various positions the part will take while in operation.

5. *Clearance layout*   The clearance layout shows the interferences, clearances, and tolerances that the design will allow under normal and abnormal operating conditions (Fig. 3-3).

6. *Tolerance accumulation layout*   This layout shows the possible results of accumulated tolerances and where tolerances may be increased to ease manufacturing. This type of layout is drawn with the parts in their tolerance extremes, either maximum or minimum, depending on the problem. Opposite extremes may be shown on either side of the centerline of a symmetrical view (Fig. 3-4).

7. *Checking layout*   The checking layout is prepared to determine the manner in which the parts of the design will be assembled and serves as a check of dimensions.

8. *Partial layout*   A partial layout is a layout of a portion of a product or unit and includes a logical reference base.

9. *Rigid bent-tube layouts for computers*   This type of layout establishes the geometry and routing of tubes, showing bends that must bypass other components, end points, intersection points, etc. The layout information is transformed, by coordinates, to a computer input data sheet. From this, the computer provides the data required to manufacture the configuration of the tube, such as straight lengths, angles of rotation, and bend angles.

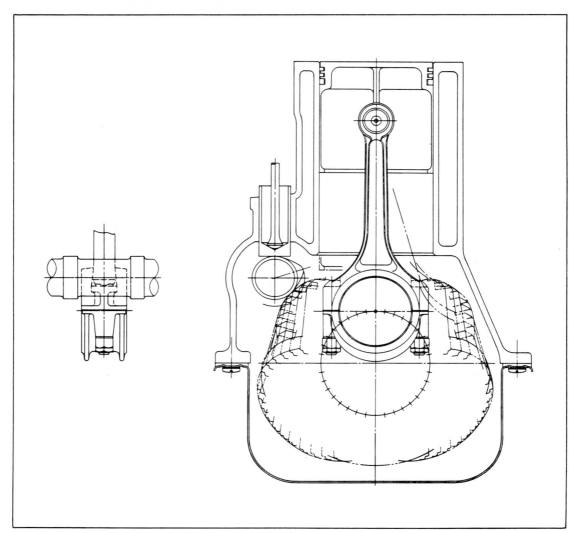

FIG. 3-2 Sample geometric layout.

LAYOUT SELECTION, PREPARATION, AND CONTROL

Suggestions on how to achieve orderly layout preparation and control at minimum cost are outlined here:

1. *Layout type selection* Layout costs can be reduced to a minimum by optimum layout type selection. This can be done by developing the design layout in the orderly stages described in this section.

2. *Responsibility* The designer should prepare and be responsible for the accuracy and completeness of the design layout. A security classification should be shown on drawings warranting a classification of "Top Secret," "Secret," or "Confidential." The designer should be responsible for documenting layout security classification.

   Drafters and checkers should be responsible for the quality of detail and assembly drawings, wiring diagrams, lists, and other documentation released to manufacturing. The detail and assembly drawings should meet such requirements as conformance with the design layout, accuracy, clarity of presentation, conformance with standards, reproduction legibility, ease of manufacture, proper assembly, adequate clearances, and proper part names and numbers. These requirements include correctness in dimensions and tolerances; specification of materials, finishes, and heat treatments; sufficient wall thicknesses; and other data required for castings, wiring diagrams, gears, sheet-metal parts, springs, and such. Quality drafting results from quality layouts.

3. *Design review* Layout drawings must be reviewed periodically by the design review committee.

4. *Accuracy* Layouts should be carefully prepared and accurate in every respect.

5. *Scale* Whenever practical, the scale of the layout should be full size. Increased or reduced scales may be used when the part or parts are too small to be conveniently shown, or too large to be drawn on standard sheet sizes.

6. *Drawing space requirements* Consideration should be given to drawing area space required for layout information, supplementary views, and listing of design requirements, as well as anticipating the need for additional information. The designer should understand the problem so that he or she can arrange an orderly layout. Overlapping views, views that interfere with the title and revision block, insufficient room to make the necessary projections, or wasted space between views, etc., should be avoided.

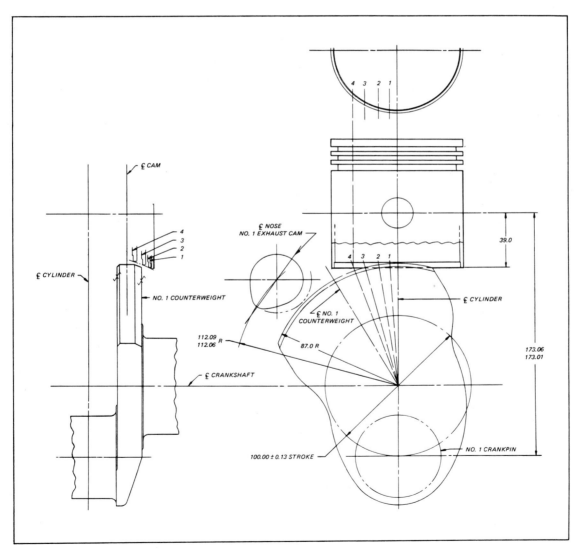

FIG. 3-3 Illustration of a clearance layout.

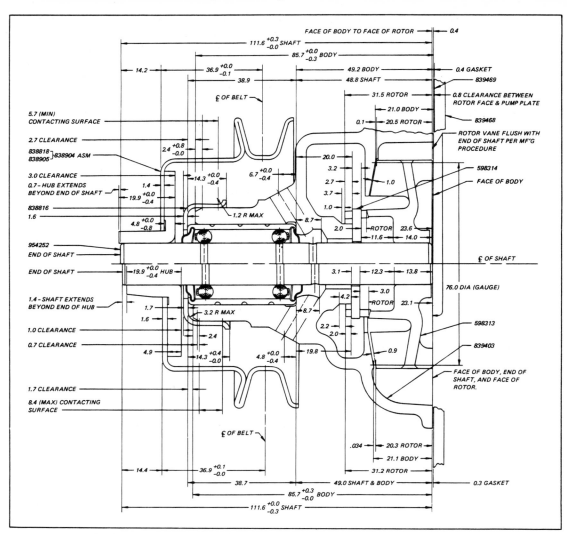

FIG. 3-4   Example of a tolerance accumulation layout.

7. *Completeness*   Layouts need not be so complete that nothing remains but to trace and add dimensions and notes to create detail drawings. (*a*) The representation of a design, with sufficient portrayal and essential information, is very important. An excellent design, if not clearly and accurately shown, could be incorrectly detailed. (*b*) Sufficient portions of the envelope and adjacent parts should be delineated by the use of phantom lines to show interrelationships and to avoid interferences. (*c*) Design conditions requiring special consideration by the drafter should be noted. Among these special conditions are all special clearances, adjustments, equipment provisions, critical dimensions and tolerances, strength and hardness requirements, and other special requirements not readily discernible by the drafter. (*d*) A list of materials should be included, listing materials, heat treatments, part names, reference numbers, etc. (*e*) Required data for springs, gears, welding, surface texture, and the like should be specified. (*f*) Each vendor-supplied item and the classification of the drawing (specification control drawing or source control drawing) should be specified. The name(s) of the vendor(s) should be listed, as well as their part identification, if available. (*g*) The sequence of assembly of parts should be noted on a layout when this sequence is not obvious. (*h*) When detail occurs many times in a regular pattern, such detail need be shown only once, with a suitable note defining number, location, and so forth.

8. *Linework* The linework should be in conformance with the specifications set forth by the design/drafting manual (D/DM). When reproductions are required, the linework should be of sufficient density to produce legible copy. It is helpful to those who must interpret layouts that the essential parts of the design be emphasized and the background subdued. This contrast can be accomplished by indicating in phantom view those parts which merely show the relationship of the design to other adjacent parts of structure. In some cases, it is preferable to draw the reference parts and structure in their opposite view and use the reverse side of the sheet for the basic layout. Erasures may then be conveniently made to either the reference structure or the basic layout without disrupting the other.

9. *Cross-sectioning* In the interest of economy, it is generally not advisable to cross-section, owing to the extra effort involved and the difficulty in making revisions.

10. *Multisheet layouts* Parts, assemblies, or installations that are too large or too complicated to be shown on a single standard-size sheet may be drawn on multiple sheets. When subsequent sheets are a continuation of a main layout, all applicable sheets should contain common reference lines to coordinate the relationship of views between the related sheets.

11. *Cross-references* Layouts should be cross-referenced in such a manner that the most current information is easily obtainable.

12. *Reference data* Design information and identifying reference data, such as drawing numbers, item names, and other pertinent information should be indicated.

13. *Standardization* Standard parts, materials, and practices should be used except when functional or design limitations dictate exceptions. Consideration should also be given to the utilization of parts or approaches used in previous designs, etc.

14. *Patents* It is important that patentable devices and ideas be submitted to the patent section for review. Such information should include prints and other data helpful in disclosing the subject. The first sketches or layout drawings of devices designed, and incorporating ideas that may possibly be patentable, should be signed and dated by the inventor, and also signed by at least one witness, preferably a witness who fully understands the design. The date the drawing was begun should appear in the title block.

15. *Revisions* It is necessary to keep layouts up to date for a certain period, and in some instances, continually. Noting progress, with records of revisions in design together with dates, should be recorded in the revision block. The method for recording layout revisions is generally the same as that used on detail drawings.

16. *Retention period* Design layouts are recorded and filed to retain a record of the design development. After use, or the expiration of a predetermined retention period, layouts may be destroyed, after they have been microfilmed.

DESIGN REVIEW AND CHECK QUESTIONS

A checklist of the items that generally require consideration in the review and checking of layouts are presented herewith. They are grouped in eight classifications, namely: (1) drawing considerations, (2) design considerations, (3) electrical and electronic considerations, (4) human factors considerations, (5) environmental considerations, (6) fabrication considerations, (7) installation and maintenance considerations, and (8) procurement considerations. Items in each classification are listed in alphabetical order.

1. Drawing considerations

   Appearance Does the general appearance of the layout conform to what was approved at the design review meeting? Is the layout clear and complete?

   Changes Are changes fully and clearly listed?

Comparison of layouts How does the layout compare with previously prepared layouts of similar designs?

Delineation Are figures, letters, and lines correctly formed, uniform, and clean? Are they sharp and dense enough to ensure legible copy?

Layout conformance Are the design requirements authentically portrayed? Does the layout require change, resulting from design review discussion?

Numbers Are the correct drawing numbers and reference numbers given?

Out-of-scale dimensions Are out-of-scale dimensions underlined to indicate that they are out of scale? Will they create trouble? Have they been approved by supervision?

Parts lists Are necessary parts listed in the list of materials?

Scale Is the layout to scale?

Signatures Have the required approval signatures been obtained?

Titles Has the title block been completed, and is the information correct?

Views Are sufficient views shown, and are they in proper relation to each other? Are directional arrows on cutting planes properly shown?

2. Design considerations

Adjustment Are proper adjustments provided? Are the adjusting mechanisms accessible? Do they cover the range required? Are positive locks provided? Are factory-only adjustments protected against tampering? Are adjustments identified where required? Is the range sufficient to provide the adjustment required under adverse tolerance conditions?

Applied surface finish Are applied surface finish requirements stated?

Balance Do any parts require a balancing control note?

Center of gravity Is the location of parts such that the equipment center of gravity is too high (or too far forward), resulting in a resonance in the vibration test range?

Clearance Do parts clear each other for assembly or operation?

Customer Are customer requirements met?

Data Are the required data for gears, splines, serrations, springs, bellows, cams, etc., specified? Are the data correct?

Design Is the principle of the design correct and practical? Are functional requirements met?

Dissimilar metals Are dissimilar metals in contact in such a way that corrosion can be expected?

Drawer slides Are drawer slides adequate for the load, and are they the type that will not bind? Do they include automatically operated locks that hold the unit or chassis in maintenance position?

Duplication of parts Could parts with slight differences be made identical?

Expansion Are dimensions and tolerances adjusted for thermal or hygroscopic expansion or contraction during operation?

Gasketing Is adequate gasketing provided?

Grain flow If the part is made from sheet metal, has the direction of grain flow been indicated where this is important to increase the durability?

Grommets Are wires and cables running through holes in metal partitions protected from mechanical damage by the use of grommets or other suitable means?

Guide pins Are guide pins provided where required?

Interchangeability Have requirements for interchangeability been considered?

Limit accumulations Are undesirable limit accumulations avoided?

Locking feature If required, does the part include a feature for ensuring its retention (wire hold, tab washer slots, or similar feature)?

Lube, drain, vent, bleed   Are holes or other features required for any of these or similar features?

Material   Are the required materials and heat treatments specified?

Motions   Have motions and their sequence been analyzed? Is the timing correct?

Patents   Have devices or ideas that may be patentable been submitted to the patent section for evaluation?

Pressure test   Does any part require pressure testing?

Protective devices   Are protective devices necessary to eliminate overstressing of delicate parts due to rough or careless handling or overloading?

Register marking   Are timing marks or some equivalent registration feature required? Are they specified?

Rigidity   Do parts have sufficient rigidity? Will structural instabilities result from mechanical parts of insufficient rigidity?

Seals   Are seals provided where needed, and is the correct type used? Can seals and O rings be assembled without damage?

Secured   Are parts secured, preventing damage from shock and vibration? For example, are terminal boards or other assemblies improperly cantilever-mounted?

Seizure   Are materials and surface treatments of contacting parts resistant to seizing, galvanic action, and similar effects?

Shear strength   Are the diameters of screws, bolts, rivets, pins, etc. large enough to withstand the applied loads? Are massive parts secured with screws of sufficient strength?

Shielding   Is sufficient shielding provided?

Shock mounts   Are the selection and location of shock mounts correct?

Standards   Are standard parts, designs, materials, and processes used wherever applicable?

Strength   Is design adequate from a strength standpoint?

Sufficient material   Is there sufficient wall thickness? Are there any possibilities of break-throughs or such?

Surface texture (surface roughness)   Are surface texture (roughness) values specified where required? Is the correct symbol used?

Test requirement   Are special inspection and functional test requirement notes or specifications stated?

Tolerances   Are the critical clearances and tolerances specified?

Weight   Are weight requirements met?

3. Electrical and electronic considerations

Arcing   Are electrical arcing hazards avoided?

Conductor width   Are conductor widths in conformance with accepted design standards?

Harnesses   Are cable harnesses adequately supported to prevent fatigue of wires at rigid termination points? Is sufficient room provided for the harnesses?

Heat-sink provisions   Are heat-sink provisions provided?

Printed-circuit patterns   Are printed-circuit patterns in conformance with accepted design standards?

Relays   Will a relay open under vibration?

4. Human factors considerations

Controls   Are controls located within the normal reach of an average or small operator? Are controls designed so that an operator can adequately grip for turning, twisting, or pushing? Are the controls designed to an operator's expectations, as derived from previous experience, as well as certain handiness factors?

Illumination   Has adequate illumination been provided? Does the design eliminate glare hazards, such as brightly polished bezels, glossy enamel finishes, or highly reflective instrument covers?

Indicators   Are visual indicators mounted so that operators can adequately see scales, indexes, pointers, and numbers?

Multiple functions   Does the design require an operator to perform too many functions simultaneously?

Operator capabilities   Does the design require an operator to perceive and handle information at faster rates than is humanly possible?

Physical strength   Does the design require an operator to exceed reasonable limits of physical strength?

Safety   Are provisions made to prevent personnel from coming into contact with voltages in excess of 70 volts rms? Are adjustment screws or other commonly worked-on parts located near unprotected high voltages or hot spots? Are guards, safety covers, and warning plates provided for electrical potentials in excess of 350 volts rms on contacts, terminals, and similar devices? Are all external metal parts, control shafts, and bushings grounded? Are personnel protected from moving mechanical parts? Are locking mechanisms for doors and drawers designed to prevent injury to the operator when the lock is released? Do doors or hinged covers have rounded corners and positive hold-open devices? Are personnel protected from imploding cathode-ray tubes? Are voltage dividers provided with test points for measurement of voltages in excess of 1000 volts? Are interlocks provided where potentials exceed 70 volts rms?

Toxic or hygroscopic materials   Has the use of toxic or hygroscopic materials or materials that will support combustion or fungus been avoided?

Variability   Does the design require an operator to make judgments more accurately than is possible within the normal range of human capabilities?

Warning lights   Are warning lights compatible with the ambient illumination levels that are expected? For example, a dim light cannot be seen in bright sunlight, and a bright light might be detrimental in a dark environment.

5. Environmental considerations

Condensation   Will the design prevent condensation of moisture in the equipment?

Temperature control   Has the necessary temperature control been provided?

Ventilation   Is ventilation adequate?

Waterproof   Is the equipment dripproof, splashproof, or watertight, as required?

6. Fabrication considerations

Brazing, soldering, or welding   Are notes or symbols included, as applicable, for brazing, soldering, or welding?

Economy   Can the design be manufactured and inspected economically? Would full or partial redesign facilitate production?

Inspection processes   Are any magnetic-particle, fluorescent-penetrant, x-ray, or similar inspection processes required, and if so, are they noted?

Welding and riveting   Has accessibility been provided for welding and riveting tools?

7. Installation and maintenance considerations

Assembly   Can parts be assembled or disassembled for servicing without unreasonable complications? Should a dowel, elongated hole, or similar feature be provided as a guide to aid in assembly?

Built-in test equipment   Is built-in test equipment provided where required? For example, where frequent observations are necessary, where measurement is critical to part life, or where testing requires disassembly of equipment or a transmission line.

Clearance   Does the part clear other parts in operation or for assembly?

Connectors   Are connectors accessible for testing and service?

Downtime   Has potential downtime been minimized by the use of replaceable assemblies?

Envelope limits   Are the envelope limits clearly shown? Are all parts within the envelope?

Puller feature   If a part has a tight fit, does it require a puller lip, a jackscrew thread, a knockout hole, or some similar extraction feature?

Removability   Is the unit or chassis completely removable without a need for extensive disassembly? Are handles or bales provided for removing units or chassis from enclosures?

Rivets   Are rivets used for mounting parts that may be subject to replacement for electrical continuity?

Room   Is sufficient room provided to remove and replace parts?

Special tools   Has the need for special tools been minimized?

Stacking assemblies   Is maintenance complicated by stacking assemblies and parts so that many of them must be removed to repair or replace one or a few?

Test points   Are test points systematically arranged, making maintenance less difficult and time-consuming?

Tool clearance   Has adequate clearance been provided for wrenches or other assembly tools?

Torque values   Have required wrench torque values been indicated where critical parts are assembled by bolts, screws, or nuts?

Visual inspection   Is the equipment designed to permit thorough visual inspection of all parts so that obvious failures, such as open tube heaters, leaking capacitors, burned resistors, and broken wires and terminals can be located quickly?

8. Procurement considerstions

Procurability   When a part is a vendor-supplied item, or includes vendor-controlled features such as materials used or process operational devices, is that part readily available to these specifications?

Restricted procurement   Is procurement that is restricted to a particular vendor or vendors specified? Have such restrictions been approved?

Vendor-supplied items   Is any classification specified (in specifications or source-control drawings) for vendor-supplied items?

SUMMARY

A diagnosis of the design-drafting operation may possibly be achieved by asking the following questions:

1. Is the problem defined?
2. Are adequate layouts prepared?
3. Are sketches prepared in advance of layouts?
4. Is there sufficient consultation on design?
5. Is maximum benefit derived from the designer's initiative, resources, and creativity?
6. Is layout preparation carried to the point at which costly comprehensive design detail has been completed, only to have the layout rejected by a customer or at the design review meeting?
7. Has the preparation of drawings been carried too far into detail, only to have checking reject the drawings or require extensive redraws or revisions? Have errors been prevented by competent supervision?

8. Has the preparation of detail drawings been carried all the way through, only to have the manufacturing costs involved in the particular design negate the product's profitability?

9. Have the items included in the checklist been considered?

10. Have producibility and possible cost reductions received adequate attention?

11. Have ample space allocations and clearances been provided.

12. Has sufficient structural strength been provided for all parts and assemblies?

13. Has adequate attention been given to service problems?

14. Have basic dimensions and other information required for checking been included?

15. Has the feasibility of incorporating any major or minor changes been considered?

Dimensioning and Tolerancing

Gary Whitmire

INTRODUCTION TO DIMENSIONING AND TOLERANCING

The objective of all dimensioning and tolerancing on drawings of fabricated parts is the specification and control of *size, form,* and *position*. Achieving this objective in a logical, economical way is frequently a puzzling problem.

Size Control

All tolerances of size define a tolerance zone. Control of the size of a machine component is frequently determined by consideration of weight and strength or rigidity requirements. Tolerances of size should be determined by:

1. Dynamic constraints
2. Problems of assembly

4-1

3. Economics

4. Any other functional requirement that the designer is able to identify

There are many cases in which the successful assembly and function of a machine are not dependent on precise tolerances of size. For these cases the effort put into tolerancing should be minimized.

Form Control

All tolerances of form define a tolerance zone. Control of the form or geometry of a machine component is made necessary by dynamic considerations, such as the need to balance a high-speed rotating member or the need to maintain a functional relationship among several parts of the same machine, as would be the case for a lathe or a planer. Tolerances of form are:

1. Implied by tolerances of size.

2. Specifically stated by explanatory notes.

3. Specified by a set of symbols adopted in recent years by the American National Standards Institute and set forth in the standard ANSI Y14.5. These symbols are intended to minimize any ambiguity concerning the designer's intentions.

Some form tolerances control an entire surface, whereas others control one line at a time (a single element) of a surface. In the second instance, each line is judged independently of every other line in the surface.

Position Control

All tolerances of position define a tolerance zone. Control of the position of holes, slots, shafts, or any other feature of a fabricated part is frequently required to ensure that the part can be assembled without rework or preselection. One very common example involves the production of matching hole patterns on mating parts that are required to line up interchangeably regardless of the distribution of hole sizes and hole positions within their respective tolerance zones. Position dimensions are always being generated from a point, line, or plane. This combination of points, lines, and planes is part of a reference datum system of three planes, each one perpendicular to the other two.

GENERAL DISCUSSION

Datums

A *datum* is the origin, or beginning place, for any measurement. Datums are points, lines, or planes that are considered perfect in every respect. A datum point, for example, has a position but no measurable size. Datum lines have length but no lateral dimension and are considered to be ideally straight. Datum planes have length and width but no thickness. They are thought of as being ideally flat. All of these are abstract mathematical concepts that provide a useful basis for devising a system of measurement.

Datum Features

A *datum feature* is a real thing that approximates the ideal. The edge of a well-made steel scale is a good approximation to a straight line, but a well-collimated beam of light is an even better one. The surface of a micrometer anvil, toolmaker's flats, gage blocks, and high-grade surface plates are good approximations to an ideal plane. All of these are used as datum features. It is important to understand that, even though we specify datum features on the drawing of a fabricated part, we actually are measuring from the gaging surfaces in contact with the actual part.

Maximum Material Condition (MMC)

This expression describes a workpiece at its heaviest weight. Two examples are given for better understanding. The first is a simple cylindrical shaft as shown in Fig. 4-1. The maximum material condition (MMC) of the 1.875 ± 0.005 diameter is 1.880. For the 4.25 ± 0.01 length, the MMC would be 4.26. The second example, shown in Fig. 4-2, is a block with a hole in it. The MMC for each of the three given dimensions shown in Fig. 4-2 is:

4.25 ± 0.01	4.26
1.12 ± 0.03	1.15
1.865 ± 0.005	1.860

With respect to the hole, the workpiece will be heaviest when the least material has been removed from it—in other words, when the hole is at its smallest diameter.

Maximum material condition is most frequently specified in its abbreviated form, MMC, or as the symbol Ⓜ.

Virtual Condition or Virtual Size

The *virtual condition* of any feature is the effective size of that feature. The virtual condition combines the influence of MMC with the position tolerance *or* the geometry tolerance. The concept of virtual condition is extremely important in the determination of tolerances that will guarantee the ability to assemble mating components even when the most adverse permissible combinations of size and position or size and form deviations from the ideal are present. Figure 4-3 shows a simple assembly of two parts that illustrates how this concept may be used in a tolerance analysis. It is easy to determine the designer's intentions for this assembly. The designer wanted the pin of workpiece no. 2 to enter the hole of workpiece no. 1 when both parts were assembled such that datum features A, B, and C were in contact. If the production of these parts were to be limited to small quantities that could be selectively assembled, then rework (filing or reaming) during the assembly procedure would be acceptable in order to satisfy the designer's intentions. There are a great many situations when this is the case, and a mismatch at assembly does not present a serious problem. There are also a great many

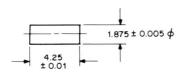

FIG. 4-1 Maximum material condition of a shaft.

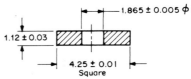

FIG. 4-2 Maximum material condition of a square.

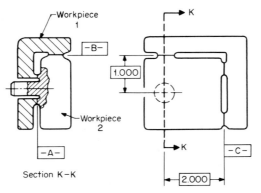

FIG. 4-3 Virtual condition assembly.

situations when production quantities are large and rework during assembly, or, alternatively, selective assembly, is highly undesirable and possibly very expensive. Such circumstances call for interchangeable assembly under the worst case of adverse tolerance combination. We can visualize one configuration of the assembly that represents a worst-case combination:

1. The hole is at its smallest size (i.e., MMC size).
2. The pin is at its largest size (i.e., MMC size).
3. The hole has been produced with its axis shifted upward by the maximum amount permitted by its position tolerance.
4. The pin has been produced with its axis shifted downward by the maximum amount permitted by its position tolerance.

Another possible worst case would occur, for example, if the shift of axes were reversed: the hole shifting downward, and the pin shifting upward. This configuration and the calculation of the virtual condition of the hole and the pin are shown in Fig. 4-4. Note that the virtual condition for both the hole and the pin must be the same in order to guarantee interchangeable assembly. This virtual condition also represents a gage size. That is, a gage for the hole can be a pin equal to the MMC of the hole minus the hole positional tolerance, and the gage for the pin will be a hole equal to the MMC of the pin plus the positional tolerance for the pin. The designer may want to make the hole a little larger or the pin a little smaller so that a line-to-line fit will not exist at the worst condition.

Note the following points about Fig. 4-4: (1) The space available in the hole for passage of the pin at assembly has been reduced by the shift of the hole; and (2) the space required in the hole for passage of the pin at assembly has been increased by the shift of the pin.

It is the designer's job to arrange matters so that the space available equals the space required under the worst-case condition. From here on, space available will be referred to as the *virtual size hole*, and space required as the *virtual size pin*. This provides us with a very powerful and useful general design equation that may be applied to any assembly situation analogous to the one shown. This equation, that virtual size hole equals virtual size pin, may be expressed as:

$$\text{MMC}_{\text{hole radius}} - \tfrac{1}{2} \text{ hole tolerance} = \text{MMC}_{\text{pin radius}} + \tfrac{1}{2} \text{ pin tolerance}$$

Multiplying each term in the equation by 2 yields:

$$\text{MMC}_{\text{hole diameter}} - \text{hole tolerance} = \text{MMC}_{\text{pin diameter}} + \text{pin tolerance}$$

Later, we will analyze many different situations using this fundamental relationship. For now, it is good to repeat the essential ideas the relationship expresses:

- The effective, or virtual, size of a *hole* is *reduced* because of position or form tolerances.

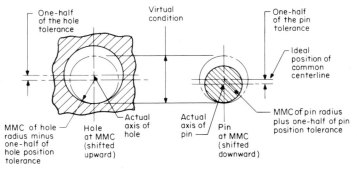

Hole − position tolerance = virtual condition
Pin + position tolerance = virtual condition

FIG. 4-4 Virtual condition definition.

• The effective, or virtual, size of a *pin* is *increased* because of position or form tolerances.

Coordinate Dimensioning and Tolerancing

In a later discussion (see Datum Reference Frame), a reference system of three planes, each one perpendicular to the other two, is explained. These planes intersect, forming three lines, each one being perpendicular to the other two. Through long years of custom, people in industry have come to think of these lines as *axes* known as the X, Y, and Z axes. It is conventional to place two of these axes, the X and Y, on shop fabrication drawings. Furthermore, by convention, the X axis is parallel to the horizontal edge of the drawing and the Y axis is parallel to the vertical edge, as shown in Fig. 4-5. These lines are called *coordinate axes*, and in most cases all dimensions represent distances measured parallel to one or the other of these two axes. Some circumstances require dimensions to be shown at an angle to these coordinate axes. The practice in such a case has been to imagine an auxiliary pair of coordinate axes, again perpendicular to one another, with the pair rotated through the proper angle (see Fig. 4-6). The tolerance applied to *each* dimension is considered to be effective in a direction parallel to the appropriate reference axis and in no other direction. This kind of tolerance is known as a *coordinate tolerance*. If the tolerance is to be applied to only one side of the ideal position specified by the dimension, it is known as a *unilateral coordinate tolerance*. If the tolerance is to be applied on both sides of the ideal position, it is called a *bilateral coordinate tolerance*. Examples of both kinds and the tolerance zones formed by them are shown in Fig. 4-7.

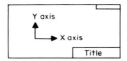

FIG. 4-5 Coordinate dimensioning drawing relationship.

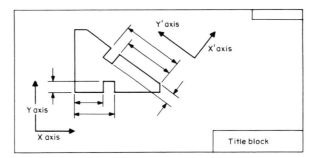

FIG. 4-6 Coordinate dimensioning, oblique.

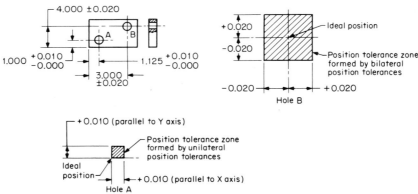

FIG. 4-7 Coordinate tolerance zone, unilateral and bilateral.

In each case, the *total coordinate tolerance* (TCT) is the width (*X* direction) and height (*Y* direction) of the tolerance zone formed by the given tolerances. For hole A there is a TCT in the *X* direction of 0.010 and a TCT of 0.010 in the *Y* direction. For hole B the TCT in the *X* direction is 0.040 (0.020 + 0.020). In the *Y* direction, again, the TCT is 0.040. It is, of course, possible to specify bilateral coordinate tolerances that are not the same on both sides of the ideal position (see Fig. 4-8). In this case the TCT in the *X* direction is 0.035 (0.005 + 0.030). The TCT in the *Y* direction is 0.025 (0.010 + 0.015). It is universally accepted practice to say that if the axis of the hole falls entirely within the tolerance zone, the position of the hole is acceptable. It is also accepted practice to say that because the tolerance zone exists throughout the full thickness of the workpiece, the perpendicularity of the axis (and therefore the hole) is governed by the size of the tolerance zone, as shown in Fig. 4-9. An even greater enlargement of the tolerance zone is shown in Fig. 4-10. In this illustration, the position shown for the actual axis falls entirely within the tolerance zone and must be considered to be acceptable even though the hole is "tipped."

Basic Tolerance

We have referred several times to the "ideal dimension." This is the dimension we would like the shops to produce. The tolerance applied to the ideal dimension is the deviation from the ideal which we think may be safely tolerated. Another name for the ideal dimension could well have been the *basic dimension*. This term has been adopted

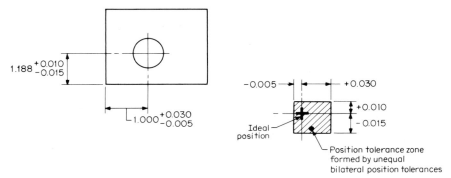

FIG. 4-8   Coordinate tolerance zone, unequal bilateral.

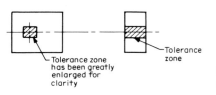

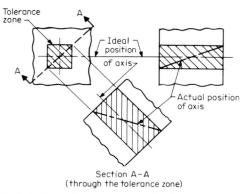

FIG. 4-9   Tolerance zone is full thickness of metal.

Section A–A
(through the tolerance zone)

FIG. 4-10   Tolerance zone at the hypotenuse.

in the ANSI Y14.5 standard to indicate the ideal dimension. The recommended practice is to enclose the basic dimension in a rectangular box: [1.875]. The use of this symbol also indicates that the tolerances shown on the drawing are to be interpreted according to the rules of ANSI Y14.5. These rules have come to be known by two names, both meaning the same thing:

1. True position dimensioning and tolerancing
2. Geometric dimensioning and tolerancing

True Position Dimensioning and Tolerancing

This notable system introduced several changes to the ways of thinking about dimensions and the tolerances associated with them. The intention was to make the job of manufacturing and inspection easier and more economical without any sacrifice in the quality of the product. These changes have already been successfully and profitably implemented in many large manufacturing plants. Future users must, however, understand these changes for and by themselves before they, too, can do the same. Let's examine the old and new systems in detail.

Old System	Geometric Dimensioning and Tolerancing
1. Position tolerance zones were square, rectangular, or angular.	1. Position tolerance zones are cylindrical.
2. The tolerance was fixed on the drawing and could not be changed except by a rather expensive administrative procedure.	2. Designers have two choices: (*a*) they can make the tolerance fixed in value, as in the old way, or (*b*) they can make the tolerance variable within limits that are determined by the actual size of the controlled feature. This variability is communicated by proper placement of the symbol Ⓜ on the drawing and requires no further administrative review.
3. Both the size and position of the feature had to be independently measured and evaluated for acceptability.	3. (*a*) The size and position *may* still be independently measured in the old way, or (*b*) acceptability *may* be evaluated by using a simple gage, with no other measurements being required.
4. For any sequence of features controlled by a "daisy chain" of dimension, the actual axis of any single feature became the datum for the next feature in line. This often led to the accumulation of tolerances additively, which in turn created many problems upon assembly of mating parts.	4. The ideal or basic position of any feature is used, as before, to evaluate the acceptability of the actual axis of that feature. The ideal or basic position (not the actual axis) is used as the datum for the next feature in line. This eliminates the possibility of accumulating tolerances. Base-line dimensioning and point-to-point dimensions are, therefore, both interpreted in the same way.
5. For any related pattern of features, such as a series of holes, *each* position dimension had its tolerance stated right next to the position dimension.	5. For any pattern of related features the position tolerance is *not* stated right next to the position dimensions. It is specified only once, along with the size and size tolerance of the feature. It applies to each and every feature in the pattern.
6. The desired control of the form of a feature was specified by an auxiliary note (or set of notes) on the face of the drawing. The language varied from place to place and from drawing to drawing according to the language skills of the drafter. As a result, so did the interpretations of these notes.	6. A unified set of symbols, defined by words and pictorial representations in ANSI Y14.5, is used to specify the desired control of form. The opportunity for variable interpretation of any given symbol is virtually eliminated.

Changes Authorized by ANSI Y14.5

These changes are few in number, but they are quite significant. Some elaboration will help in the understanding of just how useful they are.

Change 1: Square to Round Tolerance Zone

The square or rectangular total coordinate tolerance zone is replaced by a round tolerance zone by circumscribing a circle through the corners of the old total coordinate tolerance zone.

Figures 4-11 and 4-12 illustrate the old system. According to the old rule, any position that is not outside the total coordinate tolerance zone is acceptable; any position that is outside of the total coordinate tolerance zone is unacceptable. Note that in Fig. 4-11, the "acceptable position" shown is farther away from the basic position than the unacceptable ones are.

Is this logical? Why are deviations along the diagonal of the total coordinate tolerance zone less damaging to proper function than deviations along the coordinate axes? The answer is: This is not logical! Deviations along the diagonal are as damaging to ease of assembly as are deviations in any other direction. A more logical approach would be to evaluate deviations from the basic position in *any direction* on an equal basis. The only shape which permits this evaluation is the circle. It therefore is the recommended shape for position tolerance zones. Shown in Fig. 4-13 is the new tolerance zone superimposed on the old. The benefits are:

1. The positions which were unacceptable in the old way are acceptable in the new.

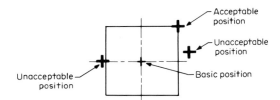

FIG. 4-11 Tolerance zone showing rejectable parts.

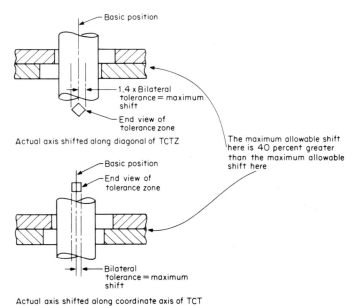

FIG. 4-12 Feature placement within tolerance zone.

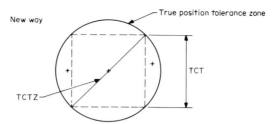

FIG. 4-13 Diameter tolerance zone.

2. The size of the "target area" has been increased by 57 percent:

$$\frac{\text{Area of circle}}{\text{Area of square}} = \frac{\frac{\pi}{4}(\sqrt{2} \times \text{TCT})^2}{(\text{TCT})^2}$$

$$= \frac{\frac{\pi}{4} \times 2 \times (\text{TCT})^2}{(\text{TCT})^2}$$

$$= \frac{\pi}{2}$$

$$= 1.57$$

3. In process and quality control, position measurements may be made along the X and Y axes according to the old way. There is no need to buy or make new and expensive tooling. Three ways are available to verify that a position measured along the X and Y axes does fall within the specified cylindrical tolerance zone.

The first way to verify the position is to use a paper gage on which the actual X and Y coordinates of position are plotted to some enlarged scale. Figure 4-14 shows a paper gage. The X and Y measurements were made in a convenient and conventional way involving no change in practice. Evaluation of these measurements—both sets show acceptable positions—was quickly and easily made using readily available paper, pencil, and ruler.

The second way is to calculate the diameter of a circle with its center at the basic position and passing through the measured position. This circle should be equal to or less than the size of the position tolerance zone (see Fig. 4-15a). This is less than the diameter of the position tolerance zone, and so the position is acceptable. For this technique it is not necessary to sketch the X and Y axes with the actual position represented to some scale. A simple right-triangle calculation will do the job.

The third way is to precalculate the diameter of the circle represented by combinations of X and Y coordinate measurements. Using the equation of the second method, we could prepare a chart making future calculations unnecessary. This chart may be quickly prepared using a calculator equipped with a polar-conversion key (see Fig. 4-15b). Find the X and Y actual measurements on the outer scale. The diameter of the circle is found on the body of the table as shown in Fig. 4-15b. Notice that mea-

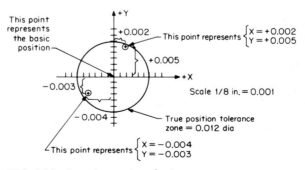

FIG. 4-14 Open inspection plotting.

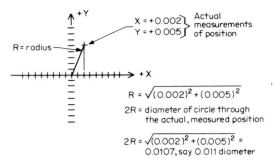

$$R = \sqrt{(0.002)^2 + (0.005)^2}$$

2R = diameter of circle through the actual, measured position

$$2R = \sqrt{(0.002)^2 + (0.005)^2} = 0.0107, \text{say } 0.011 \text{ diameter}$$

FIG. 4-15a Paper gage.

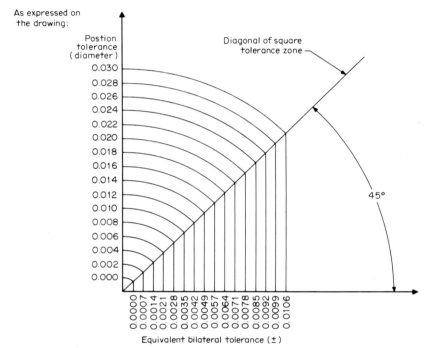

FIG. 4-15b Polar conversion chart.

surements which are recorded as "minus" do *not* require any special treatment. For our purposes they are used as though they were plus measurements.

Change 2: ⓢ versus Ⓜ

Tolerances may be made variable within very well-defined limits, or they may be fixed at the value shown on the face of the drawing. The choice is left to the designer on the basis of functional requirements. A typical tolerance symbol, which contains a lot of other information, may well look like the example in Fig. 4-16.

For the moment, our concern with the total symbol shown in Fig. 4-16 centers on the ⓢ, which stands for "regardless of feature size" (RFS). This is called a *modifier*. Its job is to indicate that the stated tolerance must not be exceeded.

There is another symbol that differs only in the modifier, but which makes a world of difference to the manufacturing and quality control people (see Fig. 4-17). The use of the symbol Ⓜ means that:

1. The given tolerance of 0.004 must be satisfied only when the controlled feature is at its MMC size.

2. If the feature has deviated from its MMC size, it would be permitted, automatically, to use a tolerance zone that is larger than specified. This extra tolerance (always welcomed by the shop) is equal to the deviation from the MMC size of the feature.

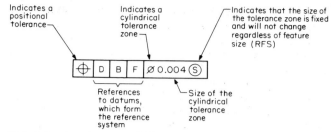

FIG. 4-16   Feature control block.

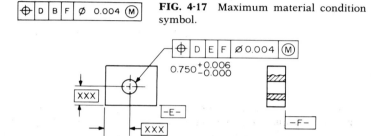

FIG. 4-17   Maximum material condition symbol.

FIG. 4-18   Drawing showing MMC and limit of size.

There is a limit, however, to the total amount of this bonus tolerance. The limit is equal to the size tolerance of the controlled feature (see Fig. 4-18). For this case, if the hole were produced at exactly 0.750, the position tolerance zone would be 0.004 diameter. If the hole were produced at 0.751, the position tolerance would be 0.005 diameter. We *add* the deviation from the MMC size to the position tolerance called out in the position tolerance symbol. A simple table for this is shown in Fig. 4-19.

Actual hole size	Permissible tolerance zone diameter
0.750 (MMC)	0.004
0.751	0.005
0.752	0.006
0.753	0.007
0.754	0.008
0.755	0.009
0.756 (LMC)	0.010

FIG. 4-19   MMC table. (LMC = least material condition, when the workpiece is at its lightest weight, that is, largest hole size, smallest pin size.)

Another way of expressing this would be as an equation:

Permissible tolerance = basic tolerance + deviation from MMC

This equation is applicable to internal and external features.

3. When one is dealing with a pattern of feature, all controlled by a common position tolerance symbol, each individual feature is independently evaluated using the above equation.

Change 3: Gaging

Acceptability of the positions of a pattern of features may be evaluated by using a functional gage with fixed gaging elements. No actual position measurement is required. Verification of the size for each feature must, however, be done. This application of go/no-go gaging techniques to a single feature or to many features simultaneously is based on the idea of *virtual size*, which gives no numerical information about the position of any feature in the pattern, but does tell unequivocally whether mating parts can be assembled interchangeably even if both mating parts have the worst case of tolerance buildup. Each gage is built to simulate the general appearance of the mating part. The positions of the gaging elements are, to gagemaker's tolerances, an exact replica of the basic positions of the mating part. The size of the gaging element is made equal to the virtual size of the controlled feature as determined by the MMC size and the true position tolerance of the feature being gaged. A numerical example will be useful in helping to understand the principles involved. See Fig. 4-20, which shows two parts that go together. This assembly is similar to that in Fig. 4-3 but has a more complex pattern of mating features. The designer has assumed the following information to begin with:

$$MMC_{hole} = 0.750 \text{ in. diameter}$$
$$TP_{hole} = 0.012$$
$$MMC_{pin} = 0.730$$
$$TP_{pin} = ?$$

Further, at assembly, the designer would like to have in the worst case a minimum radial clearance between the hole and the pin of 0.001. Note that this minimum radial clearance will occur when a mating hole and pin each are shifted in opposite directions by the maximum amount permitted by their respective tolerances, and that simultaneously the hole and pin are each at their MMC size.

The problem is first to calculate the true position (TP) of the pin so that the workpieces will assemble in the worst case. Then it is necessary to design a functional gage that will guarantee this ability to assemble in the worst case. The basic concept involving virtual size is used:

$$\text{Virtual size hole} = \text{virtual size pin}$$
$$MMC_{hole \ diameter} - TP \text{ hole tolerance}$$
$$= MMC_{pin \ diameter} + TP \text{ pin tolerance} + 2(\text{minimum radial clearance})$$
$$0.750 - 0.012 = 0.730 + TP_{pin} + 2(0.001)$$
$$0.750 - 0.012 - 0.730 - 0.002 = TP_{pin}$$
$$0.006 = TP_{pin}$$

Now let's examine the problem of designing the functional gages required for the inspection of these workpieces. The gage shown in Fig. 4-21 would be used to inspect the workpiece that has the holes. The positions of the gage pins duplicate the ideal positions of the workpiece holes with respect to the datum features. In the real world, gagemaker's tolerances will apply to all relevant gage features, but we ignore them here for the sake of simplicity. The gage shown in Fig. 4-22 would be used to inspect the workpiece that has the pins. In Fig. 4-22,

$$\text{Gage hole diameter} = MMC_{pin} + TP_{pin}$$
$$= 0.730 + 0.006$$
$$= 0.736 \text{ diameter}$$

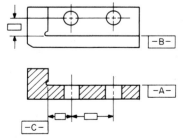 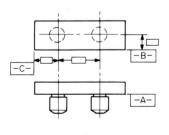

FIG. 4-20 Mating parts.

The gage shown in Fig. 4-21 guarantees that none of the workpiece holes will fit inside a virtual size cylinder that is 0.738 diameter. The gage shown in Fig. 4-22 guarantees that none of the workpiece pins will fit outside a virtual size cylinder that is 0.736 diameter. Since only workpieces with holes or pins of acceptable size will pass inspection by these gages, the gages guarantee assembly of all these parts under the worst case while also guaranteeing the desired minimum radial clearance. It is appropriate to emphasize the fact that any parts which are passed by the gages will assemble interchangeably.

It is also good to emphasize that the functional gage will automatically yield the proper amount of bonus tolerance for any individual feature that has deviated from its MMC size by a permissible amount.

Change 4: Basic Dimensioning Patterns of Holes

In order to explain the value of basic dimensioning we will discuss four different ways (Figs. 4-23 through 4-26) of dimensioning the simple hole pattern of our previous illustration. The holes in Figs. 4-23 through 4-26 are labeled I and II to distinguish them from each other.

For point-to-point dimensioning (sometimes called chain dimensioning under the old way), it is universally understood that the actual position of axis I serves as the datum for axis II. This means that the tolerance zone for axis II is shifted in the identical amount by which axis I has deviated from its ideal position. One possible configuration of tolerance zones is shown in Fig. 4-27. We can readily see that more complex patterns could become difficult to analyze and could also lead to an objectionable "tolerance accumulation."

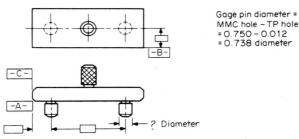

Gage pin diameter =
MMC hole − TP hole
= 0.750 − 0.012
= 0.738 diameter

FIG. 4-21 Functional gage for holes.

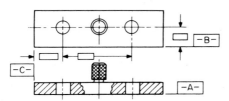

FIG. 4-22 Functional gage for pins.

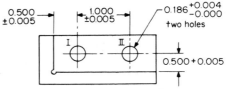

FIG. 4-23 Point-to-point dimensioning under the old system.

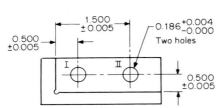

FIG. 4-24 Base-line dimensioning under the old system.

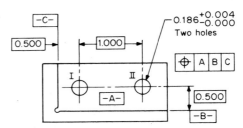

FIG. 4-25 Basic point-to-point dimensioning.

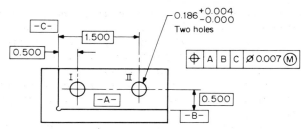

FIG. 4-26 Basic baseline dimensioning.

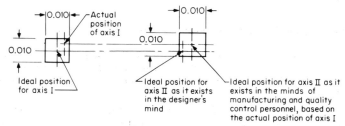

FIG. 4-27 Point-to-point tolerance zones.

Base-line dimensioning relieves the problem somewhat by having each position referred to a common datum feature whose position does not vary. It is also consistent with some types of manufacturing equipment that display the position of the cutting tool with respect to a "zero" position. This tolerance zone configuration is shown in Fig. 4-28.

There is a costly flaw, however, in the base-line dimensioning technique. In the point-to-point scheme, the tolerance between number I and number II is ± 0.005. In the base-line scheme, there is a tolerance of ± 0.010 between number I and number II. If this is not permissible, it would be necessary to change the tolerance as shown in Fig. 4-29. In this figure, two tolerances have been changed by cutting each one in half. Instead of producing one tolerance of ± 0.005, it is now required to produce two tolerances of ± 0.0025 each in order to achieve the same functional result. This is certainly a change in the wrong direction! Figure 4-30 shows what geometric dimensioning accomplishes in this case. The tolerance zones for number I and number II would be identical for *both* styles of dimensioning. This is shown in Fig. 4-31.

The possible deviation in position between number I and number II is not altered by changing from point-to-point dimensioning to base-line dimensioning. A designer therefore is free to use whichever style of dimensioning best suits production methods without worrying about degrading function or increasing manufacturing costs.

Change 5: Feature Control Symbol Application for ⊕

In the past, the tolerance assigned to each and every position dimension was shown right next to the dimension (see Fig. 4-32). This technique was used even for features that formed a pattern in which the tolerance was the same for each and every position dimension in the pattern. This practice is inefficient. In the new way, the position tol-

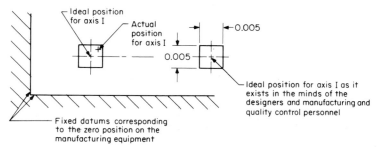

FIG. 4-28 Base-line tolerance zones.

0.186 +0.004 −0.000
Two holes
I
II
0.500 ± 0.005
0.5000 ± 0.0025
1.5000 ±0.0025

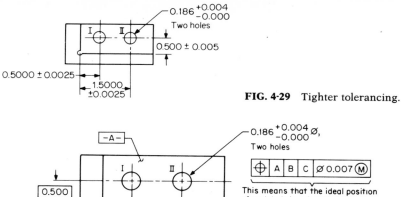

FIG. 4-29 Tighter tolerancing.

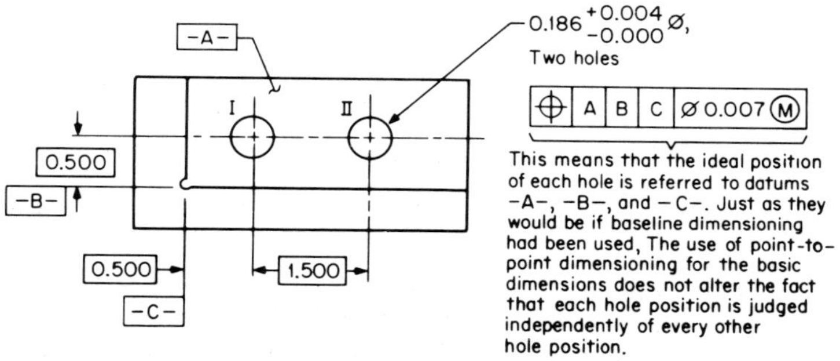

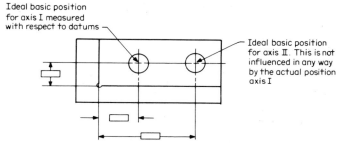

−A−
0.186 +0.004 −0.000 Ø,
Two holes
I
II
0.500
−B−
0.500
1.500
−C−
A B C Ø 0.007 M
This means that the ideal position of each hole is referred to datums −A−, −B−, and −C−. Just as they would be if baseline dimensioning had been used, The use of point-to-point dimensioning for the basic dimensions does not alter the fact that each hole position is judged independently of every other hole position.

FIG. 4-30 Geometric dimensioning and tolerancing method.

Ideal basic position for axis I measured with respect to datums
Ideal basic position for axis II. This is not influenced in any way by the actual position axis I

FIG. 4-31 Basic pattern.

0.005 ±0.004
1.000 ±0.006

FIG. 4-32 Tolerance next to all dimensions.

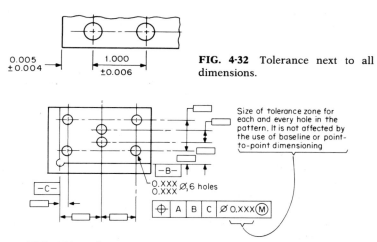

Size of tolerance zone for each and every hole in the pattern. It is not affected by the use of baseline or point-to-point dimensioning
−B−
−C−
0.XXX 0.XXX Ø,6 holes
A B C Ø 0.XXX M

FIG. 4-33 Tolerance is only stated once.

erance is specified for features in a particular pattern only once, thereby eliminating much repetitive, unnecessary work (see Fig. 4-33).

Change 6: Feature Control Symbol Application for Control of Geometry

Control of geometry has normally been accomplished by means of a note such as: "This surface must be parallel to surface marked with * to within 0.002 inches." Notes of

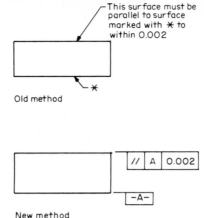

Old method

New method

FIG. 4-34   Symbol versus note.

this type are wordy (and take up a great deal of space on the drawing) and may be subject to a surprising number of different interpretations. The ANSI specification recommends the use of a set of symbols that are concise and unequivocal in their meaning. Each symbol is explained in words and by a picture so that varying interpretations of what is intended are minimized or eliminated. Figure 4-34 uses the parallelism control note above and converts it to the new system.

DATUM SYSTEM

The purpose of this section is to establish guidelines for the application of the datum reference system to engineering drawings.

Datum Reference System

The datum reference system can be used on engineering drawings to show those features which are critical to functions and to parts relationships. The datum reference system allows the coordination of related features of the parts that are intended to mate together.

Advantages

One of the most important advantages of the datum concept is the consistency that can be achieved between the manufacturing and inspection functions. One of the biggest and most costly problems that has long existed in the production cycle has been inconsistent interpretation of drawings by the machinist and the inspector. For example, when setting up a part on the machine, a machinist may choose a particular surface or feature for orientation of the part during machining. On the other hand, the inspector may choose a different surface or feature when orienting the part during the inspection process. The question becomes one of who is right and who is wrong. If the machinist is right, the inspector may be rejecting good parts. If the inspector is right, the machinist may consistently be making bad parts. The time and money wasted because of this problem can be staggering and, worst of all, needless. The problem can be avoided by the use of the datum reference system. With this system, the readers of a drawing can immediately see which features are critical to the function of the part and to that part's relationship to other parts or part features.

Definition

According to ANSI Y14.5, datums are "points, lines, planes, or cylinders and other geometric shapes assumed to be exact for purposes of computation, from which the location or geometric relationship (form) of features of a part may be established." Datums are not assumed to exist on the part itself, but are closely simulated by very precise manufacturing and inspection equipment. Examples of a datum are the surface plates used for the inspection of parts or the table of a machine such as a drill press, a milling machine, or a jig fixture.

Datum Feature

A datum feature is the actual part feature that is used to establish the datum. This relationship can be seen in Fig. 4-35.

Datum Identification Symbol

A datum is specified on a drawing by means of a datum identification symbol. An example of a datum identification symbol callout is shown in Fig. 4-36. Shown in Fig. 4-37 are different applications and corresponding descriptions of various datum identification symbol callouts.

Datum Reference Frame

In the real world there is nothing perfect, and therefore we must use approximations of perfection for measurements. It is useful to distinguish between the ideal things that can exist only in our minds and the real things that must be accepted and worked with. The ideal in the datum world is called a *datum reference frame*, and the real material thing that can be seen and felt is the *datum feature*. Sometimes these datum features are identified simply by the way in which position dimensions are shown on the drawing. These are called *implied datum features*. At other times, the features are specifically identified by *datum identification symbols*, for example, ⟨-A-⟩, properly placed on the drawing.

The datum reference frame may be represented by a machine table with vertical side plates, by a jig fixture, or by a surface plate. The datum reference frame has three mutually perpendicular planes, as shown in Fig. 4-38. A part is dimensioned and assigned datum identification symbols in order of their importance. This order of importance is generally dictated by design function, and the symbols are listed in a feature control block according to the order of importance, as primary, secondary, and tertiary. Note that the symbols need not be in alphabetical order. Figure 4-39 illustrates a typical drawing showing this concept.

The machinist sets the part on the machine according to the order of importance as indicated on the drawing. Datum C is listed first, and thus is primary, and the primary datum must have a minimum of three high points of the datum feature in contact with the datum reference frame, which is the machine table in this case, as shown in Fig. 4-40. The machinist then slides the part over to the secondary datum, A, so that a min-

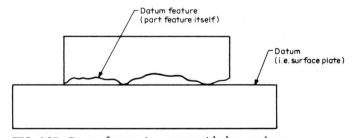

FIG. 4-35 Datum feature in contact with datum plane.

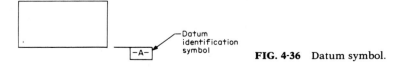

FIG. 4-36 Datum symbol.

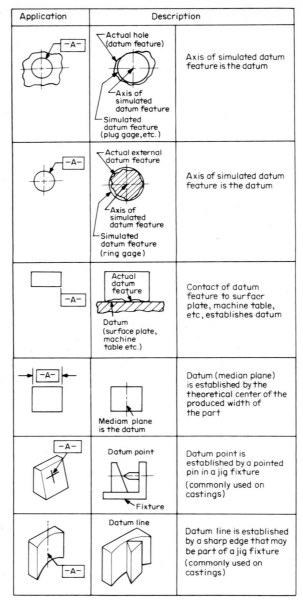

Application	Description	
	Actual hole (datum feature) — Axis of simulated datum feature — Simulated datum feature (plug gage, etc.)	Axis of simulated datum feature is the datum
	Actual external datum feature — Axis of simulated datum feature — Simulated datum feature (ring gage)	Axis of simulated datum feature is the datum
	Actual datum feature — Datum (surface plate, machine table etc.)	Contact of datum feature to surfacr plate, machine table, etc, establishes datum
	Mediam plane is the datum	Datum (median plane) is established by the theoretical center of the produced width of the part
	Datum point — Fixture	Datum point is established by a pointed pin in a jig fixture (commonly used on castings)
	Datum line	Datum line is established by a sharp edge that may be part of a jig fixture (commonly used on castings)

FIG. 4-37 Types of datums.

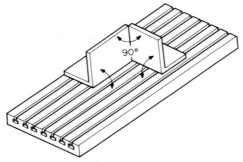

90°

FIG. 4-38 Simulated datum reference frame.

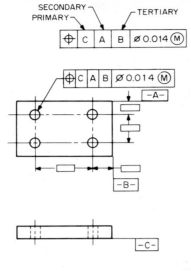

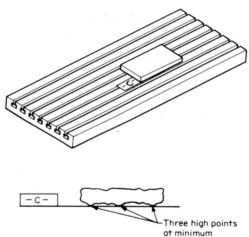

FIG. 4-39 Reading direction of datums.

FIG. 4-40 Primary datum.

imum of two points are in contact with a vertical plate that is 90° from the primary datum, as shown in Fig. 4-41. After the primary and secondary datums of the part have been established, the machinist slides the part over to the tertiary datum plane so that at least one point is in contact with it, as shown in Fig. 4-42. Next, the machinist clamps the part and makes measurements from the datum planes (not the part) to locate the placement of the holes.

At inspection and assembly, the part must be set up on a datum reference frame in exactly the same orientation used during machining.

Cylindrical Datum Features

The datum reference frame and cylindrical datum features seem to have no relationship. However, we treat the cylindrical feature as though it had two centerlines that are at 90° to each other, representing two theoretical datum planes. Theoretically, the axis of the cylindrical datum feature is the point at which the two datum planes intersect, as shown in Fig. 4-43. In the example shown in Fig. 4-44, datum A is the primary datum. The secondary and tertiary datums are represented by the two theoretical center planes, which are centered on the feature labeled -B- . In this case, one feature is used to create two datum planes.

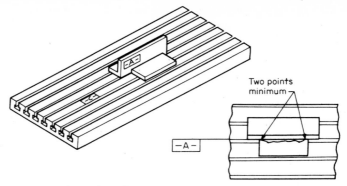

FIG. 4-41   Secondary datum.

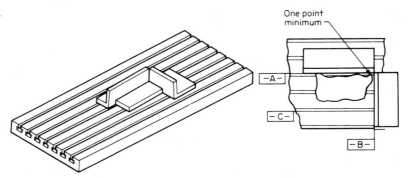

FIG. 4-42   Tertiary datum.

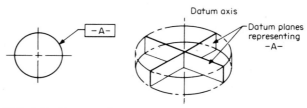

FIG. 4-43   Datum axis.

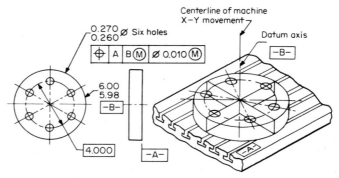

FIG. 4-44   Datum axis simulated.

Auxiliary Datum

An example of an auxiliary datum is shown in Fig. 4-45. The primary datum A must have a minimum of three points in contact with the primary datum plane (machine table); the center planes of datum B will be determined by the center of the machining head and the *X-Y* movement of the machine. The keyway is the auxiliary datum used for orientating the two holes. The two holes need not be dimensioned from the center-line, because the use of datum B automatically centers the holes.

Datums with Ⓢ (RFS) and Ⓜ (MMC) (see Fig. 4-46 and Fig. 4-47)

In *a*, the lathe chuck holds datum B securely as the primary datum. Datum A need only contact the base plate as shown. (*Note:* RFS requires that the subject feature be held or centered by touching.) In *b*, datum A is the primary datum, so a minimum of three points must be in contact with the base plate. Datum B is centered, but not with so much force as to lift datum A off the base plate. In *c*, datum B has changed to MMC. MMC requires only a fixture with a 4.000-diameter hole, representing the largest diameter of datum B (MMC). The only requirement is that the part fit into the fixture as shown. In *d* datum A is shown as primary, and datum B at MMC. Once again, a fixture and datum A must have at least three points contacting the base plate as shown. In *e* datum B is 4.000, and not perpendicular to datum A. Datum B must be remachined to a smaller diameter (within its limit of size) so that datum A will be in contact with the base plate as it was in *d*.

Datum Targets

Datum targets are most useful, in that they eliminate the ambiguity in establishing the reference system to be used by manufacturing and quality control. They are especially useful in the measurement of castings, forgings, and molded parts because of the surface irregularities frequently found on such parts. Datum targets (sometimes called tooling points) are specified on product drawings and are positioned on a tooling fixture in the same manner as a datum reference frame, with three, two, and one points for primary, secondary, and tertiary datums. An example of a tooling fixture is shown in Fig. 4-48. The datum targets shown here can represent points, lines, or areas, or any combination of the three. Datum targets may have basic or toleranced dimensions from functional surfaces and in relation to each other. Figure 4-49 shows datum target dimensioning.

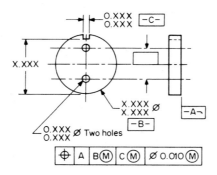

FIG. 4-45 Auxiliary datum.

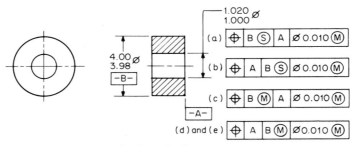

FIG. 4-46 Datums with Ⓢ and Ⓜ .

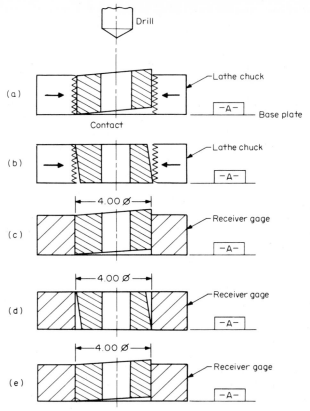

FIG. 4-47   Definition of datums with Ⓢ and Ⓜ.

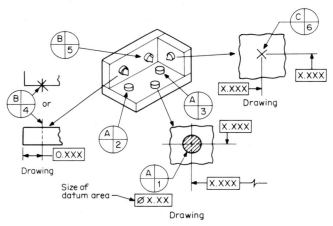

FIG. 4-48   Datum targets.

POSITIONAL TOLERANCING (TRUE POSITION)

The most common problem of positional tolerancing involves the manufacturing of *matching-hole patterns*—when two or more parts are to be joined by bolted connections. The purpose of assigning dimensions and tolerances to the hole positions is to guarantee that the bolts will assemble with the holes without the need for assembly personnel to do any rework of the holes or the bolts. The designer who attempts to solve this problem has a great variety of alternatives available. The following is an analysis of many of these alternatives. The outcome of this analysis will be a set of design equations, each one applying to a particular kind of bolted connection and style of dimensioning.

Three different kinds of matching-hole joints are in general use:

1. *Floating fastener* This is the kind of joint in which a bolt passes through clearance holes in each workpiece and is secured by a nut (see Fig. 4-50).

2. *Fixed external fastener* This is the kind of joint in which a threaded stud or a pin is assembled into one workpiece. The position tolerance for the external fastener applies to the external fastener *after* it has been assembled into its workpiece. The external fastener (or array of fasteners) must then accept one or more workpieces, each having clearance holes to accommodate the external fastener(s). This kind of joint is most commonly secured with a nut (see Fig. 4-51).

3. *Fixed internal fastener* In this joint *one* workpiece has an array of threaded holes and the mating workpieces have clearance holes. After the workpieces are placed one on another, it is necessary that a bolt be able to pass through each clearance hole and then to engage the threaded hole and be drawn tight with no interference between any bolt and its hole (see Fig. 4-52).

Each of these three different kinds of joints may be used with one of three different ways of registration.

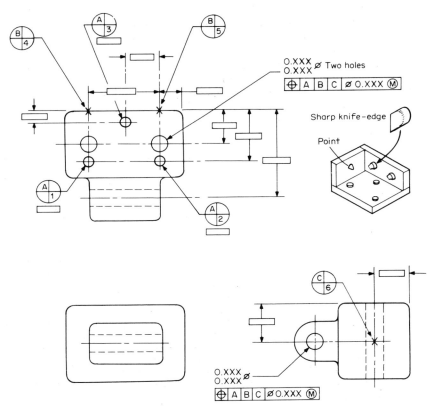

FIG. 4-49 Datum target drawing.

FIG. 4-50 Floating fastener definition.

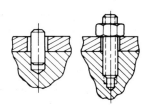

FIG. 4-51 Fixed external fastener definition.

1. The workpieces are brought into firm contact along three datum surfaces, which must remain in intimate contact to prevent any relative movement between the workpieces.

2. The workpieces are brought into registration by cylindrical pilot members. These permit limited relative movement between the mating workpieces as determined by the difference in size of the pilot members.

3. The workpieces have no physical registration means. They are free to move relative to one another until one hole pattern matches the other and assembly can be accomplished.

In addition, there is a choice among several styles of dimensioning techniques, giving a wide variety of possible combinations.

The three matching-hole joints require more than a definition to describe their characteristics. The following will describe each one on an individual basis and provide formulas so that the user can maintain 100 percent interchangeability. The following also applies to circular hole patterns.

Floating Fastener The assembly shown in Fig. 4-53a can be dimensioned in different ways. Each part has two clearance holes large enough to permit the hole location to shift or tilt in the worst condition and still allow the bolt to pass through each hole. For the assembly to be secured by a floating fastener, a nut, cotter pin, or similar fastener must be used. Figure 4-53b shows an acceptable way to dimension such a part. The two holes as a pattern may shift within a 0.06 square tolerance zone, but the hole-to-hole relationship must fall within the 0.010 diameter tolerance zone. Figure 4-54 illustrates the tolerance zones at a larger scale. As each hole departs from the MMC size, the tolerance zone becomes larger by the same amount (see Fig. 4-55). Also, note that the axis of the hole may fall outside the square tolerance zone.

When the MMC symbol, Ⓜ, is used, the tolerance zone increases as the feature departs from the MMC. For example:

Hole diameter	Diameter tolerance zone
0.510	0.010
0.511	0.011
0.512	0.012
0.513	0.013
0.514	0.014
0.515	0.015
0.516	0.016
0.517	0.017
0.518	0.018
0.519	0.019
0.520	0.020

The hole can shift within the diameter tolerance zone or, as shown in Fig. 4-55, it can tilt from one edge of the diameter tolerance zone at the top of the part to the opposite edge of the diameter tolerance zone at the bottom of the part. Figure 4-56 is very similar to Fig. 4-53 except that the ± 0.03 has been changed to a basic dimension.

A composite positional control has special rules. The larger tolerance (0.030) is the control of the two holes as a pattern to the edge of the part, and the smaller tolerance (0.010) is the control from hole to hole. In Fig. 4-57, note that the axis of the holes must always fall within both tolerance zones.

Figures 4-58 and 4-59 illustrate how datums affect the tolerance zones. Datum A establishes a perpendicular control to the bottom surface. Datum B establishes a parallelism control of the basic tolerance zone to the edge of the part. When the inspection department staff check the part, they will have to rotate it until it meets the controls stated, because the part is symmetrical. To speed up inspection and assembly of parts, it would be good company procedure to have manufacturing staff mark the parts.

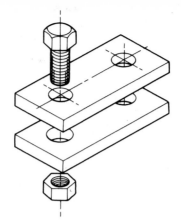

FIG. 4-52 Fixed internal fastener definition.

FIG. 4-53a Floating fastener.

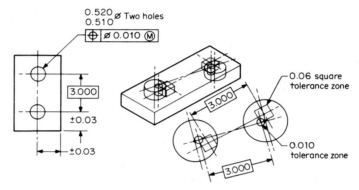

FIG. 4-53b Dimensioning floating fastener.

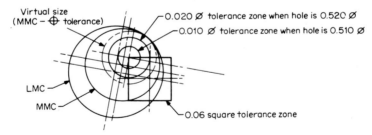

FIG. 4-54 Tolerancing zone for floating fastener.

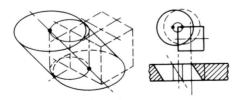

FIG. 4-55 Position of hole in tolerance zone (see table, p. 4-24).

Figures 4-53*a* through 4-59 have covered most conditions for positional control for the basic tolerance zone. That is,

1. A pattern of holes has a looser control to the edge of the part, but the pattern itself is controlled within the stated tolerance of 0.010 diameter (Fig. 4-53).

2. The tolerance zone gets larger when the Ⓜ symbol is used.

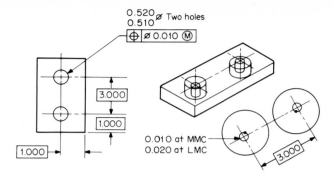

FIG. 4-56   Using basic dimensions (see table, p. 4-24).

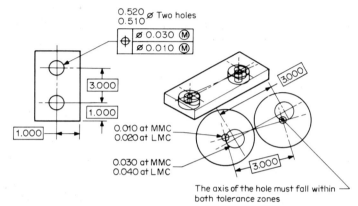

FIG. 4-57   Using a composite positional tolerance.

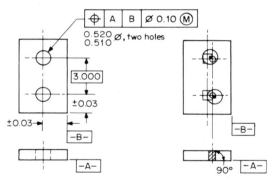

FIG. 4-58   Using datums with positional tolerance.

3. The axis of each hole may shift or tilt within the stated tolerance zone for the full thickness of the feature.

4. The pattern of holes is controlled more tightly to the edge of the part by using basic dimensions instead of a ± tolerance, which is generally looser.

5. A composite positional control provides two diameter tolerance zones, one larger than the other. This allows functional gages to be used for checking patterns of holes to the edge of the part, as well as the hole-to-hole tolerance, which is always tighter.

6. Datums are used so that the part can be oriented to its functional requirements (see Datums, discussed earlier, and also Fig. 4-58).

Fixed External Fasteners   Fixed external fasteners may be dimensioned in the same way as described for floating fasteners (see Fig. 4-60).

Fixed Internal Fasteners Fixed internal fasteners may also be dimensioned as was described for floating fasteners. The one big problem with fixed internal fasteners (see Fig. 4-61) is the tilting of the threaded hole within the positional tolerance zone. This tilt creates a very small tolerance zone for the thread or a very large clearance hole for the mating part, and the thicker the mating part, the worse the condition gets (see Fig. 4-62). Because of this condition, it is recommended that a *projected tolerance* be used (see Fig. 4-63).

Projected Tolerance Zone

A projected tolerance zone is the proper way to cut tight tolerances. Preassign a reasonable positional tolerance to the thread and use the formulas given in Fig. 4-63.

Minimum Edge Distance

Minimum edge distance can be determined by following the example shown in Fig. 4-64. Note that the axis of the hole may fall outside the square tolerance zone. Determining minimum edge distance using 0.000 tolerance is not uncommon and is permitted, but can be done only when an Ⓜ symbol is used, as shown in Fig. 4-65. The advantage of 0.000 tolerance is that it allows the machinist more latitude in selecting drill sizes.

Position of Noncircular Features

The rules are the same for noncircular features as they are for circular features. Figure 4-66 shows an example.

Circular Features

Circular features should not be treated any differently than has already been discussed. Figure 4-67 shows a typical example and how one might find minimum edge distance. Figure 4-68 shows the tolerance zones for datum A and the four holes when the datum and holes are at the LMC.

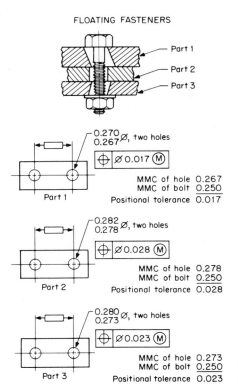

FLOATING FASTENERS

FIG. 4-59 Formula for floating fastener assembly.

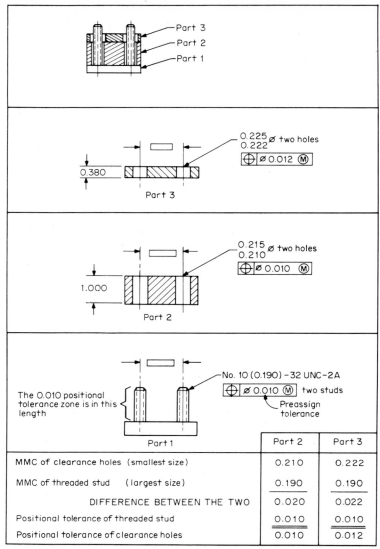

	Part 2	Part 3
MMC of clearance holes (smallest size)	0.210	0.222
MMC of threaded stud (largest size)	0.190	0.190
DIFFERENCE BETWEEN THE TWO	0.020	0.022
Positional tolerance of threaded stud	0.010	0.010
Positional tolerance of clearance holes	0.010	0.012

FIG. 4-60 Formula for fixed external fastener assembly.

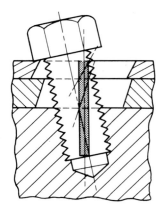

FIG. 4-61 Fixed internal fastener assembly.

Note: Preassign a positional tolerance
to A or B

0.375 – 16 UNC – 2B
four holes

0.405 – 0.437 ⌀
four holes

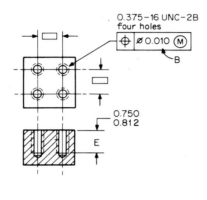

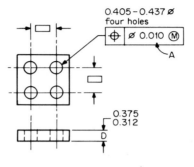

0.750
0.812

0.375
0.312

$$C = A + B \left(1 + \frac{2 \times D}{E} \right)$$

C = MMC of hole minus
MMC of bolt

To find the ⌖ tolerance of
threaded hole:

$$B = \frac{C - A}{\left(1 + \frac{2 \times D}{E}\right)}$$

To find the ⌖ tolerance of
clearance hole:

$$A = C - B \left(1 + \frac{2 \times D}{E} \right)$$

To find the MMC diameter of
the clearance hole:

C + Bolt size

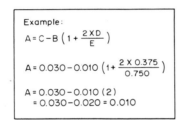

Example:

$$A = C - B \left(1 + \frac{2 \times D}{E} \right)$$

$$A = 0.030 - 0.010 \left(1 + \frac{2 \times 0.375}{0.750} \right)$$

$$A = 0.030 - 0.010 \,(2)$$
$$= 0.030 - 0.020 = 0.010$$

FIG. 4-62 Formula for fixed internal fastener assembly.

TOLERANCE OF FORM

The symbols listed below are used for control of form or geometry. They enable a designer to specify unequivocally the kind of geometry control required. This control may be applied to entire surfaces, or to the axis of a feature. The limits of size, position, and surface texture automatically play a part in form control. The symbols are used to refine these automatic controls, when the feature being controlled is a feature of size. The designer is permitted the option of applying the Ⓜ modifier to the form-control tolerance. As the feature departs from the MMC, additional form tolerance is permitted. RFS Ⓢ is understood unless otherwise stated.

Straightness —   Profile of a surface ⌓
Flatness ▱   Perpendicularity ⊥
Roundness ○   Angularity ∠
Cylindricity ⌭   Parallelism //
Profile of a line ⌒

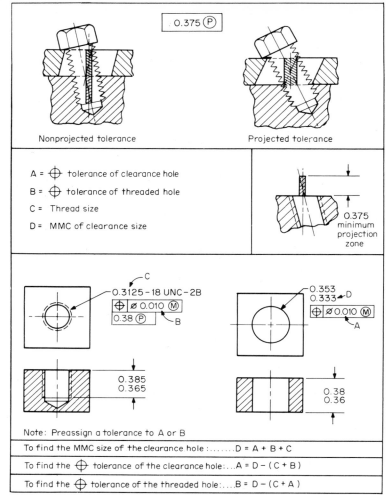

FIG. 4-63 Formulas for projected tolerance zone.

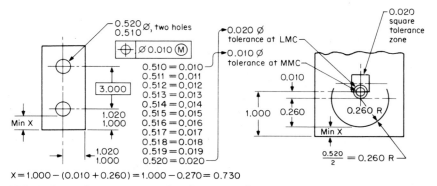

$X = 1.000 - (0.010 + 0.260) = 1.000 - 0.270 = 0.730$

FIG. 4-64 Finding minimum edge distance with square tolerance zone.

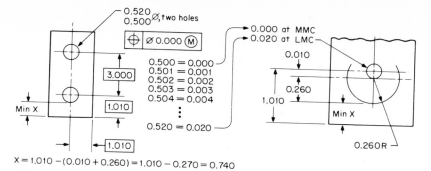

$$X = 1.010 - (0.010 + 0.260) = 1.010 - 0.270 = 0.740$$

FIG. 4-65 Finding minimum edge distance with basic dimensions.

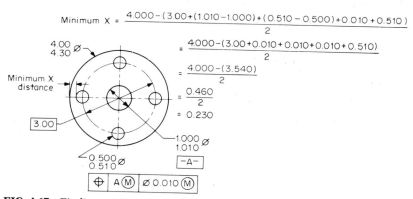

Part 1   Part 2
$$1.520 + 0.010 = 1.540 - 0.010$$
$$1.530 = 1.530$$
$$1.530 = \text{Gage size or virtual size}$$

0.010 tolerance at MMC
0.020 tolerance at LMC

FIG. 4-66 Position of noncircular features.

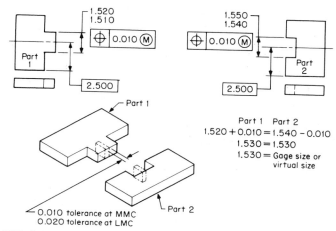

$$\text{Minimum X} = \frac{4.000 - (3.00 + (1.010 - 1.000) + (0.510 - 0.500) + 0.010 + 0.510)}{2}$$

$$= \frac{4.000 - (3.00 + 0.010 + 0.010 + 0.010 + 0.510)}{2}$$

$$= \frac{4.000 - (3.540)}{2}$$

$$= \frac{0.460}{2}$$

$$= 0.230$$

FIG. 4-67 Finding minimum edge distance with one datum.

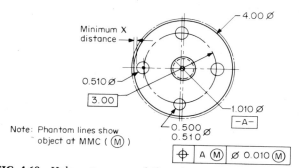

Note: Phantom lines show object at MMC (Ⓜ)

FIG. 4-68 Hole pattern may shift within datum tolerance zone.

When perfect form is required at MMC, a zero form tolerance may be specified on the product drawing, if, as the feature departs from MMC, a tolerance equal to this departure is applied. This of course may be used only when the controlling feature has size. It would not apply, for example, to the controls of flatness.

Straightness of Surface Elements — ⓢ The space between two parallel lines separated by the specified tolerance defines the straightness of surface elements. The tolerance zone must be contained within the limits of the part size in any attitude (see Figs. 4-69 and 4-70).

Comments:

1. Applicable to a round or rectangular part.
2. Cannot be referred to a datum.
3. Cannot use an ⓜ symbol.
4. Must have perfect form at MMC.
5. The two parallel lines must share a common plane with the nominal axis.
6. The stated straightness tolerance must be less than the limits of size.
7. Does not control roundness, parallelism, or flatness.
8. Applies only in the specified direction.
9. The part must be perfectly straight at MMC and may go out of straight by the amount of departure from MMC, but this departure shall not exceed the straightness tolerance. In other words,

$$1.030 = 0.000 \text{ straightness}$$
$$1.029 = 0.001 \text{ straightness}$$
$$1.028 = 0.002 \text{ straightness}$$
$$1.027 = 0.002 \text{ straightness}$$

10. A note may be added if appropriate: "Perfect form not required at MMC."
11. Cross-sectional elements cannot violate the limits of size.

Straightness of Axis — The axis or centerline of a diameter part must fall within the specified cylindrical tolerance zone (see Fig. 4-71).

Comments:

1. Applicable to a round part only.
2. Cannot be referred to a datum.

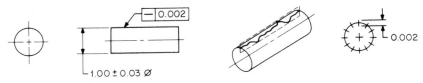

FIG. 4-69 Straightness of a diameter surface.

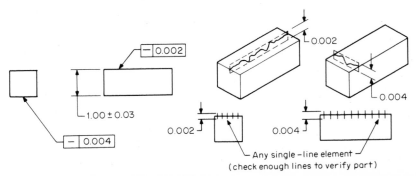

FIG. 4-70 Straightness of a flat surface.

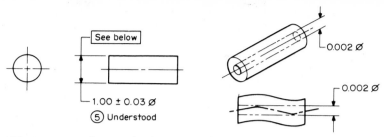

FIG. 4-71   Straightness of a diameter axis.

3. Can use an Ⓜ symbol; if Ⓜ is not used, Ⓢ is understood.

4. Perfect form at MMC can be violated.

5. The straightness feature control symbol must be attached to the diameter callout or to the dimension line.

6. Does not control roundness or parallelism.

7. The diameter symbol, Ø, must precede the stated tolerance.

8. The virtual size will be equal to the MMC size plus the straightness tolerance.

9. Cross-sectional elements cannot violate the limits of size.

		Feature size	Maximum departure from straightness
10.	— Ø 0.002	1.03	0.002
		1.00	0.002
		0.97	0.002
11.	— Ø 0.002 Ⓜ	1.03	0.002
		1.00	0.032
		0.97	0.062

12.

Ø 0.015 Ⓜ
Ø 0.002/1.00 Ⓜ

—

A composite straightness symbol may be used when the full length must be straight within a larger tolerance (0.015) and each shorter segment (1.00) in a smaller tolerance (0.002).

Flatness ⟋⟍ The space between two parallel planes separated by the specified tolerance defines flatness. The tolerance zone may be used in any attitude but must be contained within the limits of the size tolerance zone (see Fig. 4-72).

Comments:

1. Not applicable to round or curved surfaces.

2. Cannot be referred to a datum.

3. Cannot use an Ⓜ or Ⓢ symbol.

4. No virtual size or functional gaging exist.

5. Cannot apply to features of size.

6. Tolerance zone has length, depth, and thickness (0.002).

7. The stated flatness tolerance must be less than the limits of size.

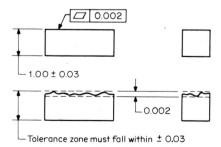

FIG. 4-72   Flatness control.

Roundness ○ Ⓢ Roundness establishes two concentric circles, creating a tolerance boundary around the circumference of the part at any cross section perpendicular to the part's axis. This tolerance boundary must not violate the limits of size and must be equidistant from its axis center, as shown in Fig. 4-73. Any and all cross-sectional elements at 90° to the axis must be round within a 0.005 wide band around the circumference of the part, as shown in Fig. 4-74.

Comments:

1. Roundness applies to the surface for a single line element only, at all cross sections.
2. Datums are not used with roundness.
3. The same definition applies to a sphere at any cross section that intersects the center.
4. Perfect form must exist at MMC.
5. Limits of size must not be violated.
6. Roundness tolerance must be smaller than its limits of size.
7. No Ⓜ symbol may be used.
8. Roundness may be used for a scribed circle.
9. A direct surface evaluation, such as the use of calipers, micrometers, indicators, profile gages, V blocks, or machine centers, is not sufficiently accurate to check true roundness. The above listing does not check lobing, waviness, elliptical section, etc. For a detailed specification on checking roundness see ANSI B89.3.1, Measurement of Out-of-Roundness.
10. Roundness is one of the most loosely defined characteristics used on drawings.
11. If the function of roundness is *not* needed for journal wear, liquid sealing, high-performance bearing diameters, or any critical application, then the following may serve the purpose:

    a. Check roundness by a radial method that involves checking variations in the distance from a center using a spindle and turntable, or by rotating the part between centers.

    b. Check by comparision to a master that is suitably positioned in relation to it.

    c. Use differently angled V blocks that might detect odd or even lobings.

12. Parts of considerable length may be a problem because their weight may cause bowing when they are suspended on V blocks or machine centers.

Free-State Variation of Roundness Free-state variation is used in geometric dimensioning and tolerancing in the same way roundness is, except that an average diameter must be specified. Free-state variation is used when the part will not stay in its round condition because of the type of material it is made of or because its cross section may be too thin. A rubber O ring and thin-wall tubing are two examples of parts that are not rigid when unassembled.

If the part is to be controlled and inspected in its free state, the drawing should use the average-diameter method as shown in Fig. 4-75, a constraining note or notes as shown in Fig. 4-76, or a π (pi) tape as shown in Fig. 4-77. The average diameter must first be found before the roundness condition can be checked. Place the part in its free state so that more than four measurements can be taken across any diametral cross section. Add all the figures and divide the sum by the number of measurements taken.

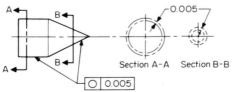

FIG. 4-73 Roundness control.

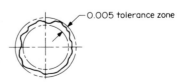

FIG. 4-74 Roundness tolerance zone.

The answer must fall within the stated average diameter shown on the drawing. The next step is to place the part on a template or similar device to determine whether any portion of the diameter falls outside the tolerance band.

Another free-state control of roundness would be to restrain the part and check it for runout, position, roundness, cylindricity, or any combination of these. An example is shown in Fig. 4-76. It is also possible to use a π (pi) tape (3.14159 linear inches per diametral inch) to control a diameter with a free-state condition. Figure 4-77 shows an example.

Comments:

1. Free-state variation can occur when distortion is caused by weight, flexibility, or internal stresses set up during manufacturing.
2. A tank will be in a free-state condition until it is presssurized.
3. A part must stay within its elastic range so that it will return to drawing tolerances when assembled.

Cylindricity ⌀ Ⓢ   Cylindricity establishes two concentric cylinders creating a tolerance boundary around the circumference of the part for its entire length. This tolerance boundary must not violate the limits of size and must be equidistant from the cylinder's axis center, as shown in Fig. 4-78.

Comments:

1. Cylindricity has all the same inherent problems that roundness has.
2. No datum may be used.
3. Cylindricity is analogous to flatness because it controls the entire cylindrical surface (flatness controls an entire flat surface).
4. Cylindricity simultaneously controls roundness, straightness, and parallelism.
5. The cylindricity tolerance times 2 should not be greater than the cylinder's limits of size.

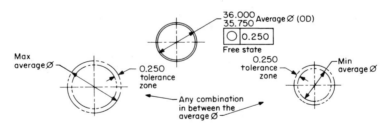

FIG. 4-75   Free-state roundness.

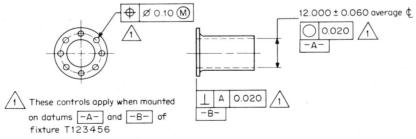

FIG. 4-76   Using a constraining note.

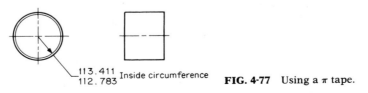

FIG. 4-77   Using a π tape.

Profile of a Surface ⌒ **and Line** ⌒ Profile tolerance establishes a bilateral or unilateral tolerance zone on or about a basic profile made up of curves, arcs, or straight lines. The part profile may vary within the stated profile tolerance (see Fig. 4-79).

Comments:

1. Profile of a line ⌒ or surface ⌒ is individually defined on the next few pages.
2. The drawing must show a phantom-line tolerance zone for unilateral conditions.
3. The tolerance zone is understood to be equally displaced from the true shape for bilateral conditions.
4. All profile dimensioning must have basic dimensions defining the profile shape.
5. Profile tolerance controls form, size, and position.
6. Other dimensional controls may be placed on a profile drawing for further control, such as a finish symbol.
7. The profile of the part must fall within its limits of size (±0.03 on Fig. 4-79).
8. Draw the true shape or section to show the desired basic profile.
9. All variations of the profile must blend together, with no sharp departures.
10. The inspection indicator must stay normal to the basic curved or angled profile.
11. Profile tolerancing may be used with other types of form controls, such as parallelism.

Profile of a Line ⌒ Ⓢ The tolerance zone is the space between two lines, each having the ideal shape of the basic profile. It is applied to individual elements of the

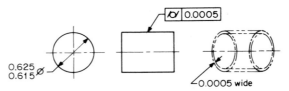

FIG. 4-78   Cylindricity control.

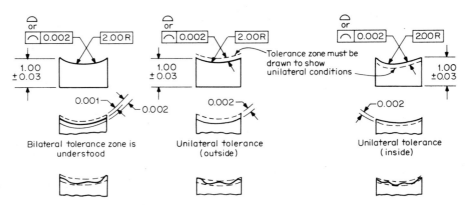

FIG. 4-79   Profile control.

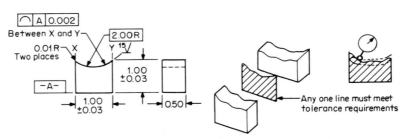

FIG. 4-80   Profile of a line.

controlled surface. The tolerance zone is established from basic dimensions as described above in the opening remarks on profile tolerance (see Fig. 4-80).

Comments:

1. The tolerance zone is 0.001 above and below the basic radius dimension.

2. The tolerance zone in this case need not be shown in the phantom on the drawing.

3. Profile of a line works well for long extruded shapes when each cross section must meet tolerance requirements or when a knife-edged cam follower tracks a single line over a cam.

4. See other comments in the opening remarks on profile tolerance.

5. Datum A establishes the relationship of the basic radius.

Profile of a Surface ⌒ Ⓢ The tolerance zone is the space between two surfaces, each having the ideal shape of the basic profile. It is applied to the entire controlled surface. The profile tolerance zone is established from basic dimensions, as shown in the opening remarks on profile tolerance (see Figs. 4-81 and 4-82).

Comments:

1. The tolerance zone is 0.001 above and below the basic radius dimension.

2. The tolerance zone in this case need not be shown in phantom on the drawing.

3. Datum A establishes an orientation of the profile tolerance zone in all directions.

4. See other comments in the opening remarks on profile tolerance.

Perpendicularity ⊥ Ⓢ Perpendicularity establishes a condition in which a surface, line, or axis must fall within a stated tolerance zone that is exactly 90° from a datum plane or datum axis, as shown in Fig. 4-83.
Comments:

1. Unless otherwise stated, RFS, or Ⓢ, is understood to apply.

2. MMC may be applied to features of size.

3. Perpendicularity must always relate to a datum.

4. Controlled perpendicularity surfaces cannot violate their limits of size.

5. The tolerance zone may apply to one direction only, as shown in Fig. 4-84.

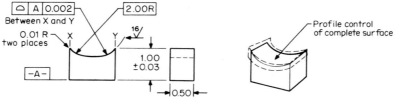

FIG. 4-81   Profile of a surface.

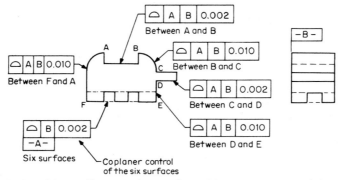

FIG. 4-82   Profile control to a datum(s). Dimensions to define shape have been omitted in this illustration for clarity.

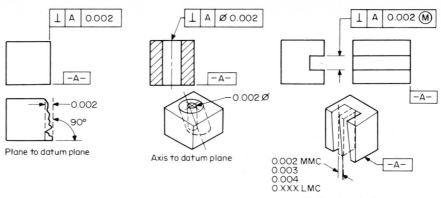

FIG. 4-83   Perpendicularity control.

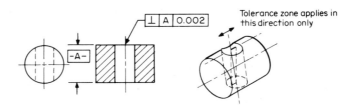

FIG. 4-84   Perpendicular control of a diameter.

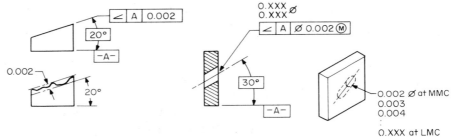

FIG. 4-85   Angularity control.

Angularity ∠ ⓢ Angularity control is used for all angles other than **90°**. It establishes a condition in which a surface, line, or axis must fall within the stated tolerance zone that is located at a basic angle (other than 90°) from a datum plane or datum axis (see Fig. 4-85).

Comments:

1. Unless otherwise stated, RFS, or ⓢ, is understood to apply.
2. MMC may be applied to features of size.
3. Angularity must always relate to a datum.
4. Angular surfaces cannot violate its limits of size.
5. The tolerance zone may apply to one direction only (see Fig. 4-84).

Parallelism // ⓢ Parallelism establishes a condition in which a surface, line, or axis must fall within the stated tolerance zone that is equidistant at all points from a datum plane or datum axis. Three parallelism tolerance conditions exist:

1. When a plane surface is to be parallel to a datum plane, the tolerance zone is two parallel planes separated by the amount of the stated tolerance (see Fig. 4-86).
2. When a cylindrical feature is to be parallel to a datum plane, the tolerance zone is two parallel planes separated by the amount of the stated tolerance (see Fig. 4-87).

3. When a cylindrical feature is to be parallel to a datum cylindrical feature, the tolerance zone is cylindrically shaped and is equidistant from the datum cylinder axis (see Fig. 4-88).

Runout ⟋ Ⓢ Runout establishes a composite control of roundness, and of position when applied to a circular element. Runout controls cylindricity, straightness, and position when applied to an entire surface. When applied to circular elements, the tolerance zone is two concentric circles radially separated by the stated tolerance on an axis coincident with the datum axis (see Fig. 4-89). When applied to an entire surface, the tolerance zone is two concentric cylinders radially separated by the stated tolerance with axis coincident with the datum axis. The part must be rotated 360° on the datum(s) for verification. Runout also provides control of surfaces at right angles to the datum axis, as shown in Fig. 4-90, or to cones or curved shapes normal to the datum axis, as shown in Fig. 4-91.

Total Runout, or Circular Runout Circular runout is very similar to roundness and to profile of a line in that it applies only to a single cross-sectional element cut through the parts, as shown in Fig. 4-92.

FIG. 4-86 Parallelism control (surface to surface).

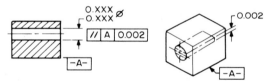

FIG. 4-87 Parallelism control (diameter to surface).

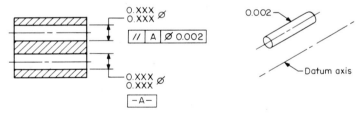

FIG. 4-88 Parallelism control (diameter to diameter).

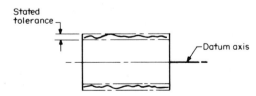

FIG. 4-89 Runout control of a diameter.

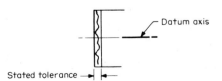

FIG. 4-90 Runout control of a perpendicular surface.

FIG. 4-91   Runout control of a cone or curved surface.

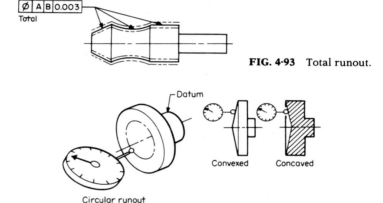

FIG. 4-92   Circular runout.

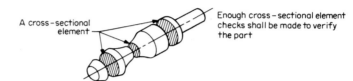

FIG. 4-93   Total runout.

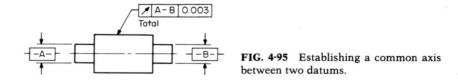

FIG. 4-94   Right-angle surfaces to the datum axis.

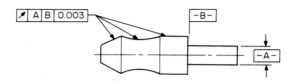

FIG. 4-95   Establishing a common axis between two datums.

Comments:

1. Runout of an element is good for automatic checking through a digital readout or a plotter because such instruments use a single pickup.

2. RFS applies to and is intended for a collective effect of deviations on the considered feature to the datum feature, regardless of feature size.

3. Choose a datum of sufficient length on the same centerline or two datums with sufficient separation on the same centerline.

4. Virtual size is the MMC size plus runout tolerance for external features, and minus runout tolerance for internal features.

5. The size of the feature must be verified independently. Total runout is the same as circular runout except that it applies to the full length of the feature (see Fig. 4-93).

Right-Angle Surfaces to the Datum Axis   This will check the perpendicularity of a surface to the datum axis and is sometimes called *wobble*. If total runout is used, it will also check flatness. Surfaces that are concaved or convexed will not be fully evaluated with circular runout, as shown in Fig. 4-94.

Common Datum Axis   Common datum axis is established when two datums of sufficient length are combined into one common datum, as shown in Fig. 4-95. Both datum A and datum B must be used to establish the common datum.

APPENDIX A   ILLUSTRATED GLOSSARY

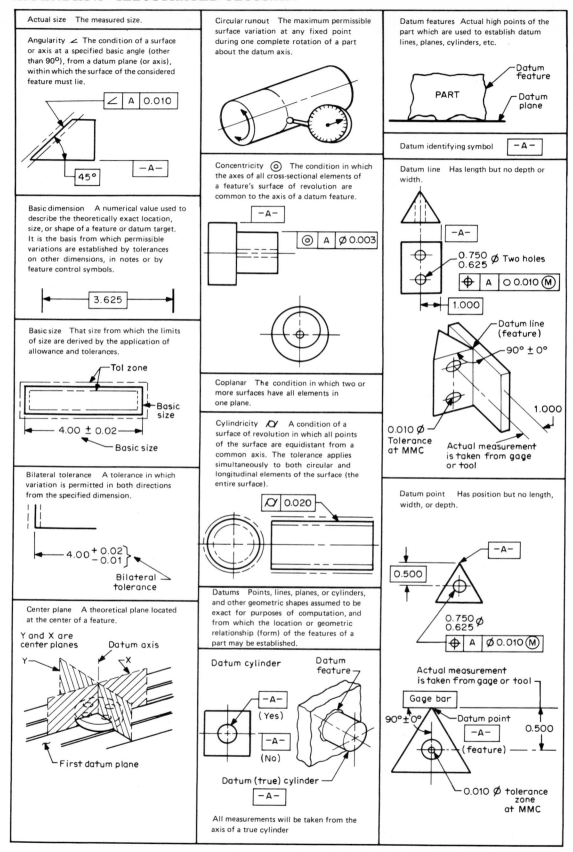

Actual size   The measured size.

Angularity ∠ The condition of a surface or axis at a specified basic angle (other than 90°), from a datum plane (or axis), within which the surface of the considered feature must lie.

| ∠ | A | 0.010 |

45°   − A −

Basic dimension   A numerical value used to describe the theoretically exact location, size, or shape of a feature or datum target. It is the basis from which permissible variations are established by tolerances on other dimensions, in notes or by feature control symbols.

3.625

Basic size   That size from which the limits of size are derived by the application of allowance and tolerances.

Tol zone
Basic size
4.00 ± 0.02
Basic size

Bilateral tolerance   A tolerance in which variation is permitted in both directions from the specified dimension.

4.00 + 0.02 / − 0.01
Bilateral tolerance

Center plane   A theoretical plane located at the center of a feature.

Y and X are center planes
Datum axis
Y
X
First datum plane

Circular runout   The maximum permissible surface variation at any fixed point during one complete rotation of a part about the datum axis.

Concentricity ⊚ The condition in which the axes of all cross-sectional elements of a feature's surface of revolution are common to the axis of a datum feature.

− A −

| ⊚ | A | ⌀ 0.003 |

Coplanar   The condition in which two or more surfaces have all elements in one plane.

Cylindricity ⌰ A condition of a surface of revolution in which all points of the surface are equidistant from a common axis. The tolerance applies simultaneously to both circular and longitudinal elements of the surface (the entire surface).

| ⌰ | 0.020 |

Datums   Points, lines, planes, or cylinders, and other geometric shapes assumed to be exact for purposes of computation, and from which the location or geometric relationship (form) of the features of a part may be established.

Datum cylinder
Datum feature
− A − (Yes)
− A − (No)
Datum (true) cylinder − A −

All measurements will be taken from the axis of a true cylinder

Datum features   Actual high points of the part which are used to establish datum lines, planes, cylinders, etc.

Datum feature
PART
Datum plane

Datum identifying symbol   − A −

Datum line   Has length but no depth or width.

− A −

0.750 / 0.625 ⌀ Two holes

| ⊕ | A | ⊚ 0.010 Ⓜ |

1.000

Datum line (feature)
90° ± 0°
0.010 ⌀ Tolerance at MMC
Actual measurement is taken from gage or tool
1.000

Datum point   Has position but no length, width, or depth.

− A −
0.500

0.750 / 0.625 ⌀

| ⊕ | A | ⌀ 0.010 Ⓜ |

Actual measurement is taken from gage or tool
Gage bar
90° ± 0°
Datum point
− A − (feature)
0.500
0.010 ⌀ tolerance zone at MMC

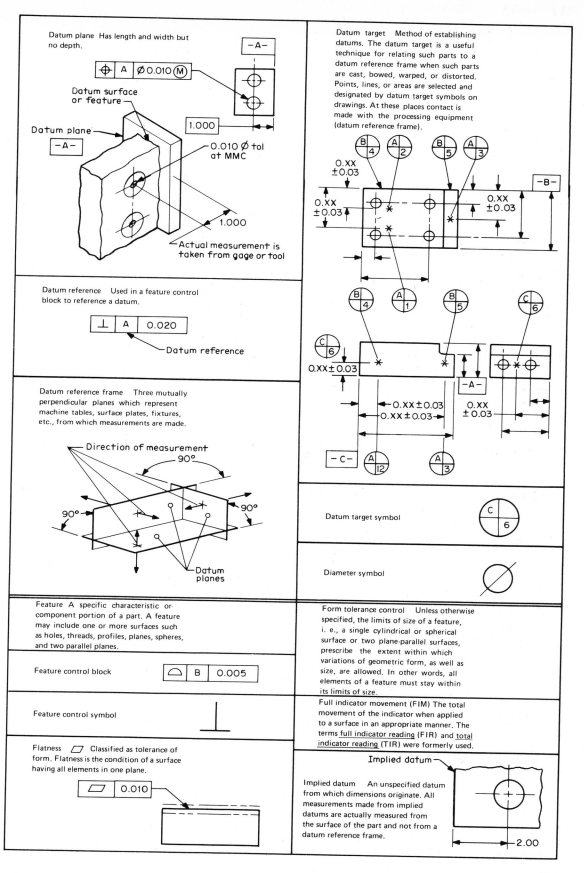

Datum plane   Has length and width but no depth.

⊕ | A | Ø0.010 Ⓜ

Datum surface or feature

Datum plane
—A—

—A—

0.010 Ø tol at MMC

1.000

1.000

Actual measurement is taken from gage or tool

Datum reference   Used in a feature control block to reference a datum.

⊥ | A | 0.020

Datum reference

Datum reference frame   Three mutually perpendicular planes which represent machine tables, surface plates, fixtures, etc., from which measurements are made.

Direction of measurement
90°
90°
90°

Datum planes

Feature   A specific characteristic or component portion of a part. A feature may include one or more surfaces such as holes, threads, profiles, planes, spheres, and two parallel planes.

Feature control block

⌒ | B | 0.005

Feature control symbol

Flatness   ▱ Classified as tolerance of form. Flatness is the condition of a surface having all elements in one plane.

▱ | 0.010

Datum target   Method of establishing datums. The datum target is a useful technique for relating such parts to a datum reference frame when such parts are cast, bowed, warped, or distorted. Points, lines, or areas are selected and designated by datum target symbols on drawings. At these places contact is made with the processing equipment (datum reference frame).

B/4   A/2   B/5   A/3

0.XX ±0.03
0.XX ±0.03
—B—
0.XX ±0.03

B/4   A/1   B/5   C/6

C/6
0.XX±0.03
—A—

0.XX ±0.03
0.XX ±0.03
0.XX ±0.03

—C—   A/12   A/3

Datum target symbol
C/6

Diameter symbol
Ø

Form tolerance control   Unless otherwise specified, the limits of size of a feature, i. e., a single cylindrical or spherical surface or two plane-parallel surfaces, prescribe the extent within which variations of geometric form, as well as size, are allowed. In other words, all elements of a feature must stay within its limits of size.

Full indicator movement (FIM)   The total movement of the indicator when applied to a surface in an appropriate manner. The terms full indicator reading (FIR) and total indicator reading (TIR) were formerly used.

Implied datum

Implied datum   An unspecified datum from which dimensions originate. All measurements made from implied datums are actually measured from the surface of the part and not from a datum reference frame.

2.00

4-43

Least material condition (LMC) The condition in which a feature of size contains the least amount of material within the stated limits of size. For example, maximum hole diameter or minimum shaft diameter.

Limit of size The applicable maximum and minimum sizes.

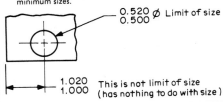

0.520 ∅ Limit of size
0.500

1.020   This is not limit of size
1.000   (has nothing to do with size)

Maximum material condition (MMC) or Ⓜ The condition in which a feature of size contains the maximum amount of material within the stated limits of size. For example, minimum hole diameter and maximum shaft diameter.

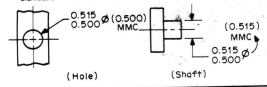

0.515 ∅ (0.500)
0.500   MMC

(0.515)
MMC

0.515 ∅
0.500

(Hole)   (Shaft)

Millimeter (mm) Metric dimensions are given in millimeters (1 in. = 25.4 mm)

Parallelism // The condition of a surface or axis equidistant at all points from a datum plane or axis.

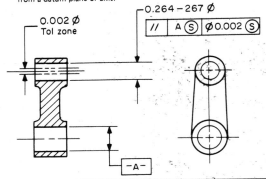

0.002 ∅
Tol zone

0.264 – 267 ∅

| // | A Ⓢ | ∅ 0.002 Ⓢ |

–A–

Perpendicularity ⊥ The condition of a surface, median plane, or axis at a right angle to a datum plane or axis.

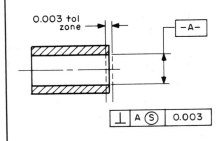

0.003 tol zone

–A–

| ⊥ | A Ⓢ | 0.003 |

Position tolerance Defines a zone within which the axis or center plane of a feature is permitted to vary from true (theoretically exact) position.

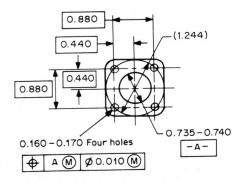

0.880
0.440
0.440
0.880
(1.244)

0.160 – 0.170 Four holes

0.735 – 0.740

–A–

| ⊕ | A Ⓜ | ∅ 0.010 Ⓜ |

Profile of a line ⌒ Establishes a tolerance, either bilateral (which is understood) or unilateral (drawn in the tolerance zone), from a true or basic profile (which is defined by basic dimensions) at all points along the profile. Profile of a line tolerance is a two-dimensional zone (at any cross section of the feature).

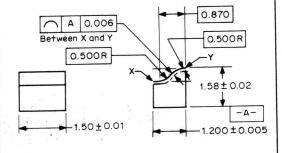

| ⌒ | A | 0.006 |

Between X and Y

0.500 R

0.870

0.500 R

X   Y

1.58 ± 0.02

–A–

1.50 ± 0.01   1.200 ± 0.005

Profile of a surface ⌒ Establishes a tolerance, either bilateral (which is understood) or unilateral (drawn in the tolerance zone), from a true or basic profile (which is defined by basic dimensions) at all points along the profile. Profile of a surface tolerance is a three-dimensional zone (the entire surface).

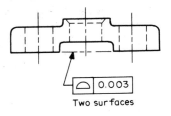

| ⌒ | 0.003 |

Two surfaces

Project tolerance zone A condition in which the cylindrical tolerance zone is projected above a threaded hole to the maximum thickness of the mating part. This should be used with all position tolerances of threaded holes. (It also provides a larger tolerance zone for 100 percent interchangeability compared to a position tolerance when a projected tolerance zone is not used.)

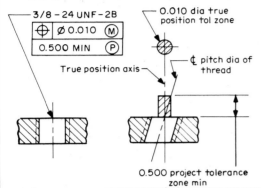

Straightness — Applied in the view in which the elements to be controlled are represented by a straight line. The axis or all points of the considered element must be within the specified tolerance.

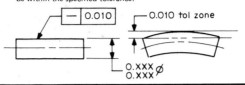

Symmetry ⌯ A condition in which a feature (or features) is symmetrically disposed about the center plane of a datum feature.

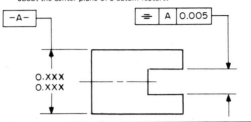

Regardless of feature size (RFS) The condition in which tolerance of position or form must be met irrespective of where the feature lies within its size tolerance.

Roundness ◯ A condition of a surface of revolution in which a cylinder or cone has all points of the surface intersected by any plane perpendicular to a common axis and equidistant from that axis. For a sphere, all points of the surface intersected by any plane passing through a common center are equidistant from the center.

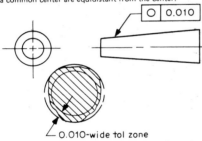

True position Denotes the theoretically exact position of a feature. A positional tolerance zone is located at the true position, which is defined by basic dimensions.

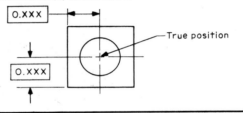

Unilateral tolerance A tolerance in which variation is permitted in only one direction from a specified dimension.

$$4.000 \; {}^{+\,0.020}_{-\,0.000}$$

Runout ⌰ A composite tolerance used to control the functional relationship of one or more features of a part constructed around, and at right angles to, a datum axis. Runout tolerance is understood to be a single circular element (part is rotated 360°). If a total runout is required, the word total must be added to the feature control block. Total runout provides composite control of all surface elements.

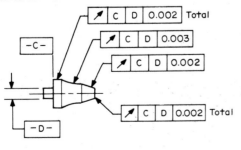

Virtual condition The boundary generated by the collective effects of the MMC limit of a feature and any applicable form or positional tolerance. The virtual condition of a hole would be the smallest hole size minus the positional tolerance. The virtual condition of a shaft would be the largest size, plus the positional tolerance.

Virtual condition for holes

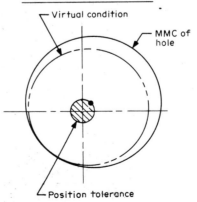

MMC of hole − position tolerance = virtual condition

Virtual condition (continued)

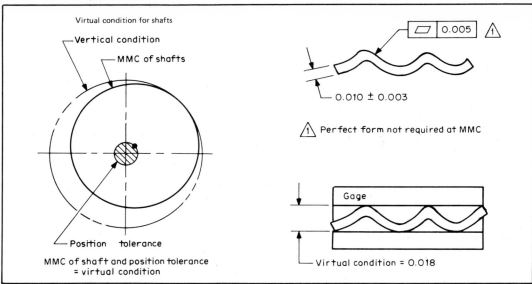

Virtual condition for shafts

Vertical condition

MMC of shafts

Position tolerance

MMC of shaft and position tolerance
= virtual condition

⌭ 0.005 ⚠1

0.010 ± 0.003

⚠1 Perfect form not required at MMC

Gage

Virtual condition = 0.018

APPENDIX B   POSITION TOLERANCING FORMULAS

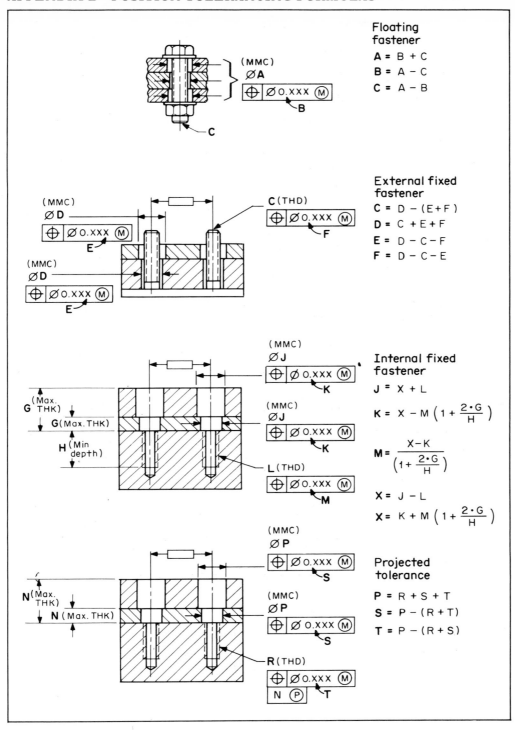

Floating fastener

$A = B + C$

$B = A - C$

$C = A - B$

External fixed fastener

$C = D - (E + F)$

$D = C + E + F$

$E = D - C - F$

$F = D - C - E$

Internal fixed fastener

$J = X + L$

$K = X - M \left(1 + \dfrac{2 \cdot G}{H} \right)$

$M = \dfrac{X - K}{\left(1 + \dfrac{2 \cdot G}{H} \right)}$

$X = J - L$

$X = K + M \left(1 + \dfrac{2 \cdot G}{H} \right)$

Projected tolerance

$P = R + S + T$

$S = P - (R + T)$

$T = P - (R + S)$

APPENDIX C TABLE OF NOMENCLATURE

	Table of nomenclature
$\varnothing$ or dia	Diameter
MMC or Ⓜ	Maximum material condition
MMC$_{hole}$	The smallest hole diameter
MMC$_{pin}$	The largest pin diameter
tol	Tolerance
min	Minimum
max	Maximum
TCT	Total coordinate tolerance
TCTZ	Total coordinate tolerance along the diagonal
Bilateral tolerance	Tolerance: ± 0.010
Unilateral tolerance	Tolerance: $^{+0.010}_{-0.000}$ or $^{+0.000}_{-0.010}$
▭	Basic dimension
TP	True position: diameter tolerance zone for tolerance of position
RFS or Ⓢ	Regardless of feature size
Modifiers	Ⓜ or Ⓢ
$T_\perp$ or $t_\perp$	A perpendicular control which is sometimes issued in conjunction with a positional tolerance to limit the amount of "tilt" an axis may have
B	In a bolted joint, the thickness or depth of the clearance hole part
A	In a bolted joint, the depth of the threaded hole
Coordinate axes	Two lines at 90° to each other from which position measurements are taken; frequently these lines are called X and Y axes or datums
R (row)	The number of rows of holes in a hole pattern
C (columns)	The number of columns of holes in a hole pattern

Section 5

Electronic Drafting

Arnold C. Noble

Since the early 1960s the electronics industries have experienced explosive and dynamic growth as a result of rapid technological advances in semiconductors, integrated circuits, and printed circuitry. The electronics industries include:

- Military and aerospace
- Telecommunications
- Computers
- Industrial controls and instruments
- Automotive
- Consumer products

This broad field of electronics has established policy and standards governing its procedural methods and techniques; therefore, it is suggested that these standards not be departed from in any way by any person who does not have a thorough understanding of the material in this section. Details of engineering standards and criteria will be presented at some length here, along with a list of military and industrial standards at the end of the section.

ELECTRICAL AND ELECTRONICS SCHEMATIC DIAGRAMS

Electrical and electronics schematic diagrams are described below, and guidelines for their preparation are provided. The accompanying illustrations clarify the text and provide examples of good drafting practices.

The electrical circuitry of any equipment may be documented with an assembly-type schematic, a set of functional schematics, a logic diagram, or a wiring schematic.

An electrical schematic may be a single- or multiple-sheet diagram; it may be one or more sheets of a detail or assembly drawing; or it may have its own drawing number.

Small schematics may be located on the same sheet on which the mechanical aspects of an assembly or part are delineated.

An electrical schematic does not necessarily show all the point-to-point wiring of a piece of equipment if the wiring data are presented with the equipment wiring dia-

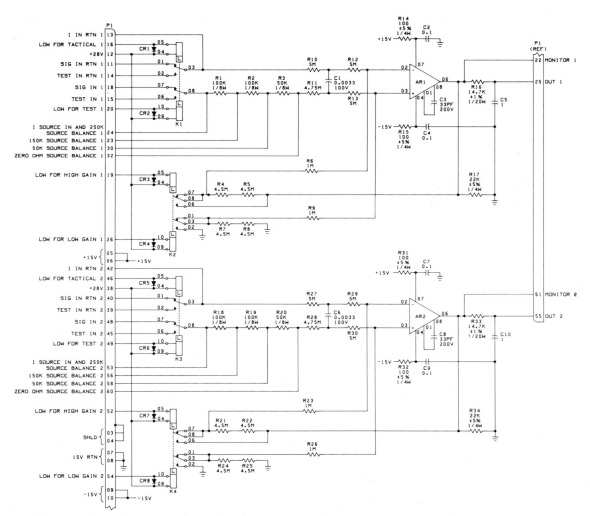

FIG. 5-1   Detail schematic. *(Interstate Electronics Corporation.)*

gram. For example, connectors used solely for interconnecting harnesses need not be shown.

Electrical schematics of parts depict, in detail, all the discrete components, electrical elements (including digital logic), and functions of the part. When these are combined to perform another function(s), a symbolic representation of that function(s) is included.

Schematic diagrams for higher-level assemblies may depict these parts in the following manners:

1. In detail, if required for clarity (see Fig. 5-1)

2. With a logic symbol or symbols depicting the function(s) performed by the part or assembly (amplifier, flip-flop, register, printed-wiring assembly, etc.; see Fig. 5-2)

3. With a circuit-element symbol (standard symbol) adequately labeled to describe the function (see Fig. 5-3)

When shown in detail, some or all of the elements or functions of lower-level assemblies may be grouped and enclosed with a broken line. When shown as circuit elements, these assemblies are depicted with the appropriate symbol and with each input and output identified. In either of the above methods, lower-level assemblies are identified by reference designation and by name or mnemonic; functions are identified by a functional name.

Assembly Electrical Schematic

An assembly schematic depicts all elements and functions of a piece of equipment (assembly, unit, or system). This type of schematic is usually selected for the documentation of simple assemblies, units, and systems (see Fig. 5-4).

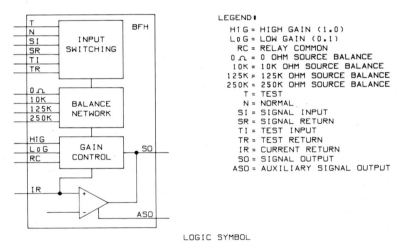

LEGEND:
HIG = HIGH GAIN (1.0)
LoG = LOW GAIN (0.1)
RC = RELAY COMMON
0 Ω = 0 OHM SOURCE BALANCE
10K = 10K OHM SOURCE BALANCE
125K = 125K OHM SOURCE BALANCE
250K = 250K OHM SOURCE BALANCE
T = TEST
N = NORMAL
SI = SIGNAL INPUT
SR = SIGNAL RETURN
TI = TEST INPUT
TR = TEST RETURN
IR = CURRENT RETURN
SO = SIGNAL OUTPUT
ASO = AUXILIARY SIGNAL OUTPUT

FIG. 5-2   Logic symbol. *(Interstate Electronics Corporation.)*

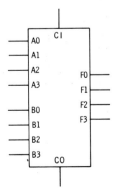

FIG. 5-3   Standard symbol. *(Interstate Electronics Corporation.)*

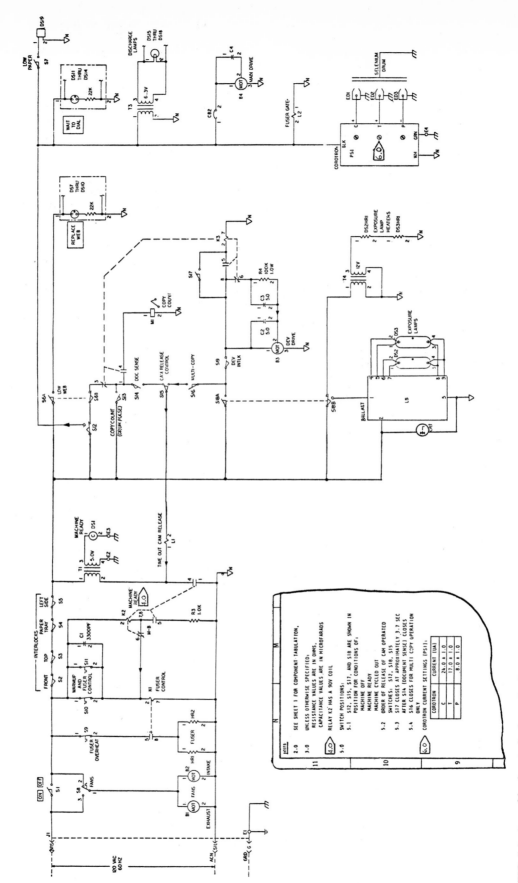

FIG. 5-4 Assembly electrical schematic. (*Xerox Corporation.*)

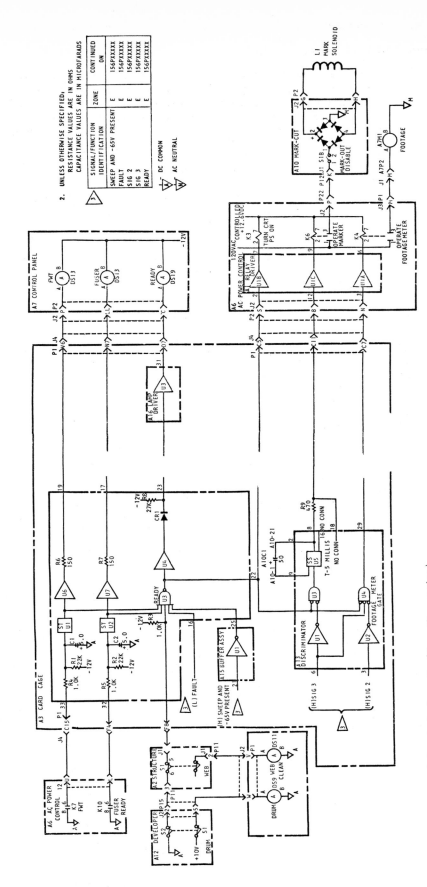

FIG. 5-5 Functional electrical schematic. *(Xerox Corporation.)*

Functional Electrical Schematic

A functional schematic depicts a complete function or operation within a piece of equipment and shows only those elements necessary to describe that function or operation. Functional schematics are usually selected for documentation of complex assemblies, units, and systems where an assembly-type schematic might prove difficult to interpret. A complete set of functional schematics for a piece of equipment shows all elements and functions of that equipment (see Fig. 5-5).

Logic Diagram

A logic diagram depicts the two-state device implementation of logic functions with logic symbols and supplementary notations, showing details of signal flow and control, but not necessarily the point-to-point wiring (see Fig. 5-6).

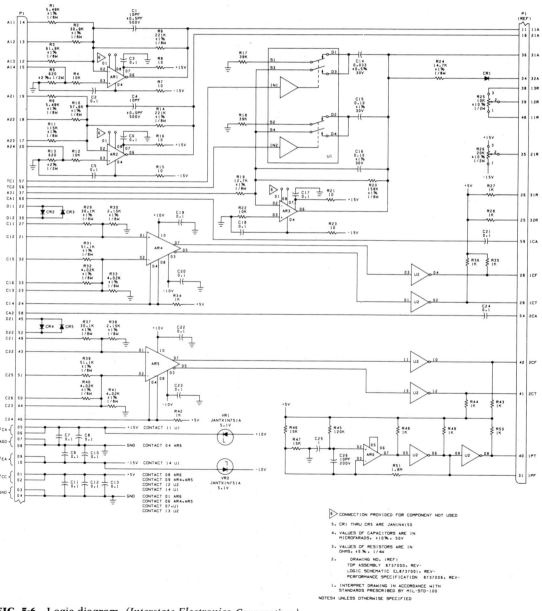

FIG. 5-6 Logic diagram. *(Interstate Electronics Corporation.)*

Wiring Schematic

A wiring schematic (see Fig. 5-7) is a combined assembly schematic and wiring diagram and is used only for assemblies containing relatively few electrical components. It may be part of the assembly drawing and is prepared as follows:

A wiring schematic is prepared as a wiring diagram using the point-to-point method, wherein electrical connections are shown using continuous lines between terminals.

Electrical components are depicted as outlines on wiring schematics. When a schematic symbol is required to permit understanding of the operation of the component, it should appear within the outline.

All symbol data, such as names, codes, reference designations, values, and ratings, are placed adjacent to or within the component outline.

All inputs and outputs of the circuit are identified.

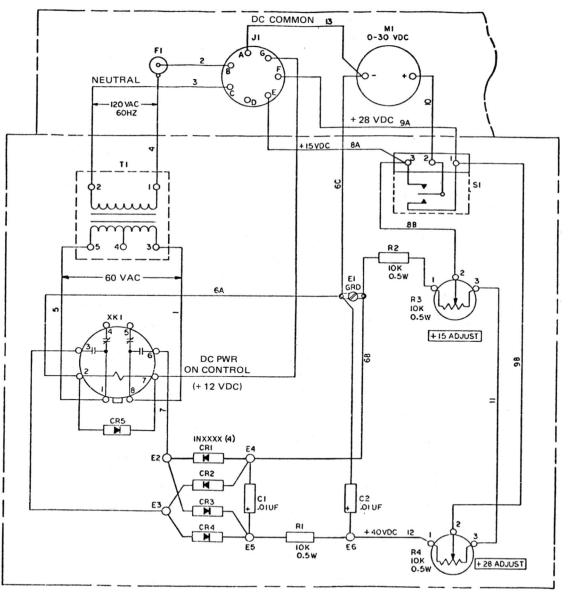

FIG. 5-7　Wiring schematic. *(Xerox Corporation.)*

Standard Conventions

Basic Electrical Circuits

Basic circuits such as multivibrators, latches, and full-wave bridges should be drawn in easily recognizable forms. Figures 5-8 through 5-10 represent a few of the established arrangements.

Series and Parallel Elements and Circuits

These should be drawn so that the series or parallel arrangement is apparent (see Figs. 5-11 and 5-12).

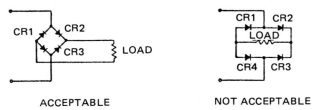

ACCEPTABLE   NOT ACCEPTABLE

FIG. 5-8   Full-wave bridge. *(Xerox Corporation.)*

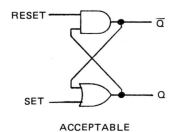
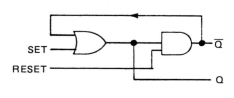

ACCEPTABLE   NOT ACCEPTABLE

FIG. 5-9   Latch. *(Xerox Corporation.)*

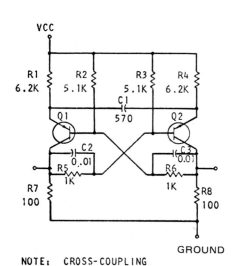
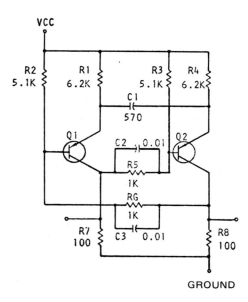

ACCEPTABLE   NOT ACCEPTABLE

FIG. 5-10   Multivibrator. *(Xerox Corporation.)*

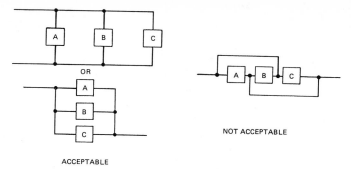

FIG. 5-11 Parallel elements and circuits. *(Xerox Corporation.)*

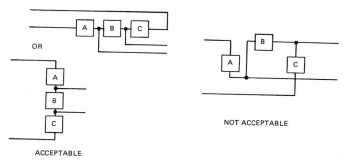

FIG. 5-12 Series elements and circuits. *(Xerox Corporation.)*

Distinctive-Shaped Symbol Orientation

It is preferred that distinctive-shaped symbols indicating signal flow, such as the amplifier, AND, and OR symbols, be oriented left to right on the diagrams.

Special Conventions

Special annotations include clocking conventions, output conventions, negation terminology, ANDed inputs, subdivisions, consolidating like devices, cascading, and standard symbol form.

Clocking Conventions

Flip-flops, counters, and registers require an external clock for synchronous operation. Storage-element types used with these devices fall into into two general classes: level-controlled, and edge-triggered. Clock inputs are defined in terms of how the outputs react.

Figure 5-13 depicts four different conventions used with the clock input. Conventions A and B are applicable to a level-controlled device such as a master/slave flip-flop. Conventions C and D are applicable to all edge-triggered devices.

Output Conventions

Three types of output drive circuits are encountered in logic circuits: totem pole, open collector, and tristate. Figure 5-14 illustrates the scheme used as part of the device symbol to define the general output characteristics. This innovation is intended to aid in troubleshooting and circuit analysis.

Negation Terminology

A small circle at the input to any element indicates that a relatively low input signal activates the element. The absence of a small circle indicates that a relatively high

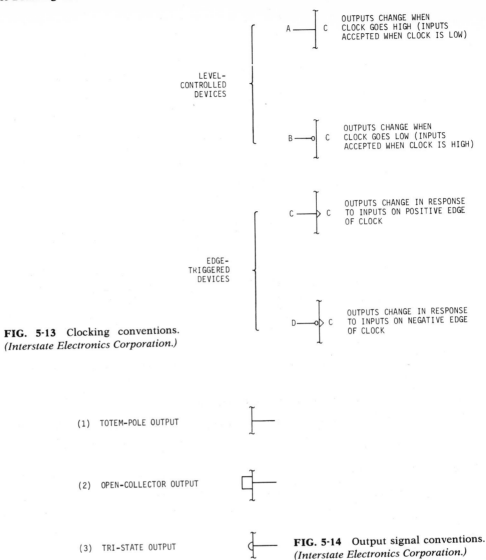

FIG. 5-13 Clocking conventions.
(Interstate Electronics Corporation.)

(1) TOTEM-POLE OUTPUT

(2) OPEN-COLLECTOR OUTPUT

(3) TRI-STATE OUTPUT

FIG. 5-14 Output signal conventions.
(Interstate Electronics Corporation.)

input signal activates the element. A small circle at the output of the symbol indicates that the output terminal is relatively low when the element is active, and the absence of a small circle indicates that the output is relatively high when the element is active.

ANDed Inputs

Some devices feature ANDed inputs for control or data. These inputs are shown with an AND gate symbol inside either the control block or the appropriate bit cell.

Subdividing Elements and Referencing

The use of control blocks and bit cells permits subdivision of a device. Subdividing often improves readability by enhancing the functional flow. For example, the control section is shown in proximity to the circuit functions that generate the control signals, and the bit cells are shown in proximity to related functions (see Fig. 5-15). As previously mentioned, a particular device is always identified by the type and location information inside the control-block symbol. Whenever a device is subdivided, both the control block and the bit cells contain type and location information. The standard engineering break line is used to denote a subdivided device.

Consolidating Like Devices

Two or more identical devices are often used to implement a particular logic function. In these instances, all symbols are combined into one large symbol to save space. When devices are consolidated, the control block sometimes becomes more complex, but the interface details are simplified because there are fewer vertical interconnect lines (see the example in Fig. 5-16).

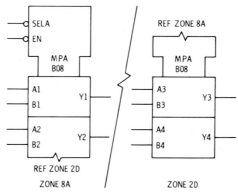

FIG. 5-15 Proper subdivision of a device. *(Interstate Electronics Corporation.)*

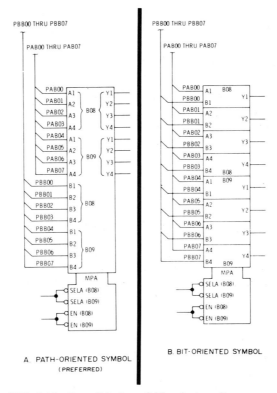

FIG. 5-16 Consolidation of like device. *(Interstate Electronics Corporation.)*

Cascading Techniques

Devices such as counters, shift registers, comparators, priority encoders, and parity generators have expansion features. Expanded element symbols are cascaded (see Fig. 5-17) rather than consolidated. Expansion outputs are interstage carry, borrow, overflow, or priority outputs. Interstage signals are more apparent when elements are cascaded. The cascading technique consists of showing the first element with the control block and bit-cell symbol to depict the initial conditioning of the function. The second and succeeding elements contain a modified control block or bit-cell array that shows only the interstage interface. All controls common to all elements are shown in the first control block.

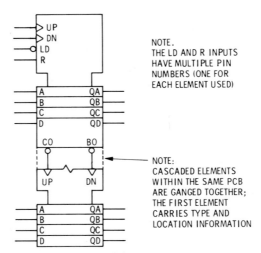

FIG. 5-17 Cascaded devices. *(Interstate Electronics Corporation.)*

Line Conventions

The selection of line thickness should take into account size reduction when it is believed that legibility may be affected. A line of medium thickness is recommended for general use on diagrams. A thin line may be used for brackets, leader lines, etc. When emphasis of special features such as a signal or transmission paths is essential, a line thickness sufficient to provide the desired contrast may be used. Line conventions, relative thickness, and suggested applications for use on electrical schematic diagrams are shown in Table 5-1.

Connecting Lines

Connecting lines may run horizontally and/or vertically on the sheet. Exceptions to this rule are made for "bridges" and "crossover functions," which are most frequently shown and understood using diagonal connecting lines. Connecting lines are drawn with a minimum number of bends and crossovers. Lengthy lines between symbols should be avoided by using an interrupted-line technique.

Connecting Line Spacing

The schematic diagram should reflect uniformity of line spacing. Minimum connecting line spacing is specified in Table 5-2.

Grouping Connecting Lines

Long parallel connecting lines should be arranged in groups of two or three (three is preferred).

TABLE 5-1   Line Thicknesses

Line application	Line types and width
For general use (symbols, connecting lines, etc.)	Medium (0.0197 in. or 0.50 mm width)
Mechanical connections	Medium (short dash) (0.0197 in. or 0.50 mm width)
Shielding and future circuits	Medium (long dash) (0.0197 in. or 0.50 mm width)
Brackets, leader lines, etc.	Thin (0.0098 in. or 0.25 mm width)
Mechanical and functional grouping boundary lines	Thick (0.0394 in. or 1.00 mm width)
For emphasis	Thick (0.0394 in. or 1.00 mm width)

TABLE 5-2   Line Spacing versus Drawing Size

Drawing size	Minimum line spacing
D, E, and J	0.30 in. (7.62 mm)
C and smaller	0.20 in. (5.08 mm)

Junctions and Crossovers

Connecting line junctions are represented by solid dots and must always appear as single junctions. The minimum dot diameter should be 0.094 in. (2.39 mm) or three times (3×) the line width, whichever is greater. Crossovers are represented by straight lines crossing without dots (see Fig. 5-18).

Line Identification

Connecting lines may be identified by the use of codes, names, voltage levels, common connection symbols, or combinations of these. These identifications provide supplementary information, aid in interpretation, and provide correlation to other electrical documentation. Identifications are placed in such a manner that it is clear which lines they identify. The identification may be "bottom-reading" for horizontal lines and "bottom-reading" or "right-reading" for vertical lines.

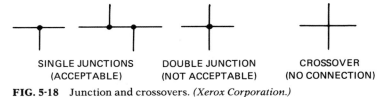

| SINGLE JUNCTIONS | DOUBLE JUNCTION | CROSSOVER |
| (ACCEPTABLE) | (NOT ACCEPTABLE) | (NO CONNECTION) |

FIG. 5-18   Junction and crossovers. *(Xerox Corporation.)*

Interrupted Connecting Lines and Addressing Requirements

Connecting lines may be interrupted as groups or individually. Interrupted groups of lines may be identified by a code and, if necessary, addressed, as shown in Example 5-1.

Individually interrupted connecting lines may be categorized as follows:

1. *Intradrawing interruptions*   Interrupted connecting lines resumed within the same drawing that are not "common connections."
2. *Interdrawing interruptions*   Interrupted connecting lines that are resumed on other drawings.
3. *Common connections*   Interrupted connecting lines depicted with a common connecting symbol.

Intradrawing Connecting Lines

Connecting lines continued to another sheet have the sheet and zone specified. Lines continued on the same sheet have zone designations specified. Interrupted connecting lines may be resumed on several sheets but can resume on any one sheet at only one point, as shown in Example 5-2.

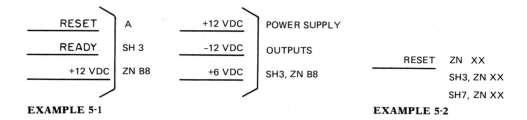

EXAMPLE 5-1   **EXAMPLE 5-2**

Numerical Values

Numerical values of resistance, capacitance, inductance, voltage, etc. are specified in a form requiring the fewest zeros. This is accomplished by using the multiplier symbols of Table 5-3 as specified in Tables 5-4 through 5-7. Numerical values of four digits are shown without commas (i.e., 4700, not 4,700). Numbers less than 1 have a zero before the decimal point (i.e., 0.014, not .014). Nonstandard (non-ANSI) abbreviations are used (i.e., μF is the ANSI abbreviation in electronic drafting but UF is actually used on drawings).

To avoid repeating symbols of units of measure that are generally applicable throughout a diagram, a drawing note may be used and only the numerical value of the units specified on the diagram.

TABLE 5-3   Multiplier Symbols

Multiplier	Prefix	Symbol
1×10^{12}	tera-	T
1×10^{9}	giga-	G
1×10^{6}	mega-	M
1×10^{3}	kilo-	K
1×10^{-3}	milli-	MILLI
1×10^{-6}	micro-	U
1×10^{-9}	nano-	N
1×10^{-12}	pico-	P
1×10^{-15}	femto-	F
1×10^{-18}	atto-	A

TABLE 5-4   Basic Unit Symbols

Quantity	Basic unit	Symbol
Resistance	ohm	ohm
Capacitance	farad	F
Inductance	henry	H
Potential	volt	V
Current	ampere	A
Power	watt	W
Time	second	SEC
	minute	MIN
	hour	HR
Frequency	hertz (cycles per second)	HZ

TABLE 5-5   Resistance Values

Range in ohms	Expressed as	Examples
Less than 1	milliohm	20 MILLIOHMS; 60 MILLIOHMS
1 to 999	ohm	50 OHMS; 75 OHMS
1000 to 999,999	kilohm	2 K; 2.25 K
1,000,000 or more	megohm	3 M; 100 M

TABLE 5-6   Capacitance Values

Range in picofarads*	Expressed as	Examples
Less than 10,000	picofarad	152.4 PF; 4700 PF
10,000 or more	microfarad	0.015 UF; 30 UF

*The basic unit of capacitance, the farad, is too large to be of practical use; therefore, the range is expressed in picofarads. The term *picofarad* has replaced the older term *micromicrofarad*.

TABLE 5-7   Inductance Values

Range in microhenries*	Expressed as	Examples
Less than 1000	Microhenry	10 UH; 100 UH; 999 UH
1000 or more	Millihenry	1 MILLIH; 50 MILLIH; 110 MILLIH

*The basic unit of inductance, the henry, is too large to be of practical use; therefore, the range is expressed in microhenries.

Diagram Layout

General Layout Guidelines

- The symbols of the diagram should be spaced to provide a balanced appearance.
- Sufficient space should be provided around symbols to avoid crowding of symbol data.
- Large blank spaces should be avoided. (*Exception:* Space provision should be made for anticipated future circuits and/or future changes.)
- Where practical, terminations for external connections should be located at the outer edges of the diagram.
- In general, diagrams should be arranged so that they can be read from left to right and from top to bottom, in the order of functional sequence or following the signal or transmission path. If this is not possible, and the direction of information flow is not obvious, then lines carrying information should be marked with an arrowhead.
- Inputs and sources should be on the left (preferred) or top, and outputs and loads on the right (preferred) or bottom.

- The layout should reflect an organizational arrangement wherein most symbols and connecting lines are in line both horizontally and vertically.

Signal-Line Emphasis   Lines of heavy thickness may be used to emphasize signal lines if they aid in the interpretation of the diagram.

Reduction of Data Density in the Field of Drawing   Use of techniques that reduce data density in the field of the diagram should be considered when making a layout. Among such techniques are the use of notes and tabulations, enclosures for repetitive circuits, and the interrupted line to reduce the number of continuous connecting lines. Judgment must be exercised in the selection of technique.

Notes are used to present information that is common to like symbols. Consider a circuit that contains several resistors with the same power rating and tolerance. Each resistor might be identified in the field of the diagram as shown in Example 5-3.

If a note such as "unless otherwise specified, resistors are 10 percent and 0.25 W, and resistance values are in ohms" were used, the only data necessary in the field of the drawing would be as shown in Example 5-4.

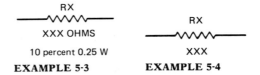

EXAMPLE 5-3   EXAMPLE 5-4

Notes that convey information unique to a piece of equipment may be devised, providing they are concise and capable of only one interpretation. Any information such as adjustment information or operating characteristics may be given in note form to eliminate the necessity of presenting this information in the field of the drawing.

Tabulations should be used to present data whenever practical and possible. These tabulations appear in the note column. Some data that can be tabulated are: relay parts location, rotary switch functions, separated components, and functional circuit location.

Enclosures are used to show repetitive circuits and mechanical or functional grouping of parts (see Fig. 5-19).

Interrupted Lines   The interrupted-line technique can be used to reduce the number of continuous connecting lines in a diagram (see Figs. 5-20 and 5-21).

Specific Layout Guidelines

The lack of dimensional guidelines is one of the chief difficulties that the drafter will encounter in circuitry work. Consequently, the drafter is forced to design or lay out

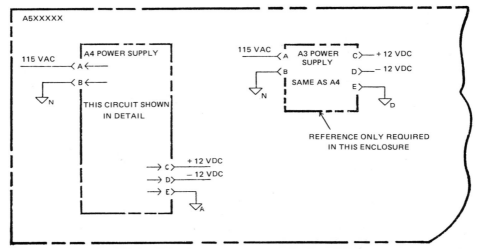

FIG. 5-19   Enclosures. *(Xerox Corporation.)*

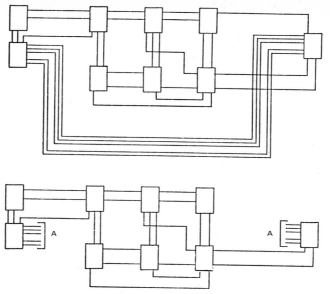

FIG. 5-20 Interrupted lines used to eliminate long interconnected lines. *(Xerox Corporation.)*

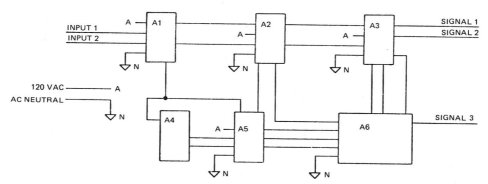

FIG. 5-21 Interrupted-line technique used to eliminate continuous connecting lines between electrically common connections; in this example, 120 V ac and ac neutral. *(Xerox Corporation.)*

the diagram on the basis of information received in the form of a freehand sketch or a revised, marked-up print of an existing diagram. It may also be necessary to form a new, more complex diagram from a compilation of circuits taken from several existing diagrams. When such diagrams are made as the design of equipment progresses, they may require considerable additions, deletions, and alterations, or possibly even a completely new approach before a satisfactory finished diagram can be drawn.

If, however, the alterations on an existing diagram are minor, the drafter's task will be quite simple because the previous diagram may be used as a guide for the layout. Then, it is only necessary to allocate sufficient additional space for the new symbols and their identification.

Component Information   Before proceeding with the diagram, the drafter should get complete and correct information about each component used on the new diagram or revision. This information may be obtained from several sources: the mechanical and electrical component drawings, the parts lists, manufacturers' catalogs, the drafting room manuals and standards, technical bulletins, the project engineer, or other similar sources.

All this data should be carefully noted on the freehand circuit sketch, a marked-up print of the existing circuit, or any other source that serves as the basis for the diagram. When space limitations prohibit such a procedure, a list of components may be prepared, giving all pertinent information and relating it to the working sketch by number, letter designation, or some other means.

Layout Considerations   After complete information has been obtained, the next step is to make one or more rough preliminary layouts to determine the space requirements of the diagram, the relative arrangement of symbols or major groups of symbols, and so forth.

Another factor to consider in diagram layouts is the width of the drawing paper, which is limited to 34 or 40 in. (76.2 or 101.6 cm). This becomes a problem in complex diagrams, which contain a number of horizontal layers of segregated component symbols. The use of the diagram is also a consideration. For example, a long diagram is more acceptable for use in instruction books than is a wide diagram that requires several horizontal folds.

Diagram Layout (Preliminary)   It is advisable to make preliminary layouts on cross-section or grid-ruled paper with 0.10-in. (2.54-mm) grid spacing. Although the initial, trial layout may not necessarily be the best one, it will serve as a guide for the next, better-proportioned layout. The drafter should strive for a well-spaced, uncluttered layout even if that means using the next larger size drawing sheet. If the diagram is made during the preliminary stages of equipment development, the drafter should always keep in mind that additional component symbols may be added later.

The complete circuit should be divided into functional segments, with each segment allotted a certain amount of diagram space. This circuit breakdown is quite helpful in making trial layouts, which can be rearranged or improved upon later.

The two-stage, resistance-coupled amplifier circuit shown in Fig. 5-22 illustrates the allotment of space requirements. Because the two stages are similar, they establish the basic overall space requirements, to which the interstage coupling components must be added.

Of the two alternative layouts shown in this sketch, that of Fig. 5-22a has a cleaner appearance and less crowding between symbols, and consequently is commonly used for electronic communication equipment diagrams. The shaded blocks represent the areas that are left for reference designations, component values, and other identifications. Space also should be allowed near the transistor envelopes for a listing of transistor pin numbers. The circuit in Fig. 5-22a is called *bottom-fed* because the feeder lines are located below the basic stage components.

Estimating Space Requirements   After a freehand trial circuit arrangement has been drawn, the grid spaces allocated for the various component symbols can be marked on the sketch between the vertical and horizontal lines that extend from the center of the symbols or connection lines (see Fig. 5-22). The grid spaces given in Table 5-8 represent average figures for the majority of schematic diagrams using 0.20-in. (5.08-mm) grid spacing.

TABLE 5-8   Estimating Space Requirements

Average diagram spacing	0.20-in. (5.08 mm) grid spaces
Component	
Capacitors	3–4
Inductors	4
Resistors	4
Diagram items	
Transistor envelope diameter	4
Resistor symbol length	3
Capacitor symbol width	1–2
Lettering height	¾ (or 0.150 in.; 3.81 mm)
Connection line spacing	1–1½
Spacing between groups of connection lines	1–2

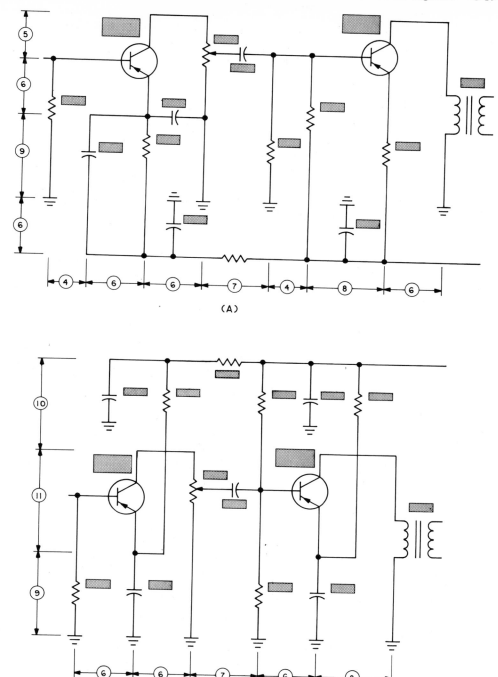

FIG. 5-22 A preliminary layout. The numbers represent the number of 20-in. (5.08-mm) grids. (*Courtesy of Nicholas M. Raskhodoff,* Electrical Drafting and Design, *3d ed., Prentice-Hall, Englewood Cliffs, N.J., 1977.*)

Although horizontal spacings between symbols have been indicated (Fig. 5-22), they are merely representative and depend upon the circuit components and arrangement. Larger or more complex component symbols, such as multiposition switches, and relays, may need more space. Terminations such as terminal boards or connectors may require additional space for identification and lettering.

Symbol Locations   Symbols should be located on grid lines with sufficient space left for lettering. It may be necessary to stagger the symbols on adjacent vertical lines to obtain space for lettering. To avoid crowding of symbols or lettering, the location of symbols should be altered as required.

Potentiometer symbols that connect to a transistor element should be drawn below the center of the symbol to obtain greater lettering area.

Symbol Lettering   It is sometimes difficult to locate space on a diagram for such lettering as component value, tolerance, type, reference designation, polarity, terminal identification, and rotation. Several methods of presenting such information are illustrated further on in this section under Reference Designations. The lettering is usually done after the circuitry details have been completed, and in some cases, ingenuity may be needed in order to eliminate the appearance of crowding or ambiguity. Although horizontally aligned symbols give the diagram a better appearance, it may be necessary in some instances to offset or stagger them to obtain sufficient lettering space.

Although vertical lettering may be used in some instances to gain space, it detracts from the readability of a diagram and therefore is not recommended.

Diagram Layout Procedures   The layout of a schematic diagram, be it complex or relatively simple, can be expedited by following a certain routine. This not only saves time, it also results in a better finished product. The beginner in circuitry drawing will benefit by following the procedures outlined here:

1. Obtain the freehand sketch or sketches or other circuit information that will serve as the basis for the schematic diagram.

2. Check this material for full information about component identification: check the subdivision of the diagram, if any, and the identification of such subdivisions; and check for possible revisions if the equipment is in the developmental stage, and allow extra drawing space.

3. Make freehand trial layouts of the diagram sections on grid paper and establish the approximate dimensions of each section, allowing symbol spaces according to the practices outlined earlier in this chapter.

4. Make freehand trial layouts of such diagram appendages as auxiliary circuits and details of complex component symbols.

5. Make a rough trial layout of tables and notes to establish the approximate space they require.

6. Make a trial assembly of the diagram section layouts and all other material to arrive at the rough complete diagram layout, allowing space for connection lines between the diagram sections and auxiliary diagram information. The shape of this rough diagram assembly should fit on a standard drawing sheet or on standard roll-size paper if it is a large diagram. The required diagram area should not impinge upon the revision and title block area of the sheet.

7. Modify the rough diagram layout as necessary.

8. Attach the rough trial layouts used in the complete rough diagram assembly to a sheet the same size as the drawing size selected.

9. Roughly sketch in the connection lines between various diagram sections and add connectors, terminal blocks, and other termination symbols.

10. Draw the finished diagram, using the rough layout assembly as a guide. The finished diagram may be drawn on 0.10- $\times$ 0.10-in. (2.54- $\times$ 2.54-mm) spaced grid paper, cut to standard drawing sheet size, or the grid paper may be slipped below the standard drawing sheet to act as a spacing guide.

Following these procedures eliminates the necessity for drawing a complete preliminary diagram after completing the first nine steps, and then drawing the finished diagram, and saves time and effort as well. The beginner may acquire proficiency and confidence, however, by drawing such a preliminary diagram before drawing the finished diagram. After the novice has gained a little experience, this should be unnecessary.

The drafter's work is simplified if the original diagram sketch, furnished by the originator, has such a circuit format that little can be gained by trying to rearrange it on the finished diagram. In such cases, it is necessary only to establish the overall dimensions of each circuit subdivision, using the dimensional practices and spacings given earlier in this section.

Improvements may be incorporated in the finished diagram as the drawing progresses. Slight modifications may be necessary to provide space for extra connection lines or additional space for lettering.

All the symbols are usually drawn first, starting with such basic symbols as logic elements. All like symbols are drawn at one time, especially if they are on the same horizontal or vertical level. If a change in symbol position is expected, it may be advisable to draw the connection lines to the approximate symbol position and fill in the symbol later. This eliminates redrawing the symbol from one position to another.

Short connection lines, such as those that extend from the transistor envelope to the immediate circuit components of the transistor, are drawn next. Ground symbols may all be drawn at the same time and vertical feeder lines brought down to the approximate position of the horizontal feeder connection lines, which are drawn next.

Junction points and crossovers should be checked carefully before lettering is added. Although there have been some attempts to omit dots at junction points, using them is more logical and results in fewer errors by the drafter or the eventual user of the diagram. The American National Standards Institute has adopted the use of junction dots in its Electrical and Electronics Diagrams standard, ANSI Y14.15-1966. The use of dots also avoids the necessity of offsetting junction connections, which would otherwise be required.

Lettering of component values and reference designations is added next. Occasionally this may require moving some of the symbols to provide additional space.

Such diagram appendages as insert diagrams, flat-pack, transistor, integrated-circuit, dual-in-line package, connector, switch, and relay-base layouts; and other similar material should be drawn next in the border of the diagram. These are followed by notes and other explanatory data.

Diagram Checklist   After the diagram has been completed, it should be checked for the following:

1. Indication of all mechanical linkages, enclosures, and shielding
2. Adequate identification of all terminations
3. Complete and clear identification and designation of components
4. Proper identification of separated symbols and circuits and of auxiliary circuits
5. All extra identification symbol markings, such as polarity, color, terminal number, transformer interconnections, switch wafer, waveform, voltage, tolerance, rotation, and so forth
6. Inclusion of all complete tabular data
7. Inclusion of special note references to symbols and other identification and, when required, the use of standard abbreviations
8. A minimum of crossovers and bends
9. Complete drawing title, identification of equipment depicted on the diagram, and other relevant data
10. Consecutive numbering of components in left-to-right and top-to-bottom sequence
11. Correct and proper identification of flat-pack, transistor, integrated circuit, dual-in-line package, auxiliary, insert, and connector layout diagrams
12. Adequate reference notes for all reference notations on the diagram
13. Selection of reference designation letters in compliance with military or industry standards
14. Adequate cross-references for related equipment or parts lists
15. Complete titles or other identification for major subdivisions on the diagram

The recommended 0.20- × 0.20-in. (5.08- × 5.08-mm) grid spacing and a 0.15-in. (3.81-mm) lettering make the diagram suitable for a 2:1 reduction for technical-manual purposes and 35-mm microfilming requirements.

GRAPHIC SYMBOLS

Because of the complexity of schematic, logic, and wiring diagrams, and also because of the need to conserve space, graphic symbols are used in electronics to denote the various electronic, electrical, and mechanical components.

Graphic symbols used for electronic diagrams have been developed over the years to portray each component in an unmistakable manner. In many cases, these graphic symbols, such as the one used for a switch, resemble the actual component or its elements, thus helping to make the diagram easier to read.

The currently accepted symbols reflect a considerable evolution since the early days of the electrical industry. At that time, the symbol used by the manufacturers had to bear a strong resemblance to the component itself. In some instances, however, manufacturers disagreed among themselves as to the choice of graphic symbols. The confusion caused by this lack of uniformity was further complicated by the emergence of the electronics industry, which, in many instances, requires the use of complex symbols for delineation of components.

Definitions and Use

A symbol is considered as the aggregate of all its parts. Graphic symbols are a shorthand used to show graphically the functions of circuits. Graphic symbols are used on schematic diagrams and as applicable on logic diagrams or wiring diagrams. These symbols are correlated with parts lists, descriptions, or instructions by means of reference designations (see Reference Designations below).

Graphic Standards

Except as noted, the symbols are from:

- Graphic Symbols for Electrical and Electronics Diagrams, ANSI Y32.2-1975 (IEEE 315-1975)
- Graphic Symbols for Logic Diagrams (Two-State Devices), ANSI Y32.14-1973 (IEEE 91-1973)

These documents have been approved by the Institute of Electrical and Electronics Engineers (IEEE), the American National Standards Institute (ANSI), the American Society of Mechanical Engineers (ASME), and the U.S. Department of Defense (DOD). In addition, IEEE 315-1975 has been approved by the Canadian Standards Association (CSA).

These standards are updated at least every 5 years by standing committees of the IEEE and ANSI. Tables 5-9 through 5-14 are partial examples of graphic symbols.

REFERENCE DESIGNATIONS

Reference designations distinguish one graphic symbol from another and correlate these identifications with actual components on the parts lists, assembly drawings, and technical manuals.

Reference designations may be placed above, below, or on either side of a graphic symbol. Other pertinent information, such as component value tolerances or ratings, terminal numbering, and other functional descriptions, may be placed around the graphic symbols (see Example 5-5). Reference designations are the main link between the electrical circuitry and the mechanical details of electronic equipment, for without them it would be impossible to differentiate between the identical components that

TABLE 5-9   Two-State Logic Devices

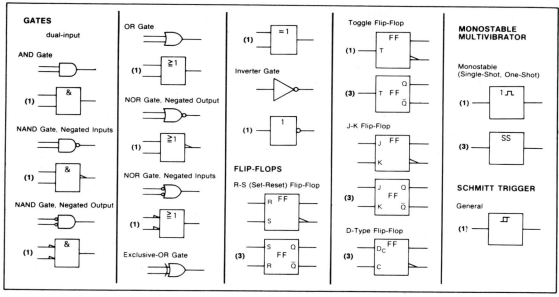

(1) = IEC Approval   **(2)** = Proposed IEEE Revision   **(3)** = Popular Industry Usage

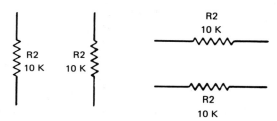

EXAMPLE 5-5   Reference designator placement.

appear throughout the equipment, or to locate the components on the mechanical drawings.

Basic Reference Designations

A basic reference designation for an item (such as a resistor, inductor, or subassembly) consists of one or more letters, which identify the class, and a number, for example: AR14 or C3. In some cases, a suffix letter may be added; for example: C7A or C7B. All letters are capitals. The designations are written with all characters the same size, on the same line, and without separation (see Fig. 5-23).

Class Designation Letter

The letters identifying the class of an item are selected in accordance with Table 5-15. For items not specifically listed, use the letters already assigned for the most similar

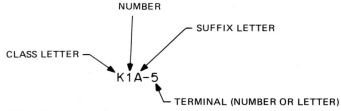

FIG. 5-23   Basic reference designation. *(Xerox Corporation.)*

TABLE 5-10 Semiconductors

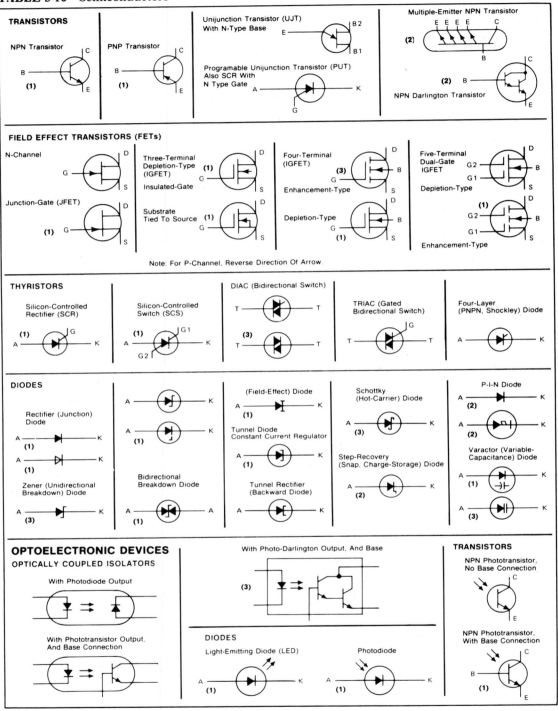

(1) = IEC Approval   **(2)** = Proposed IEEE Revision   **(3)** = Popular Industry Usage

class of items. The letters A and U (for assembly) should not be used if more specific class letters are assigned for a particular item.

Note: Certain item names and designating letters may apply to either a part or an assembly. For example, motors, relays, and printed-wiring boards can be considered a part or an assembly.

TABLE 5-11 Transmission Path

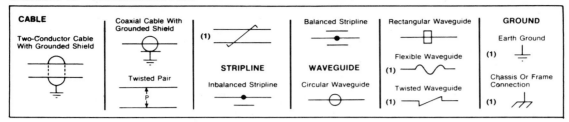

TABLE 5-12 Fundamental Circuit Components

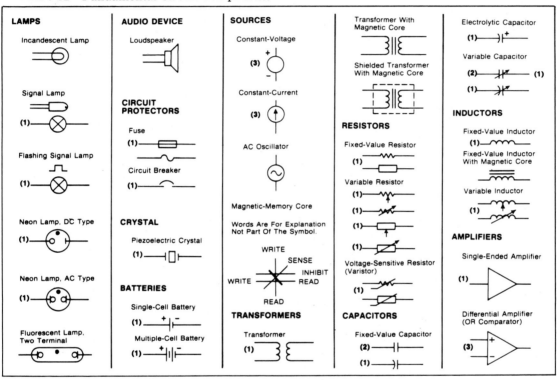

(1) = IEC Approval   (2) = Proposed IEEE Revision   (3) = Popular Industry Usage

Special Considerations For reference designation assignment purposes:

1. The term *subassembly* applies equally to an assembly.

2. A group of parts is not treated as an assembly unless it is one or more of the following:

   a. A plug-in item

   b. A significant item covered by a separate schematic

   c. A multiapplication item

   d. Likely to be handled as a replaceable item for maintenance purposes

3. Repairable assemblies not assigned a distinctive class letter in the class designation list are given the class letter A.

4. Nonrepairable assemblies not assigned a distinctive class letter (for example, an integrated circuit module) in the class designation list are assigned the class letter U. Potted, embedded, or hermetically sealed assemblies are treated as parts and considered nonrepairable. Reference designations must be assigned to elements within

TABLE 5-13 Terminals and Connectors

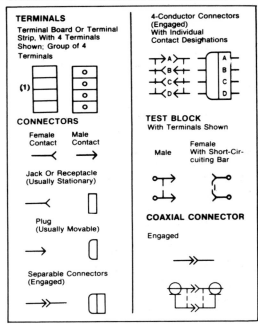

TABLE 5-14 Contacts, Switches, and Relays

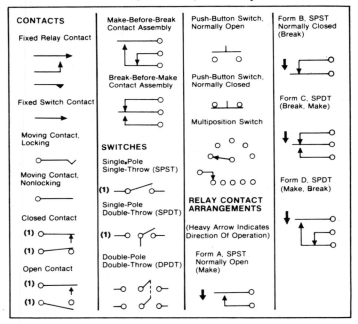

(1) = IEC Approval

such assemblies if reference to these elements is required for adequate operation and maintenance or for specification control information.

Number

The number portion of the reference designation differentiates an item from all other items identified by the same class letter. The number may be assigned sequentially as described below.

TABLE 5-15 Class Designation Letters

$A^{a,b}$ accelerometer
assembly, separable or
  repairable[c]
circuit element, general
computer
divider, electronic
facsimile set
generator, electronic
function
integrator
modulator
multiplier, electronic
recorder, sound
recording unit
reproducer, sound
servomechanism, positional
sensor (transducer to electric
  power)
subassembly, separable or
  repairable
telephone set
telephone station
teleprinter
teletypewriter

AR   amplifier (magnetic,
  operational, or summing)
repeater, telephone

AT   attenuator (fixed or variable)
bolometer
capacitive termination
inductive termination
isolator (nonreciprocal
  device)
pad
resistive termination

B   blower
fan
motor
synchro

BT   barrier photocell
battery
blocking layer cell
cell, battery
cell, solar
transducer, photovoltaic

C   capacitor
capacitor bushing

CB   circuit breaker

CP   adapter, connector
coupling (aperture, loop, or
  probe)
junction (coaxial or
  waveguide)

CR   absorber, overvoltage
current regulator
  (semiconductor device)
demodulator, diode-type ring
detector, crystal
diode (capacitive, storage, or
  tunnel)
modulator, diode-type ring

photodiode
rectifier (metallic or diode)
selenium cell (rectifier)
semiconductor device, diode
thyristor (semiconductor
  diode-type)
transducer, photoconductive
varactor
varistor, asymmetrical

DC   coupler, directional

DL   delay function
delay line
slow-wave structure

DS   alarm, visual
annunciator
audible signaling device
bell, electrical
buzzer
device, indicating (excluding
  meter or thermometer)
flasher (circuit interrupter)
indicator (excluding meter
  or thermometer)
lamp (cold cathode,
  fluorescent, glow,
  incandescent, indicating,
  pilot, signal, neon)
light source, general
ringer, telephone
signal light
siren
sounder, telegraph
vibrator, indicating
visual signaling device

E^a antenna, loop or radar
arrester, lighting
bimetallic strip
brush, electrical contact
carbon block
cell (aluminum or
  electrolytic)
cell, conductivity
contact, electrical
core (adjustable-tuning,
  electromagnetic, inductor,
  memory, transformer)
counterpoise, antenna
dipole antenna
ferrite bead rings
film element
gap (horn, protective, or
  sphere)
Hall element or generator
insulator
magnet, permanent
part, miscellaneous electrical
post, binding
protector (network, gap,
  telephone)
rotary joint (microwave)
shield (electrical or optical)
short (coaxial transmission)

spark gap
splice
terminal (individual)
terminal, circuit
termination, cable
valve element

EQ   equalizer
network, equalizing

F   fuse
fuse cutout
limiter, current (for power
  cable)

FL   filter

G   amplifier, rotating
  (regulating generator)
chopper, electronic
exciter (rotating machine)
frequency changer (rotating)
generator
magneto (ignition or
  telephone)
regulating generator
vibrator, interrupter

H^a hardware (common
  fasteners, etc.)

HP^a hydraulic part

HR   heater
lamp (heating or infrared)
resistor, heating

HS   handset
operator's set

HT   earphone
headset, electrical
receiver (hearing-aid or
  telephone)

HY   circulator
hybrid coil (telephone usage)
junction, hybrid
magic T
network, hybrid circuit

J   connector, receptable,
  electrical
disconnecting device
  (connector receptable)
jack
receptacle (connector,
  stationary portion)
waveguide flange (choke)

K   contactor (magnetically
  operated)
relay (armature, solenoid,
  reed, thermal)

L   choke coil
coil (all not classified as
  transformers)
electromagnetic actuator
field (generator or motor)
inductor
inductor, shunt

5-27

TABLE 5-15 Class Designation Letters *(Continued)*

	reactor
	saturable reactor
	solenoid, electrical
	winding
LS	horn, electrical
	loudspeaker
	loudspeaker-microphone
	reproducer, sound
	transducer, underwater sound
M	clock
	coulomb accumulator
	counter, electrical
	gage
	instrument
	meter
	oscillograph
	oscilloscope
	outdoor metering device
	recorder, elapsed-time
	strain gage
	thermometer
	timer, electric
MG	converter (rotating machine)
	dynamotor
	inverter (motor-generator)
	motor-generator
MK	hydrophone
	mocrophone
	transmitter, telephone
MP[a]	brake
	clutch
	frame
	gyroscope
	interlock, mechanical
	mechanical part
	mounting (not electrical circuit, not a socket)
	part, miscellaneous mechanical (bearing, coupling, gear, shaft)
	part, structural
	reed, vibration
	tuning fork
MT	detector, primary
	transducer (measuring or mode)
N[d]	subdivision, equipment
P	connector plug, electrical
	disconnecting device (connector, plug)
	plug (connector, movable portion)
	waveguide flange (plain)
PS	inverter, static (dc to ac)
	power supply
	rectifier (complete power-supply assembly)
	thermogenerator

PU	eraser, magnetic
	erasing head
	head (with various modifiers)
	pickup
	recording head
Q	transistor
	rectifier, semiconductor controlled
	switch, semiconductor controlled
	thyratron (semiconductor device)
	thyristor (semiconductor triode)
R	magnetoresistor
	potentiometer
	resistor (adjustable, nonlinear, variable)
	rheostat
	shunt (instrument or relay)
RE	receiver, radio
RT	ballast (lamp or tube)
	lamp, resistance
	resistor (current-regulating or thermal)
	temperature-sensing element
	thermistor
RV	resistor, voltage-sensitive
	varistor, symmetrical
S	contactor (manually, mechanically, or thermally operated)
	dial, telephone
	disconnecting device (switch)
	governor (electrical contact tape)
	interlock, safety, electrical
	key-switch (telephone usage)
	key, telegraph
	speed regulator (electrical contact type)
	switch
	switch (hook, interlock, reed)
	thermal cutout (circuit interrupter)
	thermostat
SQ	link (fusible or sensing)
	squib (electric, explosive, igniter)
SR	slip ring
	ring, electrical contact
	rotating contact
T	autotransformer
	coil (telephone induction or repeating)
	coupler, linear
	taper (coaxial or waveguide)

	transformer (current or potential)
TB	block, connecting
	strip, terminal
	terminal board
	test block
TC	thermocouple
	thermopile
TP[e]	test point
TR	transmitter, radio
U[b]	integrated circuit package
	microcircuit
	micromodule
	nonrepairable assembly
	photon-coupled isolator
V	cell (light-sensitive, photoemissive, photo sensitive)
	counter tube (Geiger-Müller or proportional)
	detector, nuclear-radiation (gas filled)
	electron tube
	ion-diffusion device
	ionization chamber
	klystron
	magnetron
	phototube
	photoelectric cell
	resonator tube (cavity-type)
	traveling-wave tube
	voltage regulator (electron tube)
VR	diode, breakdown
	regulator, voltage (excluding electron tube)
	stabistor
	voltage regulator (semiconductor device)
W	bus bar
	cable
	cable assembly (with connectors)
	cable, coaxial
	conductor
	dielectric path
	distribution line
	Goubau line
	transmission line
	transmission line, strip-type
	waveguide
	wire
WT[f]	tiepoint, wiring
X	fuse holder
	lamp holder
	socket
Y	crystal unit (piezoelectric or quartz)
	oscillator (excluding electron tube)

	oscillator, magnetostriction resonator, tuning fork	or waveguide transmission use)	resonator (tuned cavity) shifter, directional phase (nonreciprocal)
Z	artificial line (other than delay line) balun carrier-line trap cavity, tuned discontinuity (usually coaxial	gyrator mode suppressor network, general (where specific class letters do not fit) network, phase-changing	shifter, phase tuned circuit tuner (E-H, multistub, slide-screw)

[a]Designations for general classes of parts are so labeled in order to facilitate designation of parts not included in this list.

[b]The class letter A is assigned on the basis that the item is separable or repairable. The class letter U shall be used if the item is inseparable or nonrepairable.

[c]For economic reasons, assemblies that are fundamentally separable or repairable may not be so provisioned but may be supplied as complete assemblies. However, the class letter A shall be retained.

[d]Not a class letter, but used to identify a subdivision of equipment in the location numbering method.

[e]Not a class letter, but commonly used to designate test points for maintenance purposes. See ANSI Y14.15-1966.

[f]Not a class letter, but commonly used to designate a tie point on connection diagrams. See ANSI Y14.15-1966.

Sequential Assignment Within each significant item (unit, assembly, etc.), numbers are assigned in a schematic manner, preferably starting at the upper left of the schematic diagram and proceeding from left to right and top to bottom for each successive portion of the circuit. Continue from left to right until all parts of the circuit have been designated in order of input to output or in order of functional sequence.

If revisions require the deletion or addition of items, the original items are not renumbered to keep the series consecutive. Added items are assigned numbers following the highest number originally used. Numbers assigned to items that have been deleted are never reused. Because of these changes, a reference designation table is often placed on the schematic diagram so that the reader need not search for a designation that may have been removed (see Table 5-16).

TABLE 5-16 Reference Designations

Last	Not used
C7	
CR5	
R19	R17
U30	

Suffix Letter

A suffix letter, beginning with A, may be added to the basic reference designation to identify each portion of a multiple-element part or similar item (such as a multicontact relay or a multiple-unit capacitor) as follows:

1. If required for explanations in related text.

2. If portions of the part are shown separated from each other on schematic diagrams.

3. If it is necessary to identify elements and their terminals for wiring. For example, the contacts of a relay would be specified K1A-5 or K1A-6.

Terminal Identification

Component To identify terminals of parts (such as sockets, terminal boards, and transformers), the part reference designation shall be suffixed with a dash and the terminal identification. This principle may also be applied, by extension, to relays, key switches, and similar devices. For example, pin 5 of relay K7 would be assigned the reference designation K7-5. Terminal numbers when used on diagrams are placed near or adjacent to the symbol.

Isolated   The reference designation class letter E identifies isolated terminals (for example, standoffs).

Tie Points   The letters WT have been established to identify a wiring tie point on a connection diagram or wire list.

The alphabetical listing of class designation letters shown in Table 5-15 is used in the assignment of reference designations for electrical and electronics parts and equipment as described in IEEE Std 200-1975.

Parts not specifically included in Table 5-15 should be assigned to the part or class most similar in function.

REFERENCES

Military Standards

Drawing, Engineering, and Associated Lists, DOD-D-1000

Engineering Drawing Practices, DOD-STD-100

Industry Standards

Drawing Sheet Size and Format, ANSI Y14.1-1975

Electrical and Electronics Diagrams, ANSI Y14.15-1966[1]

Graphic Symbols for Electrical and Electronics Diagrams (including Reference Designation Class Designation Letters), (IEEE 315-1975)

Graphic Symbols for Logic Diagrams (Two-State Devices); ANSI Y32.14-1973 (IEEE 91-1973)

Interconnection Diagrams, ANSI Y15.15a-1970[1]

Letter Symbols for Quantities Used in Electrical Science and Electrical Engineering ANSI Y10.5-1968 (IEEE 280-1967)

Line Conventions and Lettering, ANSI Y14.2-1979

Reference Designations for Electrical and Electronics Parts and Equipment, ANSI Y32.16-1975 (IEEE 200-1975)

Copies of industry association standards are available to military personnel through the DOD Single Stock Point, U.S. Naval Supply Depot (NSD 103), 5801 Tabor Avenue, Philadelphia, PA 19120. Other organizations and private individuals may obtain copies from:

ANSI—American National Standards Institute, 1430 Broadway, New York, NY 10018

IEEE—The Institute of Electrical and Electronics Engineers, Inc., 345 E. 47th Street, New York, NY 10017

IEC—International Electrotechnical Commission, 1, rue de Varembe, Geneva, Switzerland

Some sources for military and federal standards are:

Armed Services Electro Standards Agency (ASESA), Fort Monmouth, Red Bank, NJ 07703

General Services Administration (GSA), Washington, DC 20234

National Bureau of Standards (NBS), Washington, DC 20234

Office of Technical Services (OTS), 5285 Port Royal Road, Springfield, VA 22171

[1]Supplemented by Electrical and Electronics Diagrams, ANSI Y14.15b-1973.

Bibliography

"Electrical and Electronics Diagrams," in *Modern Drafting Practices and Standards Manual*, General Electric Company, Technology Marketing Operation, 120 Erie Blvd., Schenectady, NY 12305, 1976, secs. K9; K9.1, pp. 9, 16, 25; and K9.4, pp. 4, 9.

"Electrical/Electronic Drafting Practice," in *Xerox Drafting Manual*, 2d Ed., Xerox Corporation, Joseph C. Wilson Center for Technology, Rochester, New York 14644, 1973, sec. E, pp. E1.02, E3.02—E3.04, E7.02—E7.09, E7.15—E7.20, E7.22—E7.27.

Nicholas M. Raskhodoff, *Electronic Drafting and Design*, Prentice-Hall, Englewood Cliffs, 3d Ed., 1977, chap. 10, p. 263; chap. 11, pp. 317–323; chap. 13, pp. 384, 403.

Trident Instrumentation PCB Manual, NAVORD OD 50611, vol. 1, Interstate Electronics Corporation, a subsidiary of Figgie International, Inc., 1001 E. Ball Rd., Anaheim, California, 92803, 1976, pp. 1–2, 1–3.

Printed-Wiring Drawings[1]

John C. McGlone

Today's printed circuit boards (PCBs) are much more complex than those designed a few years ago, and the trend in technology indicates that the complexity will continue to increase as more integrated circuits and packaged components are used on a given area of the board. For example, in the mid-1970s the average two-layer board had a density of about 1 in.^{2} of board area per 14-pin integrated circuit (IC). Five years later this density had become about 0.50 in.^{2} per IC. Dealing with this density is probably beyond the ability of a designer unless some type of computer assistance is available.

In addition, the availability of suitably skilled PCB designers is diminishing, making recruitment difficult. Indeed, the labor-intensive task of manual PCB layout is becoming more and more unacceptable as not only commercial pressures on reducing product development times but also labor prices continue to increase and board density complicates the design task.

The reality of these facts clearly demonstrates that there is a strong need to provide the PCB designer with a truly cost- and performance-effective tool to assist him or her

[1]All illustrations from REDAC Interactive Graphics, Inc.

in the layout of boards. The logical choice for the tool is the computer, which has the ability to carry out complex calculations very fast and the inherent ability to store vast amounts of data and to provide rapid access to individual items, and which can be programmed with all the automatic design aids and interactive features needed to design a board.

The essential makeup of a computerized PCB design system should offer the designer the following:

1. Input to the system should be easy to do and should use the same information as traditional manual methods so as to ease the transition from manual to computer-aided design (CAD).

2. The system operation should be easy to learn, easy to use, and readily remembered by the designer.

3. The interface between the computer and the designer should be highly interactive.

4. The automatic design aids supplied should encompass the entire design sequence, including component placement, track routing, dimensional checking, and postprocessing.

5. The system should make it virtually impossible for errors to be made.

6. The outputs from the system should include all engineering documentation and manufacturing aids.

THE CONCEPT OF AUTOINTERACTIVITY[2]

The harmonious relationship between the designer and the computer is a fundamental requirement for an efficient design system. Interactive graphics systems ensure that maximum effectiveness can be realized from both the designer and the computer.

To best understand the industry, it is imperative that a printed circuit (PC) designer understand the principles of PC layout and then proceed to what the role of CAD is all about.

Toward that end, an overview of the rules of manual PC design is presented first in this section; then, under Using an Interactive Graphics (IAG) System, the use of a typical CAD system is described.

RULES OF MANUAL PRINTED CIRCUIT (PC) DESIGN—AN OVERVIEW

Terms and Definitions

The following list contains a number of printed-wiring terms and definitions.

annular ring The circular strip of copper completely surrounding the holes, or extending beyond the eyelet flanges, that is required for the attachment of electronic part leads, wires, etc.

artwork An accurately scaled configuration that is used to produce the master pattern (see "printed-wiring artwork," below).

base The insulating support for the printed pattern; also referred to as the *base plate*.

bond strength A measure of the force required to separate the conductor from the adjoining surface of the base material.

circuit holes Any hole lying within the conductive area.

component A separable part of a printed-wiring assembly, such as a resistor, capacitor, microelectronic circuit device, or transistor.

conductive material A metallic foil, usually copper, that is bonded to the base material.

conductive pattern A design formed of any conductive material on an insulating base.

---

[2]*Autointeractivity* refers to an automated graphics system that can be interrupted by a designer.

conductor   A single conductive line forming an electrical connection between terminal areas.

conductor pattern   A conductive pattern that has low electrical resistance and is completely processed.

current-carrying capacity   The maximum current that can be continuously carried without degradation of electrical or mechanical properties of the printed-wiring board, or of the attached temperature-sensitive component parts.

definition   The fidelity of reproduction in the conductors relative to the original master pattern drawing.

dip soldering   A technique whereby printed-wiring boards are immersed in, or floated on the surface of, molten solder for the purpose of simultaneously soldering component parts to printed wiring.

disconnect pad   A terminal area, augmented by an affixing device, to allow replacement of a component.

etchant   The solvent used to dissolve the unwanted portion of conductive material bonded to a base.

etched printed-wiring   Printed wiring formed as a result of etching.

etching   A subtractive process consisting of chemical (or chemical and electrolytic) removal of the unwanted portion of conductive material bonded to a base.

fabrication   Any mechanical operation performed to put holes in, or cut to size, a printed-wiring board (such as drilling, piercing, routing, sawing, and blanking).

grid   A network of equally spaced, mutually perpendicular lines, the intersection of which provides the basis for an incremental location system.

indexing holes   Holes placed in a printed-wiring base to enable the base to be positioned accurately. Also referred to as "tooling holes."

indexing notch   A notch placed in the edge of a printed-wiring base and used to enable the base printed contact to be positioned accurately with respect to an interface plug-in receptacle.

interface connections   A conductor that connects conductive patterns on opposite sides of a printed-wiring base.

jumper   A point-to-point electrical connection made by hookup wire when the connection cannot be made in the printed wiring.

marking artwork   Artwork prepared as an inked composite overlay layout of the printed-wiring assembly and printed-wiring artwork to provide the master used to produce the master marking drawing and master marking pattern. The marking artwork verifies component designation clearances.

master drawing   A drawing showing the dimensional limits applicable to any or all parts of a printed-wiring board, including the base. (Data characteristics include length, width, and orientation of master pattern in relation to overall envelope, and other parameters such as circuit hole nomenclature, fabrication techniques, marking and plating requirements, material specifications, and keying-polarization of the printed-wiring board.)

master marking drawing   A drawing (Cronaflex contact print) reproduction of the marking artwork, on stabilized material, which specifies the reduction ratio to be used to obtain the master marking pattern (negative).

master marking pattern   A 1:1 scale pattern that is used to produce the markings on silk screen within the accuracy specified in the master marking drawing.

master pattern   A 1:1 scale pattern that is used to produce the printed wiring within the accuracy specified in the master pattern drawing.

master pattern drawing   A drawing (Cronaflex or Mylar contact print) reproduction of the printed-wiring artwork, on stabilized material, which specifies the grid system and reduction ratio to be used to obtain the master pattern (negative).

metal-clad base material   A laminate that consists of metallic material bonded to one or both surfaces of an insulating base.

pinholes   Small holes occurring as imperfections that penetrate entirely through the printed wiring to the base.

pits   Small holes occurring as imperfections that do not penetrate entirely through the printed wiring.

plated printed wiring   A conductive pattern formed by plating with a metallic coating.

plated-through interface connection Connection formed by the electrodeposition of metal on the sides of a hole in the base. Also referred to as a *plated-through hole*.

plating A process consisting of the electrodeposition of a metallic coating.

printed contact A portion of the printed wiring used to connect the circuit to a plug-in receptacle and which performs the function of a pin of a male plug. Also referred to as a *tab, connector contact area*, or *fingers*.

printed wiring A conductor pattern devised for the purpose of providing point-to-point electrical connection or shielding.

printed-wiring artwork Artwork that is prepared from the printed-wiring layout to provide the master used to produce the master pattern drawing and the master pattern.

printed-wiring assembly An assembly of a printed-wiring board including separately manufactured component parts that have been added.

printed-wiring board A completely processed conductor pattern or patterns, usually formed on a stiff, flat base.

printed-wiring layout A preliminary drawing prepared to locate components, to verify routing and clearances, and to provide data for the preparation of other drawings (e.g., printed-wiring assembly, printed-wiring artwork).

printing The reproduction of a pattern on a surface by any process.

reference dimension A numerical dimension on the master pattern drawing that is used to check the accuracy of the photographic reduction.

register The relative position of one or more printed-wiring patterns, or portions thereof, with respect to their desired locations on a printed-wiring base or to another pattern on the opposite side of the base.

registration The location of the printed pattern with respect to the holes.

registration-conductive pattern to board outline The location of the printed pattern with respect to the overall outline dimensions of the printed-wiring board.

registration front-to-back The location of the printed pattern on one side of the board with respect to the printed pattern on the opposite side.

reproduction targets Predetermined points that disclose the necessary photographic reduction for a particular piece of printed-wiring artwork.

resist A material used to protect the desired portions of the printed wiring from the action of the etchant.

schematic diagram A diagram showing, by means of graphical symbols, the electrical connections and functions of a specific circuit arrangement. A schematic diagram facilitates tracing the circuit and its functions without regard to the actual physical size, shape, or location of the component devices or parts (see Section 5).

solderability The capability of a material to be wet by solder.

terminal area A portion of printed wiring used for making electrical connections to the conductive pattern, such as the enlarged portion of conductor material surrounding a component mounting hole. Also referred to as *boss, pad, land*, or *terminal point*.

twist The deviation from a plane surface as measured from one corner of a board to the diagonally opposite corner.

undercut The reduction of the cross section of a metal-foil conductor caused by the etchant removing metal from under the edge of the resist.

warpage The deviation from a plane surface as measured across the length or width of a board.

Sequence of Printed-Wiring Design and Drawing Preparation

The sequence of printed-wiring design, drawing preparation, checking, and approval should be in accordance with that shown in Fig. 6-1. (A summary of information flow is given below printed circuit layout, artwork, master pattern, and master drawings which generally shall be drawn to a scale of 4:1.) A scale of 2:1 may be used, provided greater accuracy is not necessary; 10:1 may be used when exceptional accuracy is required. Printed-wiring assembly, marking artwork, and the master marking drawing shall be the same scale as the printed-wiring artwork.

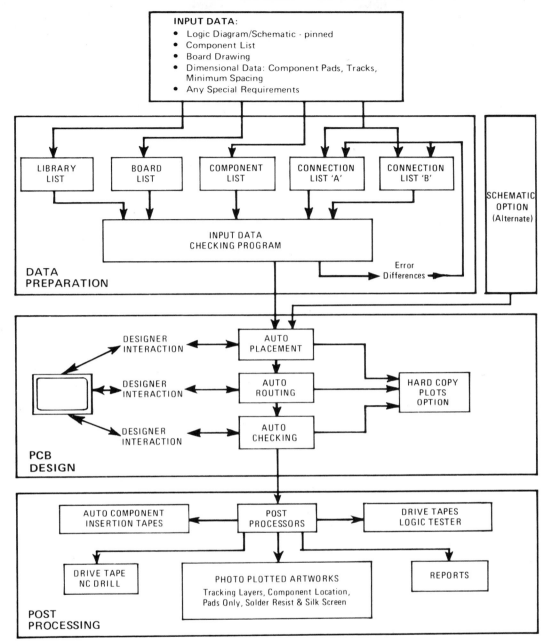

FIG. 6-1 Flowchart for PCB layout using REDAC Mini.

Design Requirements

The printed-wiring design requirements discussed in the following paragraphs are included in this guide because of their direct bearing upon drawing preparation.

Board Size and Mounting

It is recommended that for each functional surface the overall area be less than 50 in.² and the maximum dimension less than 10 in. (Boards larger than 50 in.² increase support problems.) The preferred board thickness is 0.062 in. Regardless of board size, support must be adequate to prevent the board from flexing.

Modular Grid and Dimensioning

All locations on the master pattern drawing shall be dimensioned by the use of a modular grid system. The basic modular units of length shall be 0.100, 0.050, or 0.025 in., in that order of preference. The basic unit shall be applied in the X and Y axes of cartesian coordinates. All locations shall be dimensioned or indicated by means of a grid intersection on the master drawing. This applies to the location of such things as holes; overall printed-wiring board dimensions, spacing, or component-part leads; and test point locations. All holes or other features shall be located within an 0.008-in. diameter of the true position indicated by the grid location. Wherever practical, overall board dimensions (length and width) shall coincide with lines of the 0.100-in. grid.

Indexing Holes and Notches

Indexing (or tooling) holes shall be located on printed-wiring boards to permit accurate positioning during fabrication. Two 0.062-in.-diameter holes shall be located in diagonal corners of the board, not closer than 0.1 in. to the edge of the board. These holes must be included in the printed-wiring artwork and master drawing and must be dimensioned for accuracy. If circuit density prohibits the use of the specific index holes, two existing holes, such as mounting or terminal holes, may be used as indexing holes, provided they are properly identified. Indexing notches shall be located in the edge of PCBs to assist in accurate positioning of boards using printed contacts.

Conductor Pattern

It is preferred that the conductor pattern be located on a single side of the board; if it must be located on both sides (making interface connections necessary), a terminal area shall be positioned at the interface location on each side.

1. *Width*   Preferred conductor width (final size) shall not be less than 0.031 in. Ordinarily, minimum width shall be 0.015 in.; however, 0.010 in. may be used in applications with stringent subminiaturization requirements. The current-carrying temperature-rise capacities listed in MIL-STD-275B, Printed Wiring for Electronic Equipment, should not be exceeded.
2. *Spacing*   Minimum space between conductors should be 0.015 in.; however, 0.010-in. spacing may be used in applications with stringent subminiaturization requirements. Spacing shall be in accordance with voltage between conductors as indicated in MIL-STD-275, Tables II, III, and IV.
3. *Routing*   Conductors shall be as straight as practical and run parallel to the edge of the board. Conductor length shall be as short as possible.
4. *Clearance*   Minimum clearance between conductors and board edges shall be 0.125 in. wherever possible. In special cases, zero clearance may be allowed. Conductors shall clear mounting hardware by at least 0.062 in. (This paragraph does not apply to ground planes, shielding, or printed heat sinks.)

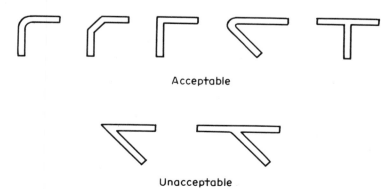

FIG. 6-2   Conductor intersections.

5. *Conductor intersections and corners*   All conductors having exterior and interior corners less than 90° included angle shall be rounded and filleted (see Fig. 6-2).

6. *Terminal holes and terminal areas*   Terminal holes and terminal area design shall be based on Table 6-1 and Fig. 6-3. The number of different hole sizes and terminal area shapes used on a board or on a related series of boards shall be minimized.

Minimum terminal area for unsupported holes shall be 0.040 in. greater than the

TABLE 6-1

Hole diameter	Terminal area	Wire diameter
0.026*	0.078	0.017 (#25)
0.031	0.094	0.020 (#24)
0.040	0.100 and 0.094†	0.025 (#22)
0.046	0.109 and 0.100†	0.032 (#20)
0.052	0.125 and 0.109†	0.040 (#18)
0.062‡	0.125	0.050 (#16)
0.093	0.156	0.080 (#12)

*For 6 leaded transistors using plated-through holes.
†For unsupported holes.
‡To be used for test points with 0.049.

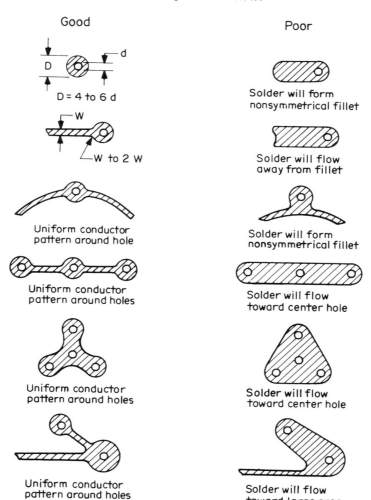

Good   Poor

D = 4 to 6 d

W to 2 W

Solder will form nonsymmetrical fillet

Solder will flow away from fillet

Uniform conductor pattern around hole

Solder will form nonsymmetrical fillet

Uniform conductor pattern around holes

Solder will flow toward center hole

Uniform conductor pattern around holes

Solder will flow toward center hole

Uniform conductor pattern around holes

Solder will flow toward large area

FIG. 6-3   Terminal area patterns.

hole diameter. The diameter of unsupported holes shall not exceed by more than 0.028 in. the diameter of the lead to be inserted, unless the leads are to be clinched and soldered to the foil.

The diameter of the plated-through holes shall be no more than 0.035 in. greater than the diameter of the lead to be inserted, unless the leads are to be clinched and soldered to the foil.

Table 6-1 is a list of hole and terminal area sizes allowable for minimum design requirements for both single-sided and double-sided boards. Table 6-2 is a list of preferred hole and terminal area sizes for both single-sided and double-sided boards.

TABLE 6-2

Hole diameter +0.000/−0.004	Terminal area	Wire diameter
0.028 ± 0.003*	0.054	Max. 0.021
0.031	0.062	0.020 (#24)
0.040	0.078	0.025 (#22)
0.046	0.078	0.032 (#20)
0.052	0.094	0.040 (#18)
0.062	0.094	0.050 (#16)
0.093	0.125	0.080 (#12)

*To be used only when necessary; TO-18 spacing, 8 lead TO-5 micrologic, etc.

Printed Connectors

It is essential that printed contacts or connector mounting pads be accurately located. Maximum contact width or pad diameter consistent with minimum spacing requirements shall be used. When male connector contacts are printed on the board, the edge of the board must be chamfered to prevent peeling of the contact when the board is inserted into the female connector.

Component Mounting

In general, components shall be mounted on one side of the board only. Components shall be mounted on the side opposite from the circuitry on single-sided boards.

1. *Terminal areas*   A terminal solder pad shall be provided at each point at which an electrical connection to the circuit pattern is to be made. Only one lead shall be inserted into a single hole.

2. *Reinforced terminal area*   All terminal areas that might be subjected to unusual attachment forces (such as from heavy components) or to frequent disconnections (from part replacement) shall be reinforced with an affixing device such as an eyelet, a terminal, or a plated-through hole.

3. *Bending of component leads*   The minimum distance from a component end seal to the start of a bend shall be 0.062 in. For components with welded leads, such as tantalum capacitors, the start of a bend shall be 0.062 in. minimum from the weld. The bend radius of component leads shall not be less than 0.062 in.

Distances between centers of terminal areas for mounting components (see Fig. 6-4) shall be computed by the following general formula:

$$S = L_{max} + 0.200 \text{ in.} - W$$

where S = lead hole spacing
  L = body length
  W = lead wire diameter

Reference Mounting Chart

L_{max}	S
0.126–0.225	0.400
0.226–0.325	0.500
0.326–0.425	0.600
0.426–0.525	0.700
0.526–0.625	0.800
0.626–0.725	0.900
0.726–0.825	1.000
0.826–0.925	1.100
0.926–1.025	1.200
etc.	

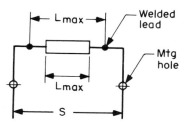

FIG. 6-4 Bending component leads.

4. *Support* Components weighing more than ¼ oz per lead require additional support, such as a clamp, bracket, or embedment, that the solder joints will not be relied on for mechanical support.

5. *Miniature components* Leads of transistors in TO-18 and other miniature cases are too close together to permit direct insertion into solder pads of acceptable size; therefore, these and similar components shall be mounted by either method shown in Fig. 6-5.

6. *Interface connections* Interface connections shall be made by one of the following methods:

a. Interface connections made by the use of a well-formed uninsulated solid copper lead (jumper) extending through a hole and clinched and soldered to the foil on each side of the board so as to contact the conductor or terminal area directly (see Fig. 6-6).

b. Interface connections made with funnel-flanged copper or brass eyelets that are well soldered to the terminal areas on each side of the board, or by means of a plated-through hole containing a continuous solder plug from one side of the board to the other side (see Fig. 6-7). Other types of eyelets may be used, provided the above requirements are met.

Standoff terminals shall not be used to provide interface connections.

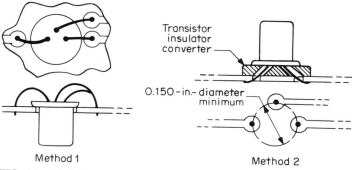

FIG. 6-5 Miniature component mounting.

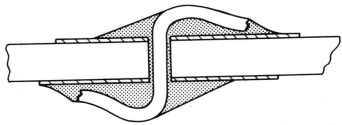

FIG. 6-6 Type I interface connection (direction of clinch optional).

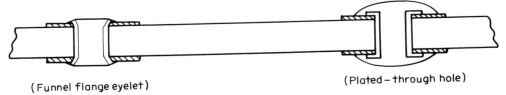

(Funnel flange eyelet)   (Plated – through hole)

FIG. 6-7 Type II interface connection.

Plating Requirements

Plating of printed-wiring boards according to IEC Technical Standards 378-152-006 is preferred.

Materials

The copper-clad laminates used to fabricate printed-wiring boards shall be in accordance with MIL-P-13949, unless otherwise specified on the controlling document.

Component Designations

Standard symbol designations (for complete list see MIL-STD-16) are as follows:

A	Assembly or subassembly
AR	Amplifier
BT	Battery
C	Capacitor
CR	Diode
DS	Lamp or indicator
E	Terminal, binding, post, electrical contact
F	Fuse
FL	Filter
G	Electrical chopper
J	Connector (receptacle, stationary)
K	Relay
L	Choke (coil)
P	Connector (plug, disconnecting)
PS	Power supply
Q	Transistor
R	Resistor, potentiometer, etc.
RT	Thermistor
RV	Varistor
S	Switch
T	Transformer
TB	Terminal board or block
TP	Test point
VR	Voltage regulator
W	Wire, cable, busbar
X	Socket
Y	Crystal unit
Z	General network (where specific reference designations do not fit)

Printed-Wiring Layouts

A printed-wiring layout drawing is a preliminary drawing prepared to locate components, verify routing and clearances, and provide data for the preparation of other drawings. Layouts must be accurately drawn, except that conductor paths may be drawn freehand. They shall contain all information necessary for design evaluation and completion of other drawings, including dimensions, hole sizes, tolerances, and board configuration and its relationship to mating structures, materials, processes, etc. All components must be drawn using maximum envelope dimensions. The general requirements given earlier under Sequence of Printed-Wiring Design and Drawing Preparation and under Design Requirements apply.

Drafting Materials

The layout shall be prepared on vellum or some other transparent material using a grid underlay in accordance with the procedure outlined earlier under Modular Grid and Dimensioning.

Procedure

The following step-by-step procedure should be used as a guide in layout preparation:

1. Study the schematic for circuit flow and electrical and thermal requirements.
2. Determine polarity and maximum envelope dimensions of components from schematic and parts list.
3. Determine critical circuits (e.g., amplifiers); determine possible electronic interface areas.
4. Lay out components. (This may be done by making heavy paper cutouts of components and moving them to determine correct placement, and then penciling in conductor lines; or by laying out components and terminal pads on a thin sheet of plastic and using this as a template for drawing components on the layout drawings.) For double-sided or multilayer boards, different-colored pencils may be used to identify the sides or layers.
5. Check the layout for electrical continuity, conformance to design requirements, and completeness.
6. Obtain a design check and approvals from the cognizant electrical and mechanical project engineers.

Preparing and Marking Printed-Wiring Artwork

Printed-wiring artwork is prepared to provide the master used to print the printed wiring on the board; it is not the master itself. Artwork shall be made using either black matte–printed circuit tape or Pelikan TN ink on Mylar, or by the "cut-and-strip" method. The artwork shall be superimposed on a standard drawing form when the Cronaflex contact print (master pattern drawing) is made.

Contrast and Definition   Artwork must have high contrast and sharp definition without splits, ragged edges, or pinholes in the conductor paths. Gray areas in inking are not acceptable.

Arrangement   Artwork shall be drawn as viewed from the printed-wiring side of the board using a grid underlay in accordance with the procedures outlined earlier under Modular Grid and Dimensioning. If the board is to be printed on both sides, a separate sheet shall be used for each side.

Reduction Dimensions   Reduction dimensions and their targets shall be specified. It is imperative that the distance between the targets be accurately drawn. Reduction dimensions shall be given in whole-, half-, or quarter-inch increments specified in thousandths. Minimum reduction tolerance is ± 0.003 in.

Indexing   Indexing holes shall be accurately shown on each sheet of printed-wiring artwork. For multiple-sheet artwork, reduction targets shall be located so that they are aligned when indexing holes are correctly positioned.

Part Identification Marking   The printed-wiring board part number (located on the circuitry side of the board, on which components are not mounted, if possible) must be printed on the printed-wiring board; therefore it must appear on the printed-wiring artwork. Preferred part number character height shall be 0.062 in. minimum. Space must be allowed for a revision letter.

Master Pattern Identification Marking   The master pattern and master marking pattern identification number and latest revision letter affecting the individual master shall be located on each piece of artwork so that it will appear on the negative from which the board or silk screen will be made. Preferred character height is 0.125 in. after reduction.

Hole Centers   Hole centers shall be in the form of circular etched-out dots. The final diameter of these dots shall be 0.020 in., minimum, or 0.031 in., preferred, to ensure proper etching. (If smaller dots are used, the acid-resistant coating will flow enough to completely cover the terminal area, preventing the etchant from removing the copper in this area.)

Master Pattern Drawing

Printed-wiring master pattern drawings provide an accurate, stable master from which reduced master patterns for board fabrication are made. Master pattern drawings shall be made from Cronaflex or Mylar contact prints of original artwork.

Quality and Accuracy   The drafter shall make a preliminary check of the contact print for reproduction quality and accuracy, retouching any pinholes or other irregularities with Pelikan TN ink. Major discrepancies such as distortion of the pattern or scale shall be cause for rejection of the contact print.

General Notes   All applicable grid information and reduction ratios shall be specified in the drawing.

Master Drawings

Printed-wiring master drawings provide the complete engineering description of printed-wiring boards. The detail drawing is made from a reproducible autopositive of the printed-wiring master pattern drawing. The conductor pattern is shown for reference, but the primary objective of the detail drawing is board design and fabrication.

Dimensioning   All dimensions required to establish board size, index hole location, terminal hole size, and all notches and cutouts shall be included.

Materials   All materials shall be specified, including copper-clad epoxy glass laminate. If terminals or eyelets are to be installed, they shall be specified on the drawing.

General Notes   All applicable process specifications, deviations from standard processes, or other information necessary for fabrication or quality control shall be specified on the drawing. When plated-through holes are used, hole diameter after plating shall be specified.

Master Marking Drawing

When markings (reference designations, terminal numbers, etc.) are to be silk-screened on the printed-wiring board, artwork and master-marking drawings must be prepared to provide masters from which the silk screen will be fabricated. All markings must be to scale, accurately positioned, and drawn as viewed from the side on which the markings appear. If markings are to appear on both sides of the board, two views must be drawn. Lettering height should be 0.062 in. in its final reduction.

Printed-Wiring Assembly

Printed-wiring assembly drawings provide the complete engineering data necessary for assembly and testing of components and associated hardware on the printed-wiring board.

Views   The assembly drawing shall be drawn from the side of the board on which the components are mounted. If components are mounted on both sides, both views must be shown.

Reference Designations   All reference designations, terminal identification, and polarity symbols shall be shown in the field of the drawing. They shall be shown on the drawing, as required for clarity, within the component outline, adjacent to the component, or connected by a leader to the component.

Dimensioning   Any critical dimensions, such as component height after soldering (which must be held during or after assembly), shall be specified.

Jumper Wires   All jumper wires shall be identified.

General Notes   Serialization marking, electrical test requirements, reference drawing numbers, conformal coating requirements, applicable process specifications, and any other general information necessary shall be specified in the general notes.

USING AN INTERACTIVE GRAPHICS (IAG) SYSTEM

The following discussion provides the reader with a cradle-to-grave description of how a PCB is designed with an interactive graphics (IAG) system. It assumes that the rough schematic has been provided as an input, and carries the process through the three sequential phases that result in the final engineering and manufacturing documentation. These three phases are:

1. Initial data preparation, during which the schematics parts information is converted into machine-readable information for the board.
2. Design of the board, the computer process whereby the board is designed, utilizing the automatic design aids and the interactive routines.
3. Postprocessing, during which the various manufacturing and engineering aids are outputted from the system.

Input Data

The input data required to lay out a PCB on a typical IAG system are the same as those required for manual layout, i.e.,

1. Schematic
2. Board drawing
3. Component (or parts) list
4. Dimensional data relating to tracks, pads, spacing, components, and the board
5. Special requirements, e.g., fixed-component or track positions, or restricted areas

Data Preparation Phase

During the data preparation phase the schematics and parts list data are converted into machine-readable data used during the board layout phase. Four lists are prepared from the input data: library, board, component, and connection lists. The data from these lists are then punched onto tape using an ASR-33 teleprinter, or similar machine, for input to the system for data checking and editing. Alternatively, with the input data terminal (option), the data can be input, checked, and edited, after which they are transferred to the system. A third method of data input is through a "schematics" option. Using this option, a designer can create the schematic and, with an automatic transfer program, transfer the schematic data into the PCB layout program, thus avoiding the use of the data input terminal.

Library List

This list is derived from the dimensional data of the components used and gives the overall dimensions of each different physical component type, which can be used for a similar component from a different manufacturer. The example of Fig. 6-8 is shown as library shape L1 in Table 6-3.

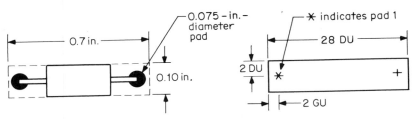

FIG. 6-8 Library shapes.

TABLE 6-3   Example of Library List

Library no.	X	Y	Pad code	No. pads
L1	28	4		2
	2	2	3	
	26	2	3	
L2	28	16		14
	2	2	1	
	6	2	2	6 [4 0]
	26	14	2	7 [−4 0]

Board List

This list is derived from the dimensional data of the board and uses the coordinates of the board, with respect to the lower-left-hand corner of the cathode-ray tube (CRT) display, in order to describe its shape. These data, if standard board designs are used, can be stored in the disk library and reused (see Fig. 6-9).

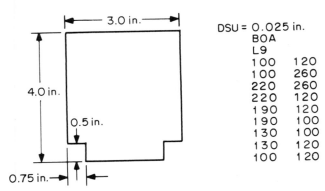

FIG. 6-9 Board list.

Parts (Component) List

This list not only cross-relates the components by their reference designation names on the schematic to the particular library shapes required, but also locates those "prefixed" components on the board.

For example, in Table 6-4 *resistor R1* uses library shape L1 and is located as position $X = 511$ and $Y = 511$. Integrated circuits (ICs) *U1* through *U10* use library shape L5 and can be positioned in the optimum location on the board as determined by the automatic placement routines.

TABLE 6-4 Example of Component List

		Position	
Name	Library no.	X	Y
R1	L1	511	511
U 1 10	L5		

Connection List

These are lists of the signal, power, and ground point-to-point connections obtained from the schematic after the designer has allocated gates to packages and pin-numbered the schematic (see Table 6-5).

The preparation of these lists, being a manual operation, could be prone to error on a large complex schematic, and therefore it is recommended that these lists be prepared by two separate people using the same schematic. A typical system equipped with input data checking software will detect and print out any differences between the two manually prepared lists. Any differences can then be checked and corrected to ensure that the PCB layout process begins with correct data.

TABLE 6-5 Example of Connection List

Component name	Pad no.	Component name	Pad no.	Component name	Pad no.
.CON					
.COD 2	Signal	Width 2			
U14	13	C3	2	U12	10
U15	8	C4	2		
U6	3	R1	1	U7	1
.CON					
.COD 4	Power	Width 4			
U 1 10	8				

Input Data Summary

In summary, the preparation of the input data has been systematized and simplified and combined with automatic error-checking software so as to (1) minimize the time required to prepare the data and (2) allow the use of "less than skilled designer" personnel. The latter consideration provides a natural career development plan from "data prep" to "designer" to "senior designer" for user personnel.

Once the input data are completed and checked, a dump is made on disk, ready for the board-design phase to start.

PCB Design Phase

The following section provides the reader with a general description of the sequence by which a PCB is designed using an IAG system. Photographics of the refresh CRT are provided for ease of understanding the automatic design aids and interactive features provided.

INPUT/OUTPUT Routine

Once the input data have been read, verified, and corrected, the design of the board can commence. The first step is for the designer to select the INPUT/OUTPUT routine, whereupon the screen picture will change to display a set INPUT/OUTPUT routine, as shown in Fig. 6-10. The designer now selects the input INITIAL option with the light pen, and the board input data dump prepared during the data prep phase is read in and displayed on the screen, as in Fig. 6-11. Note the TIME readout of ZERO minutes, indicating that the design has just started.

The initial display shows the board outline with the edge connector fixed in place; all other components are located at the origin of the working area (in the lower-left-hand corner of the CRT).

The designer then selects PICTURE, followed by EXPAND, to increase the board size to nearly fill the working area as shown in Fig. 6-12. At this point, the components are ready to be placed on the board, and the designer then selects AUTO PLACE.

AUTO PLACE Routine

A number of options are provided, as shown in Fig. 6-13, to enable the designer to obtain good component placement. These options are used in conjunction with each other and with MANUAL PLACE to obtain this placement.

The designer will first FIX the discrete components at the origin and then UNFIX the ICs. The next step is to select the MATRIX option, which places the ICs on a pre-defined grid. *Note:* The grid matrix can be defined in the data preparation stage, or can be defined before selecting MATRIX, by selecting DEFINE MATRIX and then entering the required data at the keyboard.

The "goodness" of the component placement can be evaluated at any time by the use of the CONN LENGTH routine. When CONN LENGTH is selected, the total length of connections between all components on the board is displayed on the screen. The designer uses this indicator as a guide throughout the placement process; the objective, of course, is to force the CONN LENGTH as low as possible. (The total connection

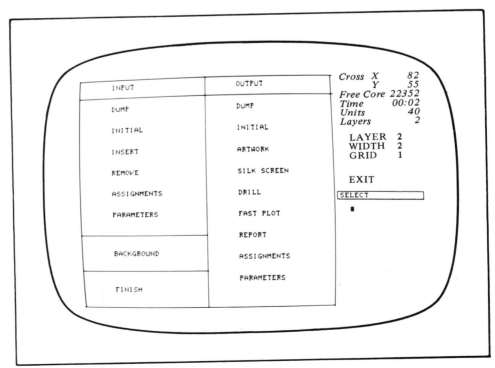

FIG. 6-10   INPUT/OUTPUT routine.

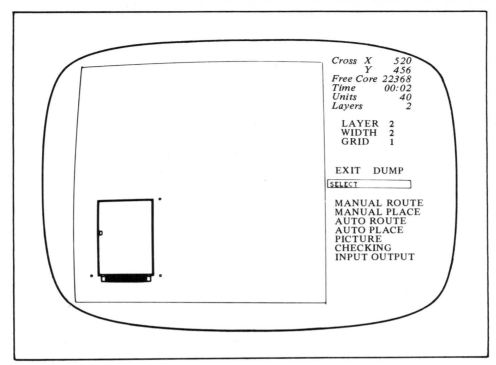

FIG. 6-11   Initial PCB display.

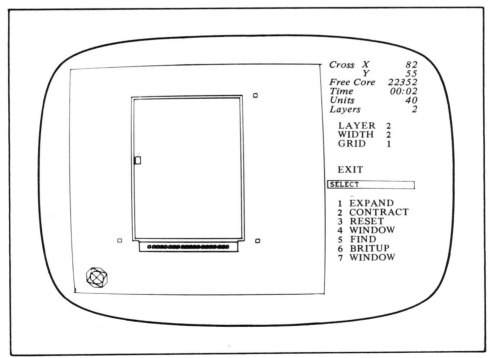

FIG. 6-12   PICTURE and EXPAND.

length of the example board, as shown in Fig. 6-13, before the optimization of component placement is 26,676 data units, a data unit in this case being 0.025 in.)

As an additional measure of "goodness," the designer observes the connection pattern at the refresh CRT to ensure that the connection density is uniform and not bunched over the board area, as is the case before placement starts. The rubber bands between the edge connector and the components represent the schematic electrical interconnections.

The designer next selects COLLAPSE, which pulls all the ICs together and down to the connector in direct proportion to the number of connections joining the components. This operation ensures that ICs that are connected to each other by a large num-

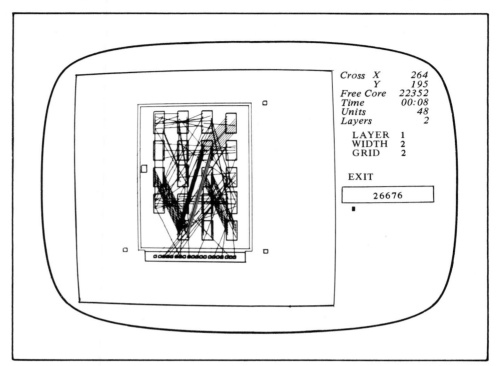

FIG. 6-13 AUTO PLACE routines (after MATRIX).

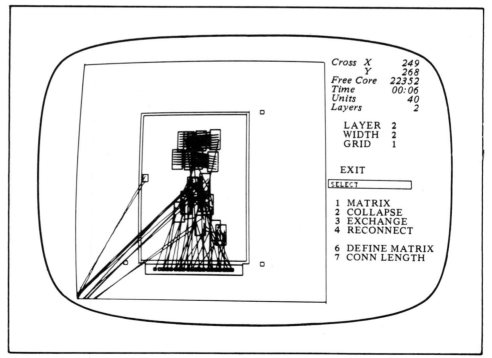

FIG. 6-14 First iteration of COLLAPSE.

ber of connections are placed close to each other. In large measure, COLLAPSE causes the connections to act like rubber bands—the greater the number of rubber bands (connections), the closer COLLAPSE will pull the components together. Note how in Figs. 6-14 and 6-15 two iterations of COLLAPSE have pulled heavily connected ICs together and, in addition, those ICs connected to the INPUT/OUTPUT connector have been pulled down close to the connector. This relative position will be retained when the ICs are moved back to the matrix, when the designer uses the routine MATRIX.

The ICs are brought back to the matrix grid using the routine MATRIX. The improvement in component placement by the previous use of COLLAPSE is retained. Following this sequence, the routine connection RECONNECT is run. RECONNECT does not relocate the components; instead, the effect of RECONNECT is to reconnect the connections so as to shorten the length of each connection on the board. The principle is shown in Fig. 6-16.

Figure 6-17 shows the connection pattern after RECONNECT has been run. The connection length at this point is 10,882 units, as compared to 26,676 at the start of component placement. Also, compare the connection pattern of Fig. 6-17 to that of Fig. 6-13, before placement was started, and note the improvement in connection density.

Next, the routine EXCHANGE is selected. EXCHANGE swaps each IC with its adja-

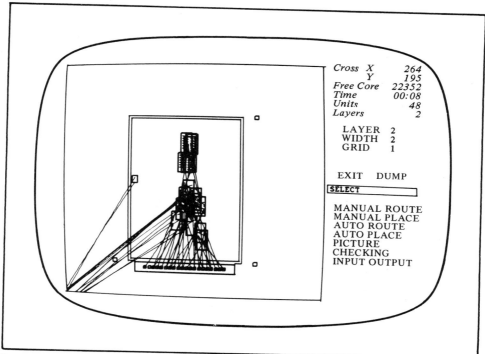

FIG. 6-15   Second iteration of COLLAPSE.

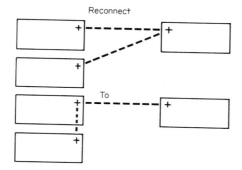

FIG. 6-16   Principle of reconnect.

cent IC in both the X and the Y directions (see Fig. 6-18) to determine whether the connection length can be further decreased.

Upon completion of the EXCHANGE routine, the designer again selects RECONNECT to further shorten the total connection length.

The placement achieved at this point should be reasonably satisfactory. If the designer feels capable of further improving placement with the light pen and the MANUAL PLACE routine, he or she may place components as desired. Figure 6-19 shows the MOVE subroutine, wherein the designer moves an IC from one location to an alternate location. Note that the connection pattern moves with the component, thus allowing the designer to evaluate the impact of the move upon the overall connection pattern. When the designer is satisfied that the ICs are in their best location, he or she uses the routine FIX to fix down the ICs in their current position. The placement of ICs is complete. Those components still to be placed at this point are the discrete components.

Concerning the placement of discretes, the reader will recall that the discrete components are off the board at location 0,0. The placement task remaining is to move these components onto the board. With the use of the automatic placement routine COLLAPSE, the discretes are pulled up by their rubber band connections and temporarily located close to the ICs to which they are attached. The results are shown in Fig. 6-20. The discretes are the square and rectilinear components shown scattered about on the board.

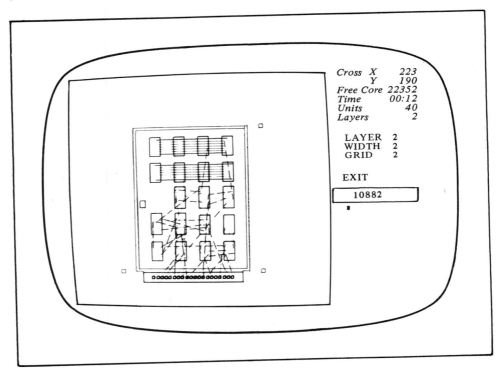

FIG. 6-17 After reconnect.

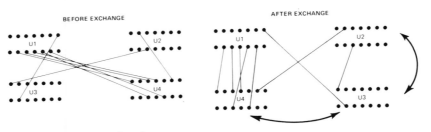

FIG. 6-18 Principle of exchange.

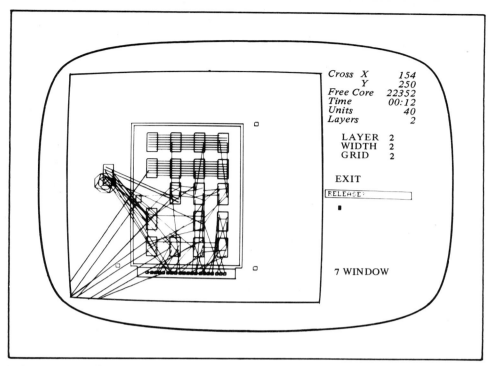

FIG. 6-19   Manual move.

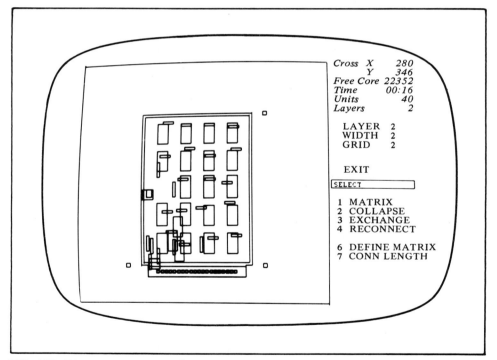

FIG. 6-20   Discrete components after COLLAPSE.

As can be seen in Fig. 6-20, the connections have disappeared from the CRT; in actuality, they have been set "invisible" by the routine PICTURE-BRITUP. This routine (Fig. 6-21) allows the various items (names, errors, layer 1–layer 8) to be set to seven levels of brightness, plus invisible. In addition, four different line types (solid, dots, etc.) can be used to display features. The designer uses this routine to help distinguish

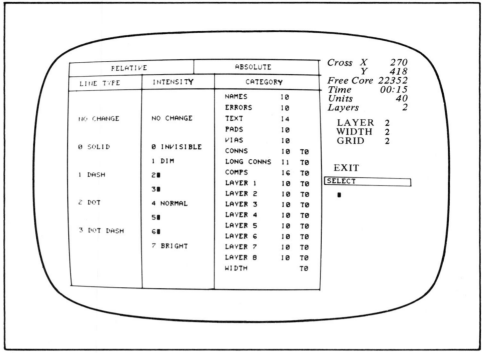

FIG. 6-21 BRITUP.

features at the CRT. For example, the tracks on layer 1 might be made very bright whereas layer 2 would be made dim. Multilayer boards might use a combination of both brightness and line type to distinguish different layers.

The discretes are moved into this final location, using the light pen and the MANUAL PLACE routine MOVE. The designer simply picks up the component with the light pen and moves and rotates it to its final position.

Figure 6-22 depicts expansion of the window of the CRT and shows how a discrete component is moved into its final position. Using the light pen, the designer can move individual components as required, either horizontally, or vertically, or, if required, the designer can rotate them in 90° steps. The designer places all discrete components in their optimum position until the total component placement is satisfactory. At this point, a final pass of the RECONNECT routine is made. This completes the final placement of all components.

The final component placement is shown in Fig. 6-23. The connections associated with the placement are shown in Fig. 6-24. The final connection length is 8,210 units, down about 70 percent from the initial placement. This very satisfactory placement was achieved in just 17 minutes, as shown by the TIME indication of Fig. 6-23.

At this point, the component placement is completed; the designer would make a dump of the database on the disk so that after the automatic routing phase (which follows placement) is completed, if it were felt that the placement could be improved, it would be possible to call back to this dump, make the adjustments to the component placement, and proceed to automatic routing again.

Automatic Routing Routine

The operator selects the AUTO ROUTE routine, and the presentation, as shown in Fig. 6-25, is presented at the CRT. There are a number of routing options available that the designer can select, depending upon the type of board involved. In the following example, we will utilize the multipass router (reentrant router) inasmuch as on this particular example board there are a memory area, power and ground connections, and general signal connections. The design strategy will be to use the different routing options within the multipass router to design the board.

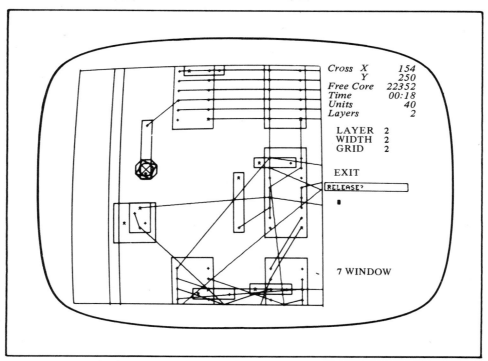

FIG. 6-22   Move of a discrete component.

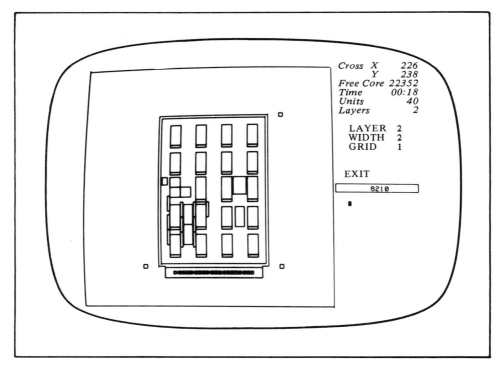

FIG. 6-23   Final component placement.

After selecting the parameters to be controlled, such as the layers to route on, the direction of the tracks on each layer, or whether feed-through minimization is desired or not, the designer activates "route" and, in this first routing pass, the router will route the power and ground connections. The results, after autorouting, are depicted in Fig. 6-26.

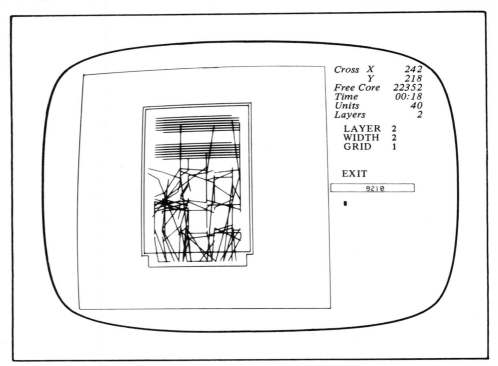

FIG. 6-24 Final connection pattern.

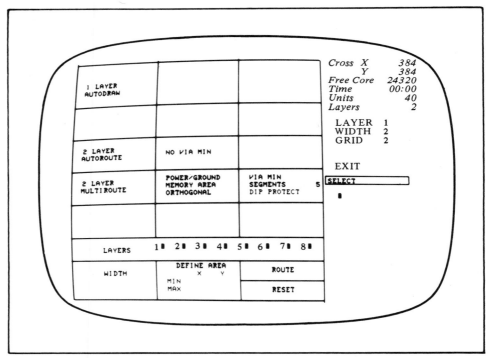

FIG. 6-25 AUTO ROUTE routines.

Following the power and ground, the designer activates the memory router, and the results of the automatic memory router together with the power and ground router are shown in Fig. 6-27. Notice the utilization of 45° angles by the memory router to achieve essentially 100 percent routing in the memory area.

The remaining connections on the board at this point are the signal connections.

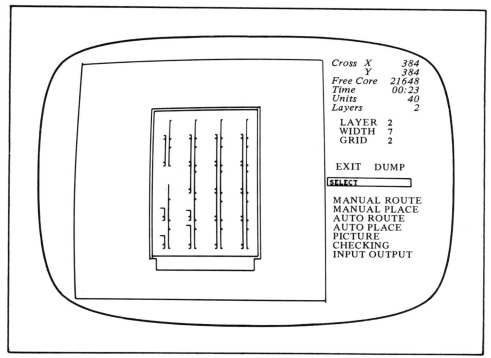

FIG. 6-26   Power and ground tracks (first pass).

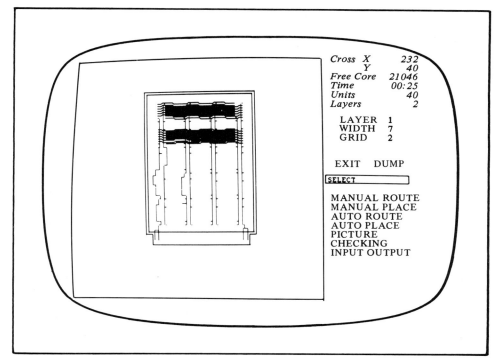

FIG. 6-27   Memory router—layer 1 (second pass).

For this pass the designer selects the orthogonal subroutine; the results of the orthogonal router on layer 1 are shown in Fig. 6-28, and layer 2 in Fig. 6-29. The unrouted signal connections at this point are as shown in Fig. 6-30.

At this point, the designer would transfer to the interactive or MANUAL ROUTE mode and rearrange some of the tracks routed by the automatic router so that the

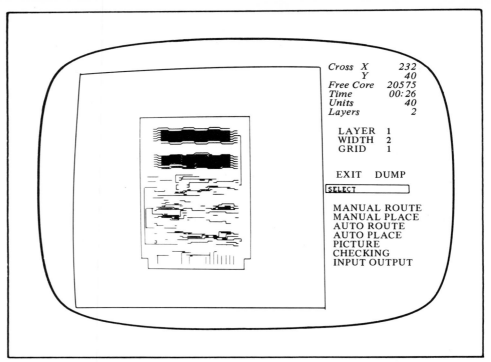

FIG. 6-28   Layer 1 tracks (third pass).

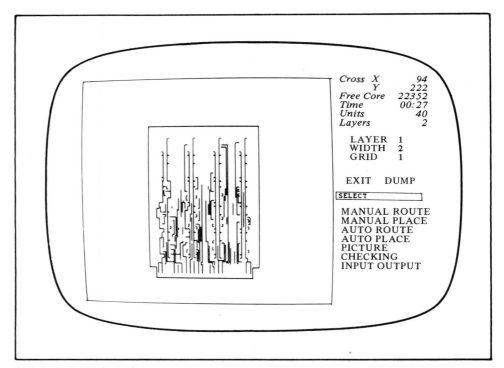

FIG. 6-29   Layer 2 tracks.

blockage that caused the unrouted connections not to route is cleared. Then the automatic ORTHOGONAL ROUTER would be turned on again (reentered), and an attempt to route the majority of the remaining connections would be made. A series of these reentrant routing/interactive sequences would be made, at the end of which any remaining unrouted connections would be interactively routed by the designer.

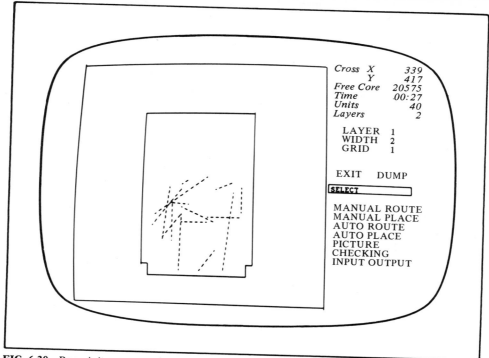

Cross X   339
*   Y   417*
Free Core   20575
Time   :00:27
Units   40
Layers   2

LAYER   1
WIDTH   2
GRID   1

EXIT   DUMP

SELECT

MANUAL ROUTE
MANUAL PLACE
AUTO ROUTE
AUTO PLACE
PICTURE
CHECKING
INPUT OUTPUT

FIG. 6-30 Remaining unrouted connections after first sequence of automatic routing.

The MANUAL ROUTE/MODIFY ROUTE used in these activities is a versatile and flexible interactive routine that allows the designer, by using the light pen, to interactively modify tracks and route connections. The routine is as simple as pointing the light pen cursor at the connection and drawing the track at the proper location to run from the first pad to the terminating pad. Note that the terminating pad is identified with the flashing butterfly and it is only to this end pad that the connection can be routed. The designer cannot make a connecting error in the process. Subroutines allow a track to be routed on any layer—a track segment can be routed at any angle or orthogonally (90°).

Figure 6-31 shows an unrouted connection in the process of being interactively routed. Figure 6-32 shows a routed track being modified.

The subroutine COPPER allows tracks to be thickened where required. It also can be used as an obstacle to the autorouter to prevent tracks and feed-throughs in selected areas of the board.

Checking

This completes the routing of all of the signal, power, and ground connections on the board, and at this time the checking activity would be made. A number of checking routines are available, as shown in Fig. 6-33.

The first check is a *connectivity check*, which compares the designed board against the original connections used at the time of start of the board design. Any differences are listed. The designer examines these differences to determine the reason and takes appropriate action. Following this, a *design rules space check* effort is accomplished, which determines whether any violations of space have occurred over the board. Specifically, every conducting surface of the board is checked against every other conducting surface of the board to determine whether the user's spacing criteria have been violated. Track-to-track, track-to-pad, pad-to-pad, track- and pad-to-copper, as well as other checks, as shown in Fig. 6-33, are made.

Any resulting violations are displayed at the CRT as flashing coded alphanumeric indicators that tell the designer precisely "what and where" the error is. If the user does not allow feed-throughs under components, a checking routine identifies any

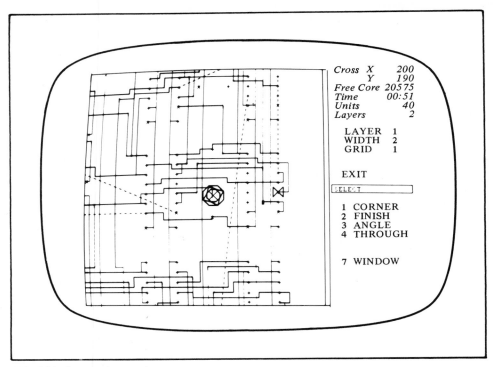

FIG. 6-31 Interactive routing.

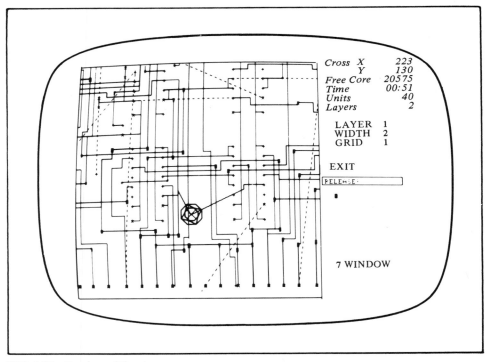

FIG. 6-32 Track modification.

feed-throughs under components. The design rule CHECK routine is extremely fast: central processing unit (CPU) time for the check on the board in Fig. 6-33 was 1 minute and 25 seconds.

Figure 6-34 shows a track-to-track violation, a track-to-pad violation, and a pad-to-pad violation. The designer examines these violations and, using the interactive TRACK MODIFY routine, makes the appropriate corrections.

Postprocessing Phase

Upon completion of the board design, it is necessary to output the board data in a form suitable for the generation of both manufacturing aids and engineering documenta-

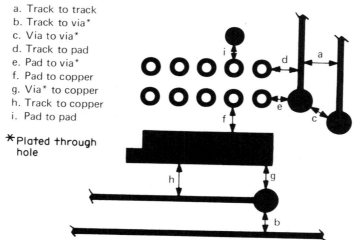

a. Track to track
b. Track to via*
c. Via to via*
d. Track to pad
e. Pad to via*
f. Pad to copper
g. Via* to copper
h. Track to copper
i. Pad to pad

*Plated through hole

FIG. 6-33 Space check parameters.

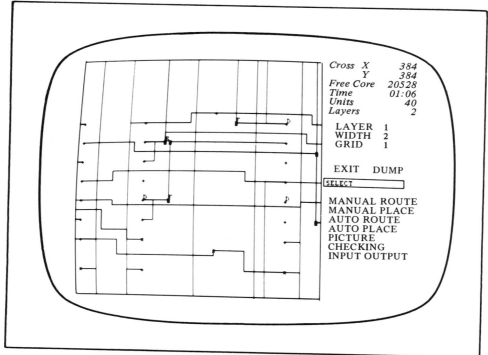

Cross	X	384
	Y	384
Free Core		20528
Time		01:06
Units		40
Layers		2

LAYER 1
WIDTH 2
GRID   1

EXIT   DUMP

SELECT

MANUAL ROUTE
MANUAL PLACE
AUTO ROUTE
AUTO PLACE
PICTURE
CHECKING
INPUT OUTPUT

FIG. 6-34 Simulated space errors.

tion. In order to do this, a number of postprocessors can be provided. These are selected by entering the INPUT/OUTPUT routines (Fig. 6-35) and selecting the OUTPUT routines as appropriate.

For Artwork The selection of the output ARTWORK results in the generation of media suitable for driving an appropriate photoplotter. Output media is produced for generating:

1. Master artwork for each layer
2. Silk screen
3. Assembly drawing
4. Solder resist artwork
5. Pads-only drilling master

(At the end of this section the 1:1 artwork, silk screen, assembly drawing, etc., of the board used in this photographic sequence is shown.)

Additional Postprocessors Additional postprocessors generate:

1. Numerical control (NC) drilling paper tape (in Excellon format)
2. NC automatic component insertion tape for both discretes and digital components (Universal and USM format)
3. Logic tester tapes (Gen Rad format)

Parts List The parts list is printed on the DECwriter.

Check Plots One-to-one or larger check plots of artwork, silk screen, etc., can be generated either off-line or on-line, depending on the system configuration.

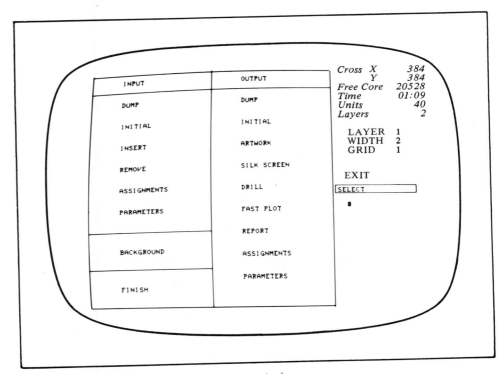

FIG. 6-35 INPUT/OUTPUT routine (postprocessing).

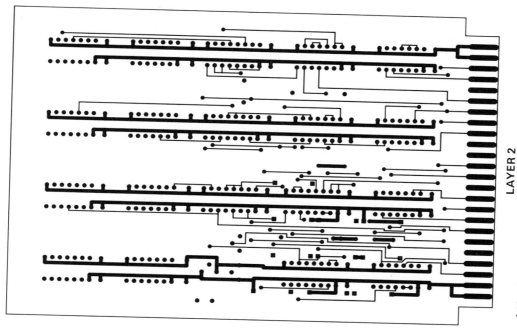

LAYER 2

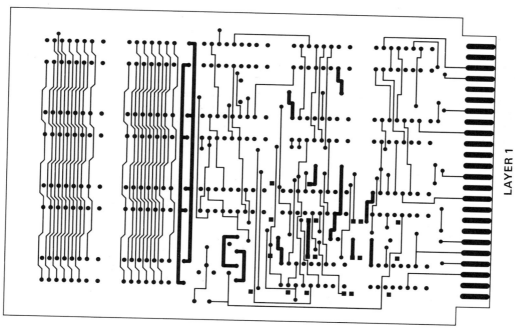

LAYER 1

FIG. 6-36 This board was designed on the REDAC Mini in a total of 1 hour and 10 minutes.

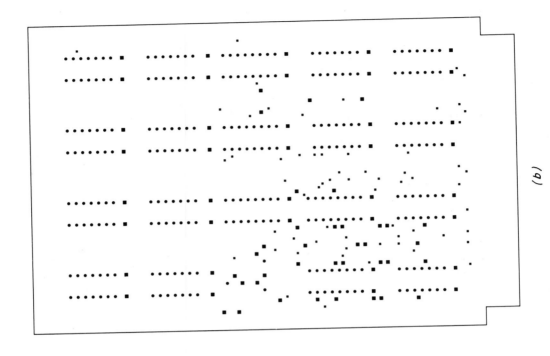

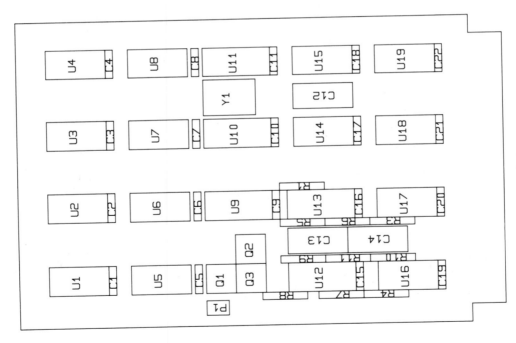

FIG. 6-37 (*a*) Assembly drawing. (*b*) Drilling drawing.

Summary

This section of the technical specifications has depicted the sequence in which a board is designed on, for example, the REDAC Mini. The total time taken on the CPU for this design was 1 hour, 10 minutes. This is further divided into:

Component placement	0 hr	17 min
Auto routing	0 hr	4 min
Clean up of tracks	0 hr	44 min
Auto check	0 hr	2 min
Clean up of "fails"	0 hr	3 min
Total time	1 hr	10 min

The plotting of the 1:1 artwork, for example, on the Gerber photoplotter is 32 minutes per layer.

Figures 6-36 and 6-37 illustrate the artwork for layers 1 and 2, and assembly and drill drawings produced by an IAG system.

Specialized Drawings

Vance D. Dutton

DEFINITIONS

specialized   Adapted to a special condition, use, or requirement. Designed for a particular purpose, occasion, person, etc.

drawing   Formal documents delineating the thinking of the design engineer in a form and format suitable for communicating to the person or persons who must convert this thinking into a finished product.

Thus specialized drawings may be defined as those documents which depict design engineering requirements in a manner that will satisfy the special needs of various users. *Specialized* is not to be interpreted as nonstandard. The drawing practices of a company should be based on the accepted industry practices of the American National Standards Institute standards for engineering drawings (ANSI-Y14), with modifications, additions, and deletions made only as necessary to suit the individual needs of a company.

FUNCTIONAL RESPONSIBILITY

One method of communication would be for the design engineer to personally contact the shop and explain to a member of the staff exactly what the requirements of the finished product are, but to do so in the majority of companies is almost as impractical as it is undesirable. In some cases there is a wide geographical separation between engineering and manufacturing, and in most cases there is a wide time separation between development of the engineering concept and the actual manufacture of the product. Close and continued personal contact is impractical for both these reasons. The practical solution is to establish a communication link between engineering and the rest of the company. This link is provided by the drafting function in its production of drawings that can be referenced at a time and place convenient to the user, can be reused for multiple production without the need for continuing personal contact, and can be expanded to include necessary user information that is not pertinent to design engineering.

Therefore, the basic function of delineating design-engineering requirements in a manner designed to satisfy the needs of the user is the responsibility of the drafting department.

DRAFTING DEPARTMENT OBLIGATION

The underlying reason for the existence of a drafting department is to provide information in a simple, clear, concise, and accurate form that will be immediately understandable to the user. In the majority of the instances the best way to convey this information is in the form of a drawing. Drafters creating drawings must recognize that their concern is one of communicating a message. They must be empathetic to their audience; they must put on paper what is in the mind of the design engineer; and they must do so with adequacy and without superfluousness. A drafter must walk the thin line between the confusion of overspecification and the inadequacy of underspecification while neither overestimating nor underestimating the intelligence and ability of the person to whom the instructions are addressed.

In the final analysis, all drawings are evaluated on the adequacy with which they satisfy the needs of the user. This basic concept cannot be compromised. Any question of inclusion versus exclusion of information should be resolved in favor of inclusion; the user should not have to guess.

The drafting department can satisfy its basic obligation only by relating drafting practices directly to the needs of the user.

DEPARTMENTAL POLICY

Traditionally the drafting department of a company has been part of the product engineering department because of drafting's functional responsibility to delineate design engineering requirements. The basic drafting obligation, however, is far more complex than to merely satisfy the needs of product engineering, so that it becomes necessary to establish a drafting department operating policy. In this respect, guidelines must be established pertaining to which information will and will not be included on drawings produced by the drafting department.

Underspecification

One theory holds that drafting output should be documents showing only the information required to describe the physical form, fit, and function of an item relative to the operation and integrity of the finished product. Intermediate configurations, operations, or procedures, and additional material requirements that may be necessary to facilitate procurement or manufacture, but which are not predictable, controllable, or known by product engineering personnel, will be the output of the manufacturing engineering department.

This type of policy is direct and to the point and may even be workable in a company engaged in the repeated manufacture of a limited product line where all drawing users are familiar with the product. This utopian atmosphere rarely exists in the real world, however, and a realistic drafting policy must take into consideration the more definitive needs of an individual company. For example:

- Product drawing information must be consistent with manufacturing capabilities, including producibility, tolerances, welding methods, and accessibility.

- Drafting may need to show shipping brackets, lifting lugs, or other features if their application or removal is critical to the integrity of the finished product, such as the addition of a handling bracket in an area where welding procedures or subsequent removal in a stress-free condition is important.

- Engineering may want to control the quality of the product by specifying necessary manufacturing information, such as the application of welding or shipping braces in a complex structure.

- Product drawings must reflect physical characteristics that will become an integral part of a component and remain as such on the finished product, such as tapped holes for handling or undercut shoulder reliefs for tool runout.

- Information should be included that will be necessary for the production of components in keeping with the physical capabilities of manufacturing equipment, such as extra stock requirements for formed plates.

Drawings must address more than form, fit, and function, and although design integrity cannot be compromised, the designer and the drafter must be concerned with shop capabilities, operations, and procedures.

Overspecification

Care must be exercised in the formulation of a drafting policy. Any attempts to specify more than form, fit, and function may result in overspecification of manufacturing detail. For example:

- A drawing should not specify manufacturing methods, tools, materials, etc., needed to accomplish a particular operation such as welding, machining, or painting.

- An engineering drawing should not specify manufacturing processes that would restrict manufacturing engineering adjustments of routing and equipment in accordance with availability. For instance, a drawing should not distinguish between weld joints to be made by manual versus semiautomatic processes if there is some question as to the availability of the semiautomatic equipment.

- Except under predefined circumstances, a drawing should not show intermediate manufacturing operations. In machining, for example, the finished dimensions alone are shown, but not the intermediate dimensions required to achieve the final dimensions and surface texture.

- Drawing information should not be more restrictive than is necessary to satisfactorily meet the functional requirements of a product. For example, one should not specify a tighter tolerance than is required simply because the manufacturing equipment is capable of holding a tighter tolerance.

- A drawing should not specify manufacturing processes unnecessarily. For instance, if a round hole is required in a steel plate, it should be specified as a diameter without indication as to whether it is to be drilled, reamed, punched, burned, or accomplished by any other means.

There are instances in which specification of the result on a drawing will automatically define the means of obtaining that result; on the other hand, the manufacturing operation that will be chosen may depend on specific results that are indicated on a drawing. For example:

- The use of certain welding equipment may dictate a specific weld-joint configuration.
- Tight tolerances and surface-finish requirements often dictate the use of specific manufacturing methods.

The Drafter's Responsibility

The product engineer must define the form, fit, and functional requirements of a product, and the manufacturing engineer must define the production processes that will be used to translate the design concept into manufactured goods. The drafter's responsibility is to combine the information supplied by both into instructions that are interpretable by others. The drafter must define the finished product and consider manufacturing process requirements while always keeping in mind the ultimate use to which the drawing will be put.

Each company should determine its own unique data requirements and establish a drafting department operating policy defining the limits of content of engineering drawings. This is not an easy task, but it is well worth the effort. A beneficial side effect of the policy will be the dissolution of any traditional boundaries that may be separating the product and manufacturing engineering functions. Drafters will find themselves in the role of a catalyst that merges the other departments into a system of increased efficiency.

COMPATIBILITY

Thus far we have addressed drawing content as determined by input from product engineering and manufacturing engineering. Such a drawing is of no intrinsic value. It must be accurate and complete, but it is a tool, an instruction, a communication device that must always be made with the idea in mind of who will use it and for what purpose.

Drawing Users

Other departments depend on the output of drafting for their effective performance, and drafting must be constantly aware of each department's functions, needs, and problems. To varying degrees, drawing users may include:

Design engineering	Inventory control
Research and development	Production control
Manufacturing engineering	Manufacturing
Purchasing	Quality control
Cost accounting	Data processing
Shipping and receiving	Customer services
Management	Publications
Marketing	Field erection
Facilities	

Drawing Uses

The most important use of drawings is in manufacturing finished products. As indicated by the number of other departments that depend on drafting output, however, drawings have many other uses, including:

- Design evaluation
- Documentation of research and development work

- Reexamination of design concepts, reference for new concepts, and understanding of how a product works
- Testing and installation of a product
- Support of spare parts and repair parts ordering
- Parts identification
- Cataloging and classifying of parts
- Competitive purchase of original items
- Substitution of interchangeable items
- Maintenance and overhaul of equipment
- Manufacture by a subcontractor, customer, or licensee
- Field modification of completed equipment
- Presentations to prospective customers
- Inspection and quality verification of in-process and completed products
- Determination of manufacturing facility needs
- Production of machine maintenance manuals and advertising literature
- Maintenance of historical information of previously manufactured machines and equipment
- Patent information
- Establishment of finished equipment cost and selling price
- Determination of shipping requirements

User-Use Compatibility

As can be seen by the above listing of users and uses, the drafting department can serve many needs, and the manner of drawing presentation depends to a great extent on the ability of users to understand and extract the information each requires. A company must recognize this user-use concept of drawing presentation and must accordingly seek to avoid the production of drawings that are designed strictly for the convenience of engineering and drafting.

The output of the drafting department must be compatible with the needs of a variety of users and, as such, must be a reasonably flexible medium subject to change of approach as the individual company grows. Given the choice between two or more methods of presentation, the final decision should be based on the needs of the user and not on the convenience of the drafter.

PLANNING

First-Priority User

Obviously, the production of engineering drawings that will satisfy the needs of such a wide variety of users requires a certain amount of commonsense judgment relative to who gets first priority in the planning of the drawing method of presentation. The highest priority should be toward manufacturing. If the product does not get manufactured, there is no need for any of the other departments, including drafting. The drafting output should therefore be designed basically to satisfy the needs of manufacturing, with modifications and additions performed as necessary to accommodate the various other departments. These modifications and additions will become evident through experience and evolution and should be documented into the design/drafting manual (D/DM) as they arise to ensure compliance and continued usage.

Design Review

Detail, assembly, subassembly, and all other drawings should be created from information supplied as a result of a coordinated preplanning effort accomplished through close cooperation among the engineering, manufacturing, and drafting operations while the designs are in the layout stage. This type of a design review procedure will produce more comprehensive drafting output for lower cost by eliminating backtracking for redesign and redraw, and by ensuring that the needs of manufacturing are inputted prior to the drafting stage. With this input, the drafter can plan the work to meet the needs of the user.

OBJECTIVES

The production of user-oriented drawings may necessitate the expenditure of more resources in the drafting department than would the production of the simpler, more straightforward textbook type of drawings. Therefore, the objectives and benefits to be realized must be defined in order to justify the expenditure. Aside from the obvious—and sometimes economically difficult to define—benefit of providing the drawing user with all the necessary information, so as to eliminate any need for further research, phone calls, or personal contact, the user-oriented drawings system

- Reduces at the manufacturing stage the lost time, scrap, and rework that may result from inadequate or confusing drawing presentation.
- Reduces overall processing time and manufacturing costs while increasing drafting costs by a minimum amount in comparison.
- Reduces the confusion, errors, and low productivity that may occur in drafting and other departments as a result of a random and inconsistent approach to drawing content and preparation techniques.
- In some instances actually simplifies drawings to improve user understanding.
- Reduces the effort within other departments, such as manufacturing engineering, quality control, and publications, that may be required to produce additional drawings designed to meet these departments' specific needs.

The Drafting Department Budget

Drafting department budget and staffing levels should be established on the basis of overall company needs and with a complete understanding of the variety of services that are actually performed. Any misunderstanding of the extent of reliance by others on the output of drafting which results in a limitation or reduction of drafting resources can cause a chain reaction of wasted time and confusion throughout the company.

DRAWING STYLES

As has been previously stated, drafting is a communication device, and therefore the language or drawing style used should be one most readily understood by the intended audience. There are many different drawing styles available, the most widely used being orthographic. Each style has its advantages and disadvantages, and the particular style chosen for an assignment depends on the use to which the drawing will be put and the print-reading abilities of the intended audience, not on the convenience to the drafter. This is not to say that the drawing user cannot be educated to read a drawing style that is more convenient to drafting. Obviously, the greatest effort on the part of the communicator will be in vain if the intended audience does not possess at least fundamental print-reading abilities.

Among the various styles or drawing languages are orthographic, diagrammatic, isometric, perspective, pictorial, photographic, and exploded assemblies.

Orthographic, or right-angle projection, drawing is the most familiar style throughout industry and may be considered as the standard approach for the communication of engineering information. Orthographic drawing should be the rule of a drafting department, with any other drawing styles being used on an exception-only basis.

Diagrammatic drawing, or the use of standard symbols and conventions, is perhaps second only to orthographic projection in acceptance throughout industry. It is the most technical and most professional method of engineering communication, and because it is the least-pictorial style, it is the most efficient method for the drafter. Because of its symbolic nature, however, this drawing style should be used only for communicating to a limited audience of skilled drawing users. The use of diagrams will be pursued further under Drawing Types later in this section.

Isometric drawing, with its three-dimensional approach to depicting what is required, is more readily understood by the drawing user and is rapidly gaining in popularity for certain applications. Piping installations, for instance, in which items seem to always be behind one another on an orthographic drawing, are much easier to understand when presented on an isometric drawing. A progressive drafting department should be using, or at least investigating and proposing the use of, isometric drawings for selected applications.

Perspective projections have very limited use in the engineering field, being mostly confined to architecture. They are usually intended for use by the advertising department to depict how a new product is going to look when it is completed.

Pictorial drawings are closely allied with perspective projections, once again the main purpose being to illustrate the appearance of the finished product. They are usually not prepared for manufacturing purposes, because they are not as applicable to the description of complex or detailed forms as an orthographic drawing is. A pictorial drawing, however, does have the advantage of simplicity and ease of understanding, which allows a person with limited print-reading ability to visualize the subject being presented. This one great advantage makes pictorial drawings worthy of consideration as a drafting style for communicating assembly and installation information. Pictorial drawing can be used when it is desirable to have a technically correct drawing that depicts the final product in an easily interpreted three-dimensional mode.

Photography can be a very efficient drafting aid in those instances in which an after-the-fact drawing is required. An object can be photographed and reproduced onto standard drawing format, and additional views, notes, and callouts can be added to produce a completed drawing. Application of photography in the majority of drafting operations is limited by the need to have the final product in hand prior to the production of such a drawing.

Exploded assemblies depict individual parts projected into space in the sequence in which these parts are to be assembled. As is the case with pictorial drawings, exploded assemblies have the advantage of simplicity and ease of understanding, but care must be taken if this drawing style is selected for use by a drafting department, as the actual relationship between assembled parts is not shown. Confusion rather than clarification may result on the assembly floor.

A company should determine which drawing styles are to be used for each type of drawing it produces, and this information, as well as the limits of drawing content, should be documented into a drafting department operating policy. There is no reason why several different drawing styles cannot be used as long as there is a clear understanding of the conditions under which each is applicable and there is agreement with the individual departments that will use the drawings.

DRAWING TYPES

In addition to drawing content and style, another consideration intended to tailor drafting practices to the needs of individual users is drawing type. Producing drawings of various types amounts to restricting information to that which is needed at any given point in the processing and manufacture of a product, and then creating, as much as is practical, separate and distinct information for each user. Practicality has its limits, however, and no drafting department can be expected to spend the time to create a separate drawing for each user unless this is absolutely necessary.

Many types of drawings are used in industry, including:

Detail	Schematic diagrams	Modification
Casting	Installation	Combination
Structural	Assembly	
Machining	Field erection	

Each drawing type has a special purpose and, when combined with effective part identification and a bill of materials, serves as an individual component within a complete package that describes the design, construction, and configuration of the finished product.

Detail Drawings

Detail drawings provide the total information required to manufacture and inspect a single item. Subcategories of detail drawings include:

Monodetail   depicting a single item

Multidetail   depicting more than one item on the same drawing by means of multiple dimensioning

Tabulated   depicting more than one item on the same drawing by means of a table that lists variable dimensions

Detail drawings provide concise and easy-to-interpret information to the drawing user when they are created in accordance with approved and documented drafting rules. Some of the required rules for detail drawings of burned and forged items include:

Detail Drawing Dimensions

Dimensions given are to be those which the drawing user needs with no addition or subtraction required to obtain necessary measurements. A typical example of how drafting-room practices must remain flexible is the advent of numerical-controlled (NC) burning equipment, which will cause drafting to change its dimensioning practices from those which are required by a shop layout worker to those required by an NC programmer. In a company that utilizes NC burning equipment, the drafter will be required to dimension actual points of intersection as opposed to work points. The coordinate axes and dimensions that had been acceptable to a shop layout worker will not be the same as the ones needed by the parts programmer. Tangent points and diagonally dimensioned points must be located in reference to the horizontal and vertical planes.

An important consideration is that if drafting supplied all the dimensions required by the parts programmer, a certain percentage of those dimensions would be redundant, a practice that should be avoided. However, the extra work required of the parts programmer can be significantly reduced by drafting's consideration of the programmer's specific needs. Figure 7-1 shows a burned plate dimensioned satisfactorily for shop layout, and Fig. 7-2 shows the same plate dimensioned for the NC programmer.

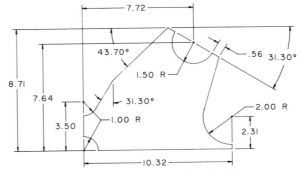

FIG. 7-1   Plate detail dimensioned for shop layout.

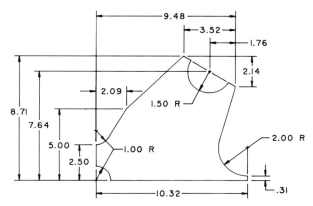

FIG. 7-2   Plate detail dimensioned for an NC programmer.

Detail Drawing Positioning

Detail items should be shown in the same relative position in which they appear on the assembly drawing, except as dictated by other considerations, such as:

- For ultimate stock-plate utilization and nesting purposes, detail items should be rotated to show their least stock requirement. Figure 7-3 shows a plate detailed in the position it occupies in the assembly drawing. Figure 7-4 shows the same plate rotated for least stock.

- For simplicity of understanding, bent plates should be shown as "bend up." This not only standardizes drafting practices, it also depicts the plate as it will be seen by the worker in the shop as the plate is placed in the press. For plates requiring multiple bends, the majority of the bends should be shown as "bend up" with a previously agreed-to code for indicating which bends are up and which are down. One system involves the shop layout worker using three punch marks to indicate "bend up" and four punch marks to indicate "bend down." A similar coding, using three or four

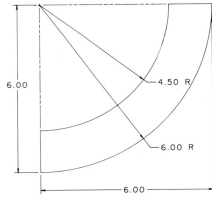

FIG. 7-3   Plate detailed in the same position as it will be assembled.

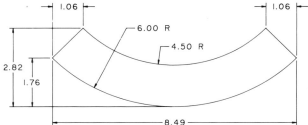

FIG. 7-4   Plate detail rotated for least stock.

short diagonal lines intersecting the bend line, is indicated on the detail drawing (see Fig. 7-5).

- Rolled plates should be shown as rolled "near-side in" for the same reasons as stated for bent plates (see Fig. 7-6).

Formed Plates

In order to reflect the information required to process and inspect formed plates, the following guidelines are suggested (see Fig. 7-6).

- A development view is required showing all detail dimensions of the flat plate, including any extra stock required because of manufacturing-equipment restrictions. Extra stock is shown by dotted lines. This view gives the information needed to cost the total material requirement and to burn the stock plate.

- A formed view is shown adjacent to the development view and reflects the finished part only. Removal of excess stock after or during forming operations is implied by the finished part dimensions and the dotted lines in the development view.

- The formed view is to be shown and dimensioned to allow both manufacture and face-plate inspection by quality control.

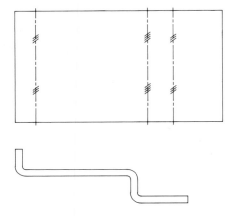

FIG. 7-5 Plate detail—multiple bends.

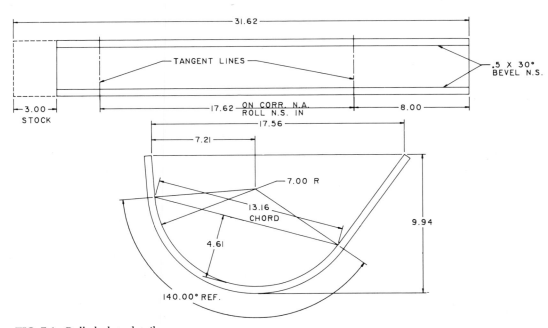

FIG. 7-6 Rolled plate detail.

An understanding of manufacturing capabilities and drawing user needs can lead to significant cost reductions, as can be illustrated by the excess-rolling-stock requirement. Most types of rolling equipment cannot produce a radius to the end of a plate, necessitating a straight portion that must be burned off after the rolling operation is complete. In many applications, the drafter can modify the part (with the approval of the designer) to show a straight portion on the finished item equal to the extra stock required for rolling (see Fig. 7-7). This design detail eliminates wasted stock and the need for two burning operations.

Drawing Scale

Detail drawings do not necessarily have to be to scale, but they must be shown in the correct proportions. The example shown in Fig. 7-8 illustrates how the drawing user can become confused by a drawing that does not at least look like the finished item. The out-of-proportion drawing is dimensioned correctly, but the 0.50 $\times$ 20° bevels appear to be required on the shorter of the two ends. This type of seemingly unimportant drafting error can result in significant cost to the company in the form of scrapped material.

Multiple Dimensioning

Multiple dimensioning of details to show more than one item on the same drawing can be a valuable practice for reducing drafting effort; however, this should only be done on the simplest of items to avoid unnecessary confusion on the part of the drawing user. Figure 7-9 illustrates the use of multiple dimensioning in an unacceptable manner that will cause more lost time in drawing interpretation than the time saved by the drafter. Figure 7-10 shows two acceptable uses for multiple dimensioning.

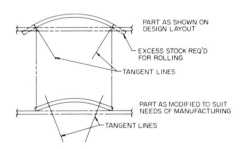

FIG. 7-7 Rolled plate modified to eliminate stock waste.

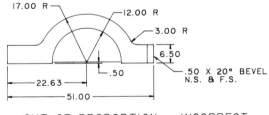

OUT OF PROPORTION – INCORRECT

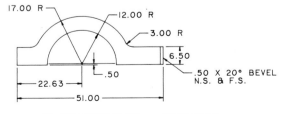

IN PROPORTION – CORRECT

FIG. 7-8 Drawing scale accuracy requirement.

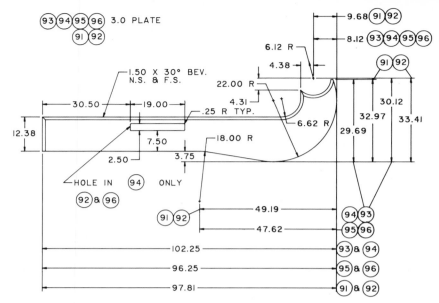

FIG. 7-9   Unacceptable multiple dimensioning.

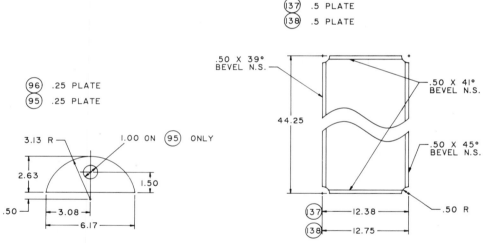

FIG. 7-10   Acceptable multiple dimensioning.

Right- and Left-Hand Details

Another accepted drafting use of multiple details is to show right-hand and left-hand items on the same detail picture. This practice is acceptable only in those instances where no more than two manufacturing operations are involved, and it should not be combined with other multidetail practices. For example, Fig. 7-11 illustrates right- and left-hand items (or shown and opposite items) that require only two operations: burn and bevel. This information should be acceptable to the drawing user. Figure 7-12 shows the same items with the addition of a bending operation. These items should be shown separately. Figure 7-13 is an attempt to combine right- and left-hand items with the practice of multidimensioning. This, too, becomes overly confusing to the drawing user and should be avoided.

Tabulated Drawings

Tabulated drawings, like multidetail drawings, can be very beneficial in increasing drafting-room efficiency, but it is important to remember that the drawing itself is not

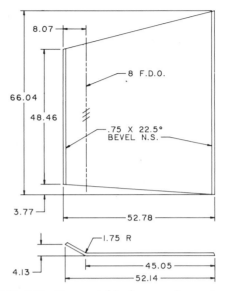

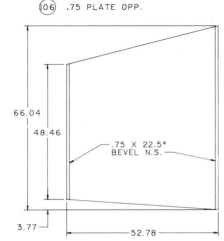

FIG. 7-11 Acceptable shown and opposite detailing.

FIG. 7-12 Unacceptable shown and opposite detailing.

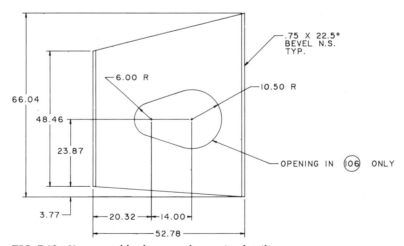

FIG. 7-13 Unacceptable shown and opposite detailing.

a finished product; it is a communication device that becomes worthless if it cannot be interpreted. Tabulations should be used only after acceptable limits have been determined, agreed to by the drawing users, and documented into the drafting manual. Figure 7-14 shows an acceptable tabulated drawing. Figure 7-15 illustrates the extremes that can come out of a drafting room when the actual needs of the user are ignored in favor of a reduction in drafting effort. Needless to say, Fig. 7-15 is totally unacceptable.

Casting Drawings

Casting-drawing requirements present what is perhaps the most convincing argument for the adoption of a form, fit, and function policy by a drafting department. Different

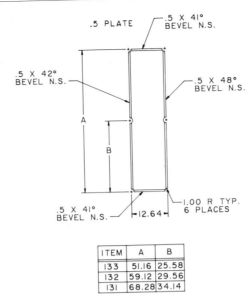

FIG. 7-14 Acceptable tabulated drawing.

foundries have different practices relative to stock allowances, location of gates and risers, special pouring techniques, and so forth, that might necessitate changes to rough-casting drawings every time the supplier was changed. The drafting department could very easily adopt a policy of showing only the finished part, with all additional information to be supplied by manufacturing engineers or established by the supplying foundry. It sounds simple enough, but the truth is, without rough-casting drawings of some kind, the following problems result:

- The purchasing department cannot obtain competitive quotes, because casting prices are based on weight and quantity. A rough-casting weight cannot be estimated without a rough-casting drawing.

- Quality control cannot inspect the rough castings without some kind of a specification for reference. A part-to-print comparison at the receiving department would be impossible without a print.

- Manufacturing engineering cannot establish the standard hours required to machine a part without some indication of the amount of stock that must be removed.

Separate Rough-Casting Drawings

Separate rough-casting drawings would satisfy the above requirements, but they can be expensive in terms of the drafting, checking, processing, and changing hours invested. Their creation and maintenance expense would have to be offset by subsequent benefits in the manufacturing and procurement process. Drafting would have to work closely with persons familiar with casting practices acting as a liaison between design engineering, manufacturing engineering, and the supplying foundry to determine necessary stock allowances that would provide a sound casting without requiring an excessive amount of machining time to remove the excess material. Then, if and when purchasing changed the supplying foundry for economic or other reasons, another drawing might have to be made incorporating the new supplier's requirements.

Drawing and Specification Combination

A logical alternative is to create a separate document (engineering specification) giving standard stock allowances based on generally accepted industry practices, with the engineering drawing showing only the finished part, special stock allowances as deemed necessary by manufacturing engineering to produce a quality casting, and special features as may be required to facilitate gates, risers, and pouring techniques, again as deemed necessary by manufacturing engineering.

This drawing and specification combination would satisfy all the benefits of a separate rough-casting drawing while at the same time providing any supplying vendor with specific information (finished-part dimensions) that can be used to determine the amount of material to be removed during machining—information that is generally needed by a foundry to identify pattern alterations that may be required to facilitate their own specific pouring techniques.

In those instances in which a supplying foundry must violate the drawing or the engineering specification to accommodate specific foundry practices, an approved deviation would be required documenting these changes. The deviation would accompany the part delivery and would be used to make any necessary adjustments to the

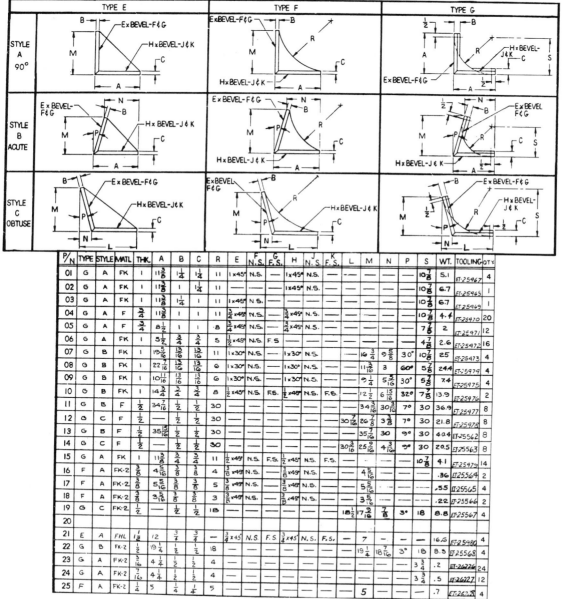

P/N	TYPE	STYLE	MATL	THK.	A	B	C	R	E	F N.S.	G F.S.	H	J N.S.	K F.S.	L	M	N	P	S	WT.	TOOLING	QTY
01	G	A	FK	1	11 3/8	1 1/4	1 1/4	11	1x45°	N.S.	—	—	1x45°	N.S.	—	—	—	—	10 7/8	5.1	ET-25967	4
02	G	A	FK	1	11 3/8	1	1 1/4	11	—	—	—	1x45°	N.S.	—	—	—	—	—	10 7/8	6.7	ET-25965	1
03	G	A	FK	1	11 3/8	1 1/4	1	11	1x45°	N.S.	—	—	—	—	—	—	—	—	10 7/8	6.7	ET-25969	1
04	G	A	F	3/4	11 3/8	1	1	11	3/4x45°	N.S.	—	3/4x45°	N.S.	—	—	—	—	—	10 7/8	4.4	ET-25970	20
05	G	A	F	3/4	8 1/8	1	1	8	3/4x45°	N.S.	—	3/4x45°	N.S.	—	—	—	—	—	7 5/8	2	ET-25971	12
06	G	A	FK	1	5 1/2	3/4	3/4	5	1/2x45°	N.S.	F.S	—	—	—	—	—	—	—	4 7/8	2.6	ET-25972	16
07	G	B	FK	1	19 5/16	13/16	13/16	11	1x30°	N.S.	—	1x30°	N.S.	—	16 3/4	9 5/8	30°	—	10 7/8	25	ET-25973	4
08	G	B	FK	1	22 7/16	13/16	13/16	6	1x30°	N.S.	—	1x30°	N.S.	—	11 3/16	5 5/8	60°	—	5 7/8	24.4	ET-25974	4
09	G	B	FK	1	10 11/16	13/16	13/16	6	1x30°	N.S.	—	1x30°	N.S.	—	9 1/4	5 5/16	30°	—	5 7/8	7.4	ET-25975	4
10	G	B	FK	1	14 3/4	3/4	3/4	8	1/2x45°	N.S.	F.S.	1/2x45°	N.S.	F.S.	12 1/2	6 15/16	32°	—	7 7/8	13.9	ET-25976	2
11	G	B	F	1/2	34 7/16	1/2	1/2	30	—	—	—	—	—	—	34 3/16	30 1/16	7°	30	36.9	ET-25977	8
12	G	C	F	1/2	—	1/2	1/2	30	—	—	—	—	—	30 7/16	26 7/8	3 3/8	7°	30	21.8	ET-25978	8
13	G	B	F	1/2	35 15/16	1/2	1/2	30	—	—	—	—	—	—	35 7/16	30	9°	30	40.4	ET-25562	8
14	G	C	F	1/2	—	1/2	1/2	30	—	—	—	—	—	30 3/16	25 9/16	4 3/16	9°	30	20.5	ET-25563	8
15	G	A	FK	1	11 3/8	3/4	3/4	11	1/2x45°	N.S.	F.S.	1/2x45°	N.S.	F.S.	—	—	—	—	10 7/8	4.1	ET-25974	14
16	F	A	FK-2	3/8	4 5/16	3/8	3/8	4	3/8x45°	N.S.	—	3/8x45°	N.S.	—	—	4 5/16	—	—	—	.36	ET-25564	2
17	F	A	FK-2	3/8	5 5/16	3/8	3/8	5	3/8x45°	N.S.	—	3/8x45°	N.S.	—	—	5 5/16	—	—	—	.55	ET-25565	4
18	F	A	FK-2	3/8	3 5/16	3/8	3/8	3	3/8x45°	N.S.	—	3/8x45°	N.S.	—	—	3 5/16	—	—	—	.22	ET-25566	2
19	G	C	FK-2	1/2	—	1/2	1/2	18	—	—	—	—	—	18 1/2	17 3/16	7/8	3°	18	8.8	ET-25567	4
20																						
21	E	A	FHL	1 1/2	12	3/4	3/4	—	3/4x45°	N.S.	F.S	3/4x45°	N.S.	F.S.	7	—	—	—	16.6	ET-25980	
22	G	B	FK-2	1/2	19 1/4	1/4	1/4	18	—	—	—	—	—	—	19 1/4	18 7/16	3°	18	8.3	ET-25568	
23	G	A	FK-2	3/16	4 1/4	1/2	1/2	4	—	—	—	—	—	—	—	—	—	3 3/4	.2	ET-26226	24
24	G	A	FK-2	7/16	4 1/4	1/2	1/2	4	—	—	—	—	—	—	—	—	—	3 3/4	.5	ET-26227	12
25	F	A	FK-2	1/4	5	1/4	1/4	5	—	—	—	—	—	5	—	—	—	—	.7	ET-26228	4

FIG. 7-15 Unacceptable tabulated drawing.

machine shop standard hours allocated for stock removal. In other words, any foundry supplying the rough casting would either cast it to the specification or document and gain approval for all variances.

Adoption of this drafting policy means that the outputted drawing will be of the finished part with any special stock allowances shown on the same drawing by the use of a dotted-line format. This combining of information intended for two audiences (pattern maker and machinist) into one drawing (Fig. 7-16) is acceptable only to the extent that the print remains easily interpretable by both persons. On more complicated parts, separate drawings should be made for the rough casting (Fig. 7-17) and the finished part (Fig. 7-18).

Structural Drawings

Structural or weldment drawings are highly specialized, and therefore the drafting rules governing their presentation are relatively easy to define. The drawings should:

- Contain all the information required by manufacturing engineering to establish weld process specifications and standard manufacturing hours.

- Provide the weld setup technician with complete information tailored to the needs of the welding process used.

- Depict each subassembly that must be manufactured as a separate unit while at the same time establishing the dimensional relationships among the separate subassemblies.

- Not employ multidrawing principles unless the resultant drawing will clearly present no interpretation problems.

- Not be combined with other drawing types, such as machining, detail, or field erection, except under strictly enforced drafting guidelines.

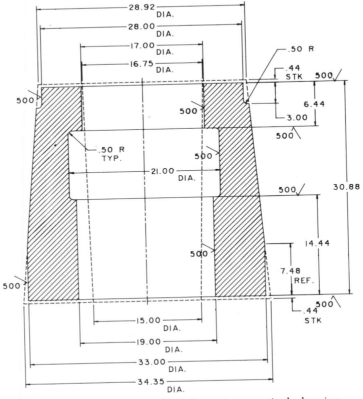

FIG. 7-16   Casting and machining information on a single drawing.

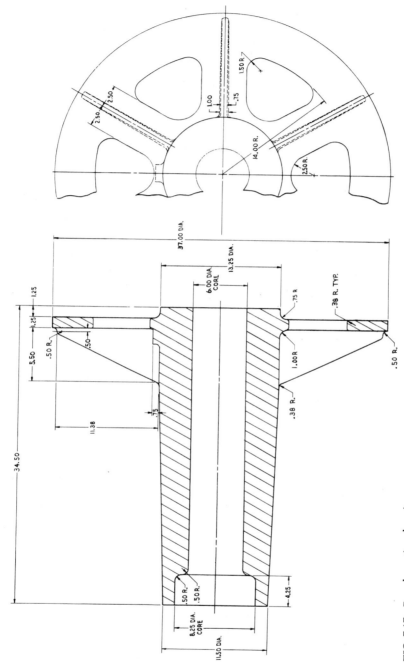

FIG. 7-17 Rough-casting drawing.

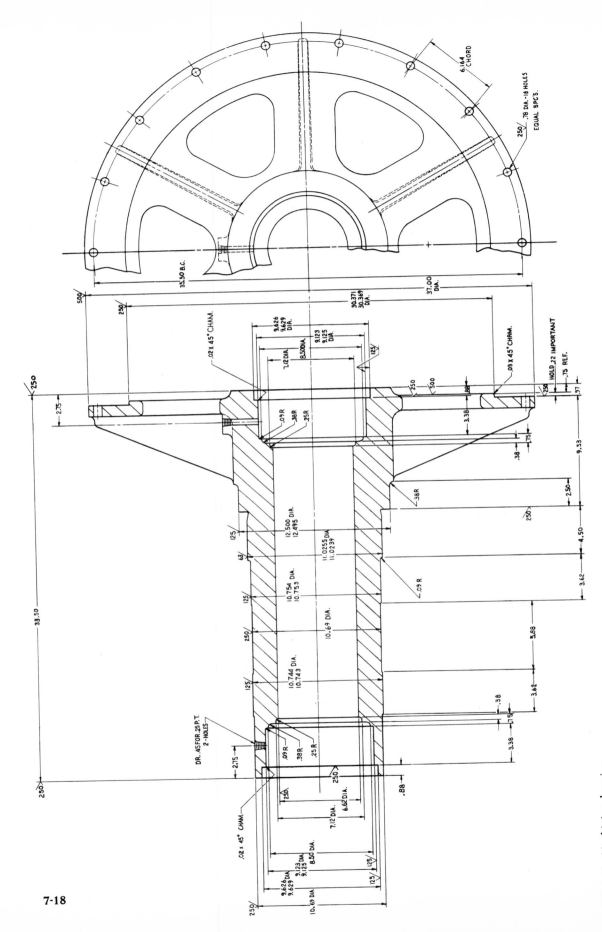

FIG. 7-18 Machining drawing.

7-18

Again, the D/DM is an important tool in the operation of the drafting department and should include guidelines for the creation of weldment drawings.

For one thing, all drawings should be oriented in accordance with preestablished rules relative to the position of the various parts in the finished product. One method of doing this would be to state that all structural drawings are to depict the unit as it would be seen while viewing the finished product from the rear and/or the left-hand side. Any drawing that deviates from this practice for reasons of clarity of presentation must include a notation that indicates how the unit is being viewed; for example, "Looking from front of machine" or "Looking from right-hand side of machine."

Also, each structural unit that must be set up and welded complete prior to being relocated for welding to another structural unit, or series of structural units, must be shown, dimensioned, and identified in a separate drawing.

Weld Setup Dimensioning

To assist the manufacturing department in obtaining tighter alignment of bulkheads to adjoining sections and to improve overall fit, a system of work line dimensioning may be used by the drafter and the setup personnel. The basic principle of this method is the location of all major setup points from work lines established on the top side of the bottom plates. Variations in overall dimensions of the bottom plates will have a minimum effect on bulkhead locations.

Figure 7-19 shows a complete structural component which, because of its size, must be manufactured in smaller units that are then shipped to the erection site for final assembly. The shaded portion of Fig. 7-19 indicates a subassembly that is shown relative to its adjacent subassemblies in Fig. 7-20. The engineering drawing for this same structure is shown in Fig. 7-21. Work lines have been established on the engineering drawing. These work lines will be duplicated by weld setup personnel on the top side of the bottom plate of the structure, and all vertical plates will be located from these reference lines. Similar work lines are employed on all adjacent structures, thus reducing misalignment and fit problems by having all setup personnel working from a series of common reference points.

Welding Accessibility

Showing the centerlines of all bulkhead openings on the plan view of structural drawings is a relatively simple drafting practice that can make the job of reviewing the prints for accessibility and weld-process specification much easier. These centerlines

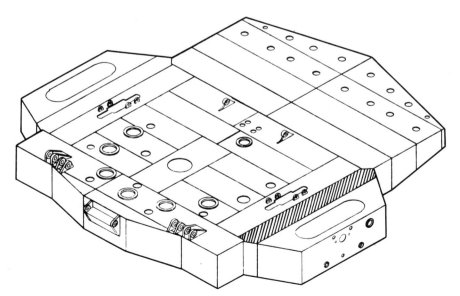

FIG. 7-19 Structural assembly.

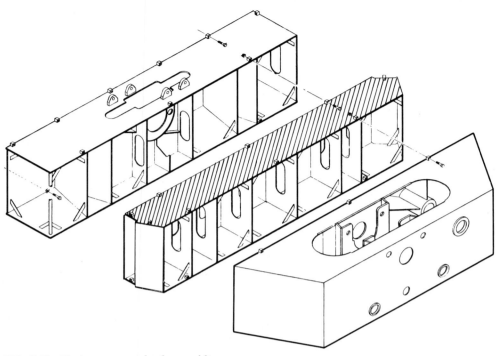

FIG. 7-20   Mating structural subassemblies.

will quickly orient drawing users to hard-to-get-at places and possible blind compartments.

One of the usual duties of weld setup personnel is to note the filler-metal specification and the required preheat temperature, both of which appear in a conspicuous place adjacent to the actual weld joint as instructions to the welder. This information should be included in each weld symbol or as a general note on the face of the engineering drawing (see Fig. 7-21).

Machining Drawings

Machining drawings should be created in accordance with accepted standard industry practices relative to dimensioning and tolerancing methods. Our concern here is how can, or should, these standard practices be supplemented to better suit the needs of the persons who are to use the prints. For the purpose of illustration, a gear has been selected, but the principles explained apply to any item that requires machining, regardless of its size or complexity.

Critical Dimensions

It is understood that the establishment of critical dimensions on a machined part must be based on the functional requirements as defined by the design engineer. It should also be noted, however, that before a drafter can determine which additional dimensions are critical to the manufacturing and inspection process, and are therefore in need of tolerancing, he or she must understand how the part is to be machined. A knowledge of the manufacturing sequence to be used will also establish datum surface requirements and other information that would not be evident from the functional requirements alone.

Manufacturing engineering, with the assistance of design engineering, should determine and categorize machined parts that are considered to be critical enough to merit special attention. Each category of parts should then be analyzed and a manufacturing procedure established sequencing the steps required to machine the parts. Drafting must then establish drawing practices based on the manufacturing procedures as well as the functional and quality-control requirements.

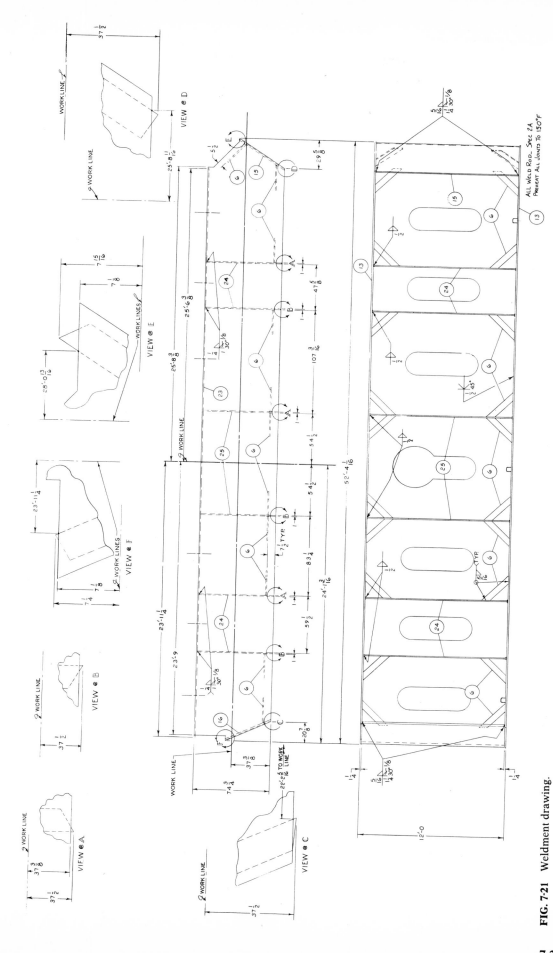

FIG. 7-21 Weldment drawing.

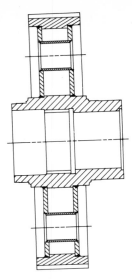

FIG. 7-22 Critical machine component.

1. Set the part on jacks with the short extension down.
2. Adjust the jacks to split the stock and level the top face. (This is the reference rim face.)
3. Center the part on the table and clamp.
4. Rough-cut the reference rim face and the hub face.
5. Rough-turn the upper hub end and rough-bore.
6. Rough-turn the outside diameter.
7. Release the clamping tension. Indicate the reference rim face for flatness. If flatness has been lost, do not reclamp. Instead, shim the gaps under the pull-downs and then reclamp.
8. Finish the reference face.
9. Finish-turn the outside diameter (500 microinches, μin.) and finish-turn the indicator path (125 μm).
10. Finish-turn the bearing journal and the bearing shoulder.
11. Release the clamping tension and check all toleranced dimensions in the free state. Get an inspection check.
12. Remove the gear blank. Put parallels in place and indicate them as okay if they are within 0.001 in.
13. Set the part on parallels on the reference rim face. Use feelers to ensure contact between the rim face and all parallels.
14. Indicate and center the part on the table.
15. Clamp and check the indicator path for runout.
16. Rough-turn the rim face, hub extension, and bore.
17. Finish the rim face and the bore.
18. Finish the bearing journal and the bearing shoulder.
19. Loosen the clamping and check all toleranced dimensions in the free state. Get an inspection check.

FIG. 7-23 Boring-mill machining sequence (gear blank, type 1).

The gear shown in Fig. 7-22 has been defined as a critical machine component, and with the functional requirements as a reference, manufacturing engineers have established a machining sequence as shown in Fig. 7-23. Note that the manufacturing engineer has requested a reference rim face (sequence steps 2, 4, 7, 8, and 13) and an indicator path on the outside diameter (sequence steps 9 and 15), neither of which is a functional requirement of the part. Figure 7-24 shows the finished part with these two features added and the remainder of the finished-surface relationships indicated in accordance with the machining sequence.

This system of tolerancing and dimensioning will result in a quality part only if all machining operations are set up in the same fashion. That is, once the tolerances have been certified in accordance with step 19 of the machining sequence, all subsequent machining operations, including keyseaters, slotters, shapers, and gear cutters, must be performed with the same gear-blank orientation. These operations must all be set up to the reference rim face and the indicator path in the outside diameter.

With the information shown in Fig. 7-24 given on the engineering drawing, the machinist and the inspector both know how to set a part up to emphasize important surfaces without prior knowledge of the part function, direct contact with engineering, or reference to the assembly drawing.

Geometric Tolerancing

A next logical step would be to initiate the use of geometric tolerancing symbols in place of the notes shown in Fig. 7-24, but remember, the drawing is a communication device, and any changes in the method of presentation must be done only with the concurrence and understanding of the drawing user. Do not initiate the use of any new language without the assurance that lost time and confusion will not result.

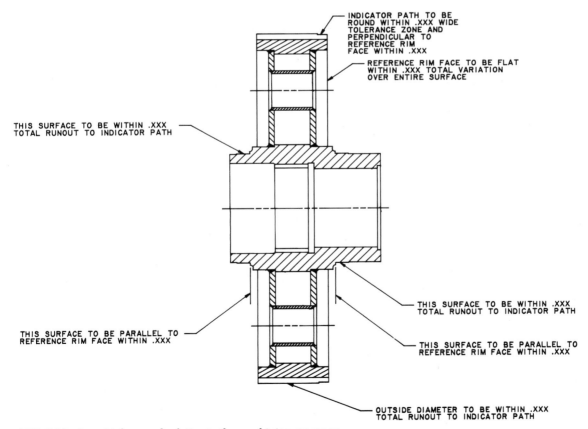

FIG. 7-24 A part toleranced relative to the machining sequence.

Additional Tips

Additional tips for machining drawings:

- When coordinate dimensions are given for hole patterns to be produced on NC equipment, add a chordal dimension as a reference for use by the inspector.
- When tolerance dimensions are given as maximum and minimum limits, place the value the machinist will reach first on top (see Fig. 7-25).

Schematic Diagrams

Schematic diagrams, employing the use of symbols and conventions known and understood throughout the industry, are used to show characteristics, connections, and relationships of items within an electrical or mechanical system. This type of drawing is used as a shorthand method of showing how a system works or is assembled without regard to its physical characteristics. Such drawings are beneficial to the assembler, machine operator, design engineer, and maintenance personnel, in that they provide a description of an overall system's construction and operation.

Installation Drawings

Installation drawings are used to show how and where an item or series of items is mounted in relation to other components and how these items are attached to their supporting structures. This type of drawing is frequently used to show the routing and physical assembly of electrical and piping systems.

Drafting department output for complex piping systems should employ the use of both schematic diagrams and installation drawings to define the full system requirements. Installation drawings, with an accompanying bill of materials listings, are required for processing through manufacturing engineering, inventory control, purchasing, manufacturing, and field erection. Total-system schematic diagrams are required for engineering, field erection, and customer interpretation and analysis.

Multiple Installation Drawings

The installation drawings are to be structured according to the method of manufacturing installation. For example, if a total piping system has components in two areas of a machine that do not assemble together until field erection, and each area is to be piped prior to field erection, two drawings will be required. Each drawing is to include all items and information necessary to install the system in the area for which it is drawn, including piping, pumps, motors, brackets, fasteners, etc.

Schematic Diagrams

Separate schematic diagrams should be made for each enclosed-circuit machine function. These drawings will not have an associated bill of material. They should contain complete system definition to allow for interpretation and analysis without referral to the individual installation drawings.

Schematic diagrams can become quite complex and difficult to interpret by even the most skilled drawing users. For this reason, they should never be combined with other

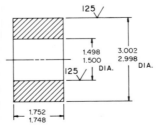

FIG. 7-25 The relative position of tolerance values.

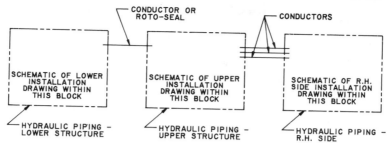

FIG. 7-26 A schematic diagram format.

types of drawings, and system requirements for new products should be shown on a separate diagram regardless of how similar it may be to an existing diagram.

In order to define a complete system, it will sometimes be necessary to transfer the information from more than one installation drawing onto a single schematic diagram while at the same time allowing for easy referral from the schematic diagram back to the individual installation drawings. This can be accomplished by diagramming each individual installation drawing as a unit on the schematic diagram, surrounding the unit with a heavy phantom line, connecting as required with applicable conductors, and noting the units with the installation drawing name or number (see Fig. 7-26).

All graphic symbols must be drawn in accordance with the accepted industry practices of the American National Standards Institute. Any deviations from these standards must be shown on each schematic diagram in the form of a legend giving a total symbol definition.

Piping Drawing Procedure

It is often difficult to develop good piping systems completely in the drafting room, for it is much easier to optimize routing and mounting methods at the time of actual assembly to the supporting structures. Good piping design, for instance, requires that the designer and the drafter work with the hardware as well as with the drawings. A company should be able to produce neat and functional piping systems at a minimum cost to all involved departments by adopting the following procedure:

1. Design and detail the required schematic diagram.

2. Lay out and detail installation drawings to the extent that impossible situations are eliminated and bulk hose, pipe, and fitting requirements are determined.

3. Custom-fit the piping to the prototype machine. The designer and detailer must be present to guide the pipe fitters, authorize on-the-spot improvements, and ensure that the system is functionally correct. They should document exact lengths and locations and then complete the installation drawings for use on future machines.

4. Photographically document the completed prototype piping for use as a guide for future machines, parts records, and service manuals.

Assembly and Field-Erection Drawings

Assembly drawings and field-erection drawings both show the configuration of two or more parts or subassemblies in their proper relative positions and include views, dimensions, joining hardware or weld requirements, special instructions, and procedures as necessary to accomplish the assembly operations. The only difference between the two drawing types is in the persons for whom the communication is intended. Orienting these drawing types to the needs of the user involves showing only the information required by the person or persons using the print. A part that is to be assembled at field erection should not be shown on a drawing intended for shop assembly; a part that is welded into a structure prior to that structure being moved to the assembly floor should not be shown on the assembly drawing; and so forth. In other words, assembly drawings and their associated bills of material should be structured in accordance with the manufacturing sequence that is to be used.

Drawing Orientation

As with structural drawings, assembly and field-erection drawings of parts should be oriented in accordance with preestablished rules relative to the position of these parts in the completed machine. Adjacent machine components should be shown in phantom to ensure understanding of the final relationships among assemblies and subassemblies.

Modification Drawings

Modification drawings show the total information required to make corrections or improvements to a completed product. They are usually used to modify a product that has already been delivered to the customer, but they can also be used to change a component that has been manufactured but not yet shipped.

Modification drawings are different from all other drawing types in that they are intended to communicate total instructions required to make the change regardless of how many different persons must use the prints. Modification drawings should include information on all work, including burning, forming, welding, machining, disassembly, assembly, etc., that must be performed to the existing product. Any items that are to be added once the modifications are complete should be detailed in accordance with normal drafting practices. For example, the modification drawing shown in Fig. 7-27 gives total instructions on what to disassemble and discard, where to locate new items and how to attach them, configuration of adjacent components, and exact location of the modification on the machine. The drawings required for the manufacture of the

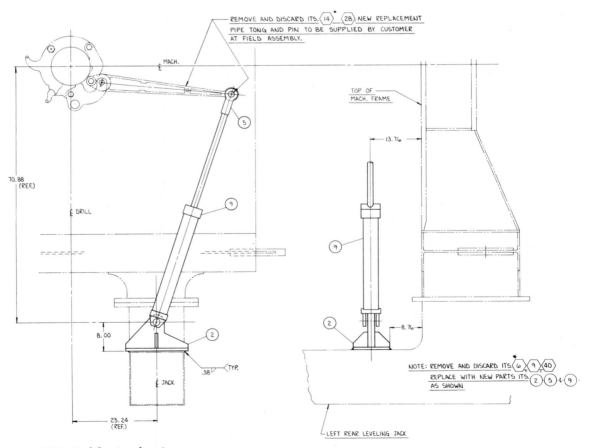

FIG. 7-27 Modification drawing.

items to be added are not included, however, because the parts are to be supplied finished complete and ready for assembly.

Modification drawings should be incorporated into the normal drawing system prior to the manufacture of any future duplicate products. They are intended to modify existing products and provide a historical record to allow determination and verification of modifications that have been made. They are not intended to be used for the manufacture of new products.

Combination Drawings

As has been previously stated, if a drafting department were required to produce separate drawings for each and every print user based on the individual needs of each, the drafting room would probably have to be as big as the rest of the company combined. Practicality dictates that certain types of information must be combined on the same drawing. This is not to say that one should ignore all the above. The drawing user is still the prime consideration for the establishment of drafting-room practices, and consequently any and all combining of information onto a single drawing must be done only under closely controlled rules and regulations. The following statements are examples of what can be done if the rules are strictly enforced.

Simplicity of presentation on an understandable drawing is the basic rule of the drafting department. Drafters, checkers, and supervisors must all contribute to commonsense control of the drawing content.

Acceptable Combining

Figure 7-28 shows a part drawing that combines burning, welding, and machining instructions onto one piece of paper. All information is sufficiently clear, and this drawing should present no problem in the processing and manufacture of the finished part. All similar parts should be drawn in the same format, with each block of information appearing in the same relative position.

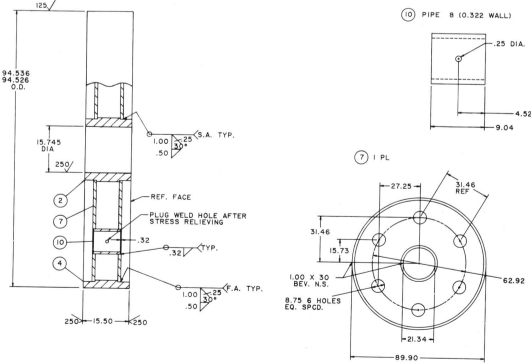

FIG. 7-28 Combination drawing—burning, welding and machining.

Unacceptable Combining

Detail drawings depicting the burning or forming of individual items should never be shown on a weldment or assembly picture regardless of their simplicity. Thus, even though the detail of items 7 and 10 in Fig. 7-28 are on the same sheet of paper as the structure, they have been pulled out of the part drawing and shown as separate details.

Multiple-Sheet Drawings

In a system of multiple-sheet drawings combining all manufacturing processes under one final finished-part number, the information should be carefully arranged in accordance with manufacturing needs—details of burned and formed items on one sheet, weldment on one sheet, machining on one sheet, etc., with all notes and supplemental information being placed on the sheet containing the picture to which they refer.

Even complicated components can sometimes have weldment and machining information combined onto one drawing. The example shown in Fig. 7-29 shows a relatively complex structure that is symmetrical about a centerline. The drafter has shown the welding information on one side and the machining information on the other, an acceptable practice if combined with a quality drawing containing explanatory notes.

Some manufacturing processes require the use of special drafting practices. The part drawing shown in Fig. 7-30 has become quite difficult to interpret because the drafter has chosen to show too many different manufacturing operations on one drawing. Included on this drawing are welding, premachining, stress relieving, flame hardening, final machining, and a series of process instructions. All the information is there, somewhere, but the drawing has been created for the convenience of the drafter, not the drawing users. Figures 7-31 and 7-32 show this same part redrawn as required to communicate effectively and understandably to the users. Note that separate informational requirements have been placed on separate drawings.

HISTORICAL INFORMATION

The practice of retaining historical information on the face of a drawing is extremely detrimental to the understanding of what must be communicated. Previous part configurations and drawing requirements as they existed prior to a drawing change are information that is seldom, if ever, required to produce a new part. This information should be retained by the creation of a historical file. All drawings should be microfilmed and recorded prior to a change, and the changed drawing should reflect only the information that is needed to process and manufacture the part. A user is expected to read and interpret every line, symbol, and word on a print, and any superfluous information will inevitably lead to confusion, wasted time, and scrapped material.

SUMMARY

The underlying reason for the existence of a drafting department is to provide information in a simple, clear, concise, and accurate form that will be immediately understandable to the user. No drafting department will meet this obligation all the time, but every drafting department should work toward the reduction of the communication gap between drawing creator and drawing user.

FEEDBACK

Every company should encourage feedback and constructive criticism from the drawing users in order to inform the drafters, checkers, and supervisors of the adequacy and various uses of the department output. Communication is, by definition, the giving *and* receiving of information. The receiving portion should be facilitated by periodic

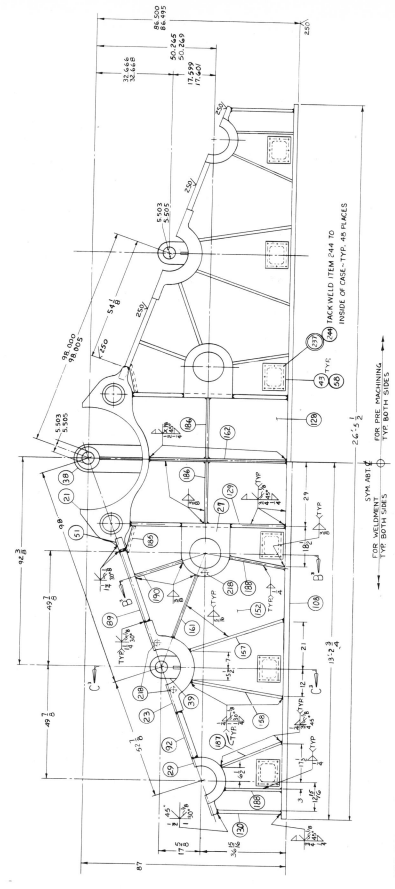

FIG. 7-29 Combination drawing—welding and machining.

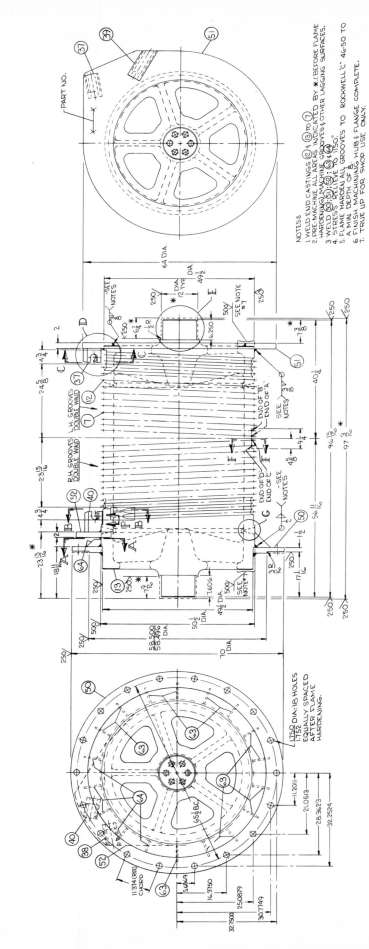

FIG. 7-30 Unacceptable combination drawing.

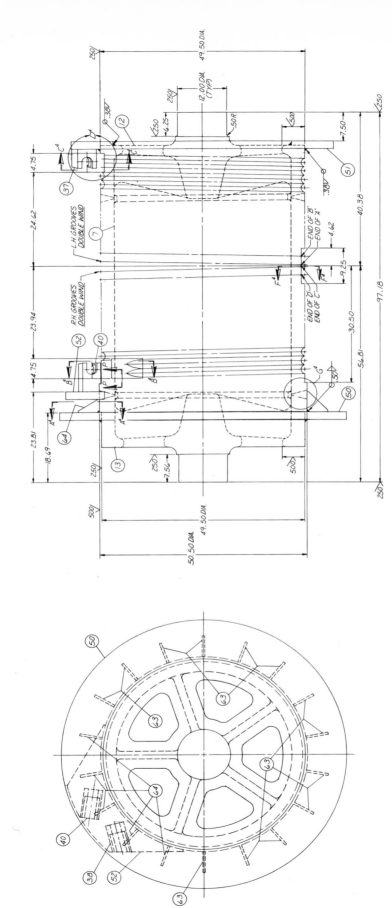

FIG. 7-31 Welding and premachining.

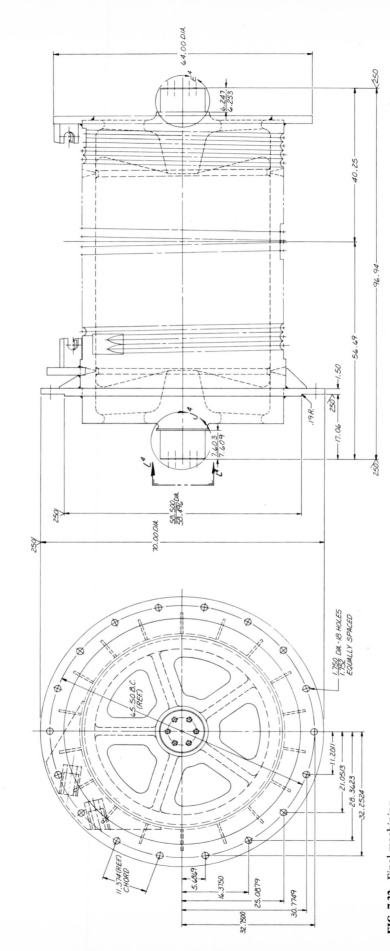

FIG. 7-32 Final machining.

personal contact between drafting personnel and drawing users, as well as by a formal and easy-to-use drawing change request system.

In an organization that isolates drafting from manufacturing, for instance, the drawing is prepared, checked, and approved by persons unfamiliar with the real needs of manufacturing personnel. The drawing is created in a manner that is readily understood by the drafter but which may be confusing to the user. The part shown in Fig. 7-30, for instance, was undoubtedly clear to the drafter when it was drawn simply because the drafter lived with it from the time it was a blank piece of paper. After putting down every line, every dimension, and every note, that drafter probably sat back and felt good about having conserved paper.

Consider the Users

Consider the plight of the users. Their first exposure to any work that must be done is when they unfold a print. A user will most likely begin by picking up a red pencil, sitting down to the workbench, and adapting the print to the real manufacturing needs. The user may add notes or merely move existing notes closer to the applicable view. He or she may cross out information that is not applicable to that particular phase of the work; delete entire views by transferring a dimension or two; add views; sketch an isometric; or pencil in little sarcastic notes about the drafting department. This is not done to show that the user knows more about drafting than the drafter, but it is instead a way of converting the given information into the information manufacturing requires.

The user then sets about the task of creating the product and, in a company that does not encourage feedback, will either throw the print away or save it for the next time the same item has to be built. Seldom are these prints routed back to drafting for drawing clarification and drafting-room practice improvements for better future presentations.

These prints are valuable; drafting can develop an appreciation of the difficulties encountered by the drawing user only by opening up a return communications channel. Drafters, checkers, and supervisors should make periodic visits to the shop and encourage constructive criticism, asking the workers how their real needs may be better served, developing a better understanding of print user problems, and bringing back the red-penciled prints for further study. The information that is gathered by these shop tours should then be reviewed, edited, and documented as future drafting-room practices.

Once a return communications channel is opened and operating, drafting is responsible to act. When the requested changes are found to be constructive and should be implemented, the drawings should be changed and a new print sent to the requestor or, at the very least, notice that the drawing will be changed. When the requested change is not to be implemented for any reason, the request should be returned to the originator with an explanation of why the suggestion is unacceptable. The results of two-way communication will invariably be beneficial to the operation of any company.

DOCUMENTATION

In many respects, drafting is an art rather than an exact science, and therefore the degree of control exercised over the actual delineation of engineering information is difficult to establish. A certain amount must be left to the discretion of the drafter, but the scope of this discretion must be limited. Every good drafting department should have a D/DM containing all standard and specialized practices that reflect the state of the art as required and acceptable to the individual company.

The manual should emphasize that all drawings, specifications, and related documents be accurately and logically presented to enable manufacturing personnel and others using the documents to produce the desired end results economically, completely, and with a minimum of error. (See Section 14.)

D/DM Review and Revision

Because drafting department practices will inevitably change with improvements and changes in manufacturing and other departmental processes, provisions for continual review and revision of the contents should be made so as to eliminate outdated material and to add useful information as needed. The manual will be meaningful only as long as it is kept up to date, as it must communicate current drafting practices to seasoned employees as well as specialized drafting practices to the new recruit.

D/DM Value

A good drafting manual will be invaluable not only in reducing costly misinterpretation by the drawing users, but also as an aid to educating newly hired drafting-room personnel to the individual practices and needs of the company. Drafting as taught in schools or as practiced in other companies usually does not conform to the specialized needs with which a new employee is confronted.

Theory versus Practice

The subject of user-oriented drawings by its very definition implies that drafting as practiced by one company does not necessarily parallel the practices of another company. For this reason, each drafting department must search out, learn from experience, document, and use those modifications, additions, and deletions to standard industry practices that are unique to its drawing users' needs. A textbook can provide, at best, only theory and generalized examples that are useful as a starting point.

Measurement with SI (Modern) Metric[1]

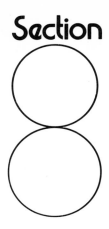

Harold E. Guetzlaff

Some people have the erroneous idea that the SI (modern) metric system of measurement is a non-American concoction, intended to belittle American technology. To reach the truth about the subject of measurement requires a bit of historical background, which is briefly given here, and the realization that American technology can actually be enhanced by the coherence of the SI metric measurement system.

[1]The author wishes to acknowledge the following sources for permitting the use of applicable information during the development of this chapter: Society of Automotive Engineers, Inc., for *SAE Recommended Practice—SAE J916c, Rules for SAE Use of SI (Metric) Units;* Clearvue, Inc., for W. T. Ryan and Paul Joe Vest, *Modern Metrics Made Easy,* Swani, Roscoe, Ill., 1976; Penton/IPC, Inc. (*Hydraulics & Pneumatics* magazine, January and February 1976), for "The Year of SI," by Walter Ernst; and the American Society for Testing and Materials for *ASTM E380-76, Metric Practice Guide.*

FIRST METRIC

The metric system is the most recent of the attempts to correct the confusion in measurement inherited from ancient times. What made the metric system a reality was the tremendous social and political upheaval of the French Revolution. In their enthusiasm to disrupt existing and accepted European traditions, the leaders of the Revolution believed that even the measurement system should be changed, because it had been devised and regulated by monarchy. A group of scientists was commissioned to establish a new measurement system based upon an unchanging, absolute standard found in the physical universe. Although work of this nature had been investigated prior to this time, the finalization of this commission was very difficult. The result established the standard unit of length, the *meter* (French spelling, metre), which when expanded or diminished by the use of decimal multiples and submultiples, is the basis for the commonly used units derived for area, volume, and weight.

The physical standard representing the meter was to be constructed so that it would equal one ten-millionth of the distance from the North Pole to the equator along the meridian of the earth running near Dunkirk, France, and Barcelona, Spain.

The metric unit of weight, the *gram*, was defined as the mass of one cubic centimeter (a cube that is $\frac{1}{100}$ of a meter on each side) of water at its temperature of maximum density. The cubic decimeter (a cube $\frac{1}{10}$ of a meter on each side) was chosen as the standard unit of volume. This measure was given the name *liter* (French spelling, litre).

CGS METRIC SYSTEM

The first official metric system, called the centimeter-gram-second (CGS) system, was proposed in 1795 and adopted in France in 1799, but it was not until 1840 that the French government, in response to a lack of public enthusiasm toward the voluntary plan, made CGS a compulsory system of measurement. Following the determination of France to make the CGS system mandatory, other countries began adopting the same system of measurement. In fact, in 1866, as the United States was beginning its advance in technological development, an act of Congress made it "lawful throughout the United States of America to employ the weights and measures of the metric system in all contracts, dealings, or court proceedings."

In 1873 the British Association for the Advancement of Science recommended the use of the CGS system, and since then it has been widely used in all branches of science throughout the world. Because of the growing use of the metric system throughout Europe, the importance of the accuracy of its base units became a matter of concern. This concern caused the French government to invite various nations to send delegates to an international conference to discuss the construction of a new prototype meter as well as a number of identical standards for the participating nations. In 1875 an international treaty, the Treaty of the Meter, set up well-defined metric standards for length and weight and established permanent procedures whereby refinements in the metric system could be recommended and adopted. This treaty was signed by 17 countries, including the United States. In 1893 the United States officially discarded the traditional English standards for the determination of length and weight (or mass), and from that point on the yard and the pound have been derived from standards of the meter and the kilogram.

MKS AND MKSA SYSTEM

During the years following the Treaty of the Meter, much discussion took place over whether the base unit of weight, the kilogram, and the term *weight* were synonymous. Finally it became necessary to put an end to the ambiguity concerning the meaning of the word weight, which was used sometimes for mass and sometimes for mechanical force.

The General Conference of Weights and Measures (CGPM, an abbreviation of the French title) declared, at its third meeting in 1901, that the kilogram is a unit of mass, and that the term weight denotes a quantity of force. The weight of a body is the product of its mass and the acceleration due to gravity. The decision to define the kilogram (and gram) in a different way from that in which they had been defined in the CGS system required a new system; this new system became known as MKS, referring to the base units meter-kilogram-second.

The International Electrotechnical Commission (IEC) accepted a recommendation in 1935 that the MKS system of units of mechanics be linked with electromagnetic units by the adoption of a fourth base unit, the ampere, as the unit of electric current. With this addition the MKSA system came into being.

DEVELOPMENT OF THE INTERNATIONAL SYSTEM OF UNITS (SI)

No standard is infallible, and therefore any standard can be considered fair game for improvement, particularly if any ambiguity develops. Such was the case with the CGS metric system of measurement. After the Treaty of the Meter and the establishment of the CGPM, there were continual attempts to improve upon the metric system by redefining the base units so that they could be reproduced in a laboratory, and there were ongoing efforts to clarify certain ambiguous details, such as the mass-force dilemma. After 160 years the CGPM decided that it was time to officially declare an improvement over the old CGS system by adopting a coherent metric system of measurement that had its own base units.

The basis of the modern metric system was the established MKSA system, with the meter redefined, the kilogram a unit of mass, and the clearly defined second and ampere. Two other base units were also included—the kelvin, for thermodynamic temperature, and the candela, for luminous intensity. The new system, which was to replace the CGS system, was called the "International System of Units," abbreviated internationally as SI, from the French *Système International d'Unités*. The major advantage of the SI metric system was the use of distinct and separate units for mass and force. With the advent of SI the force unit acquired the name *newton*, which is used wherever the former kilogram-force unit, or that entity which is commonly called weight, would have been used in the MKSA system. In 1960 the eleventh session of the CGPM formally adopted the SI system of measurement with six base units; in 1971 a seventh base unit, the *mole*, was added. The mole is the unit representing the amount of a substance.

INCH-POUND VERSUS SI

Most countries engaged in some form of industrial technology had adopted a metric system as their official measurement standard prior to the introduction of the SI system. United States' trade with these metric countries had not been noticeably hindered by the difference in measurement systems, probably because it was fairly easy, if awkward, to make a conversion when needed. With the introduction in 1960 of the modern, coherent SI metric system of measurement, which the United States' closest ally, the United Kingdom, decided to adopt in 1965, the reason for staying with the inch-pound system became less valid. To make matters worse, Canada decided in 1970 to adopt the new SI system. Canada decided to go slowly with its changeover implementation, however, until it received a more positive indication that the United States would engage in a transition program. Such a program was initiated a few years later by the U.S. Metric Conversion Act of 1975. One other complication developed in the world trade picture that made it impractical for any country to continue to hold a negative attitude toward the SI measurement system. The European Economic Community (EEC), better known as the Common Market, issued a directive in 1971 that required the nine member countries to use SI metric units in all commercial transactions by the end of 1977. The effect of this directive on the United States is that each EEC country has the privilege of deciding individually whether or not it will accept U.S. goods with measurements described in U.S. customary units, whether dual units (both U.S. customary and SI for each measurement) will be acceptable, or whether only SI units will be permitted.

Although the United States is often considered to be the world's most technologically advanced nation, world trade has a tremendous impact on all U.S. technological disciplines. When world trade might be adversely affected by the different measurement system used in the United States, it is only logical to devise some means by which U.S. technology may take a more positive approach toward working with the SI metric system.

The Metric Conversion Act of 1975 gives U.S. industry an opportunity to voluntarily change from the U.S. customary measurement system to the new SI metric; at least, the Act instructs industry as to the recommended units that can be used in conjunction with U.S. customary measurements if the dual approach is desired. The Act gives U.S. industry the opportunity to compete with foreign industry in all instances in which measurement requirements must be met. Resistance to the change to the SI system by those segments of U.S. industry which are strictly domestic, and by the public in general, is natural, because such change disrupts a way of life that has been taken for granted. The change to SI requires that eventually everyone who deals with anything involving measurement will have to be somewhat reeducated.

BECOMING FAMILIAR WITH SI

A well-organized transition program begins with familiarity with the SI base units (Table 8-1), the SI supplementary units, the SI derived units, the approved SI prefixes, and the SI prefixes that are considered to be normal usage.[2]

TABLE 8-1   SI Base Units

Quantity	Name	Symbol
Length	meter	m
Mass	kilogram	kg
Time	second	s
Electric current	ampere	A
Thermodynamic temperature	kelvin	K
Amount of substance	mole	mol
Luminous intensity	candela	cd

Each SI base unit is a separate entity, depending on no other base unit for definition or verification. Therefore, the fact that a cubic decimeter (dm^3), which has been given the special name of *liter* (L), has a mass of approximately one kilogram (kg) is coincidental. Many metric documents, tables, and charts state that there is a close relationship between volume and mass, and that one liter of water has a mass of one kilogram. This was true only from 1901 to 1964. In 1964 the 12th CGPM revoked the 1901 definition of the liter and said, in effect, that the liter shall no longer be defined in such a manner as to give the impression that the liter has any relation to mass. The 12th CGPM did allow the word liter to be employed as a special name for cubic decimeter (dm^3) but cautioned anyone who would do so not to use the word liter to describe the results of high-accuracy measurements. The reason for this precaution is that the 1901 liter and the SI cubic decimeter differ by about 28 parts in 10^6. Although such a minute difference would have little effect on most volume measurements, the knowledge that there is a difference may have an effect upon some disciplines.

Supplementary units (Table 8-2) were adopted by the 11th CGPM in 1960 at the same time the base units were adopted. Questions have been raised as to why these units were not included as additional base units, and some SI reference material actually states that they can be regarded as either base units or derived units. It

[2]Definitions for SI base units, supplementary units, and derived units with special names are provided later (see Definitions).

TABLE 8-2   SI Supplementary Units

Quantity	Name	Symbol
Plane angle	radian	rad
Solid angle	steradian	sr

appears that supplementary units should be regarded as base units when we observe that they are used in the same manner as the official base units.

For those people who have reservations about the transition to SI and who are not well acquainted with any metric system of measurement, some satisfaction should be derived from the fact that the engineering use of measurement units in nearly every metric country of the world involves a struggle to learn new and unfamiliar terms. In other words, the transition to SI affects more people than just those who are accustomed to using the U.S. customary system of measurement.

It should be noted that a number of SI units (base and derived) are no different when identified as being part of the SI system from what they were before the introduction of the SI system. The base units meter, second, and ampere, as well as the supplementary unit radian, are the same, and most of the electrical derived units are unchanged.

Derived units (Table 8-3) are expressed algebraically in terms of base units and/or supplementary units. The derived units shown in Table 8-3 were given special names by the CGPM because of the complexity of their derivation; for example, the pascal usually is shown with the formula of N/m^2, but it could be also shown as $kg/(m \cdot s^2)$ and probably will be shown this way when it becomes necessary to solve an equation that makes use of base-unit factors. On the other hand, many of the derived units with special names honor an individual who worked and contributed much to a specific

TABLE 8-3   SI Derived Units with Special Names

Quantity	Unit	Symbol	Formula
Frequency	hertz	Hz	$1/s$
Force	newton	N	$kg \cdot m/s^2$
Pressure or stress	pascal	Pa	N/m^2
Work, energy, quantity of heat	joule	J	$N \cdot m$
Power	watt	W	J/s
Electric charge	coulomb	C	$A \cdot s$
Potential difference (electromotive force)	volt	V	W/A
Electrical capacitance	farad	F	$A \cdot s/V$ or C/V
Electrical resistance	ohm	Ω	V/A
Electrical conductance	siemens	S	A/V
Magnetic flux density	tesla	T	Wb/m^2
Magnetic flux	weber	Wb	$V \cdot s$
Electrical inductance	henry	H	$V \cdot s/A$ or Wb/A
Temperature	degree Celsius	°C	For temperature interval $1°C = 1$ K; for temperature, $°C = K - 273.15$
Luminous flux	lumen	lm	$cd \cdot sr$
Illuminance	lux	lx	lm/m^2
Absorbed dose	gray	Gy	J/kg
Activity (of a radionuclide)	becquerel	Bq	$1/s$

scientific field, and these special names have existed for many years, with the result that they have been accepted as SI units without change. Additional derived units are listed later (see Definitions).

In general, a prefix is used to indicate the order of magnitude of a base unit, a supplementary unit, or any SI derived unit having a special name. Using a prefix properly eliminates the need for nonsignificant digits and leading zeros in decimal fractions, and it also provides a convenient alternative to the powers-of-ten notation preferred in computation. For example,

$$12\ 300 \text{ mm becomes } 12.3 \times 10^3 \text{ mm or } 12.3 \text{ m}$$

$$12.3 \times 10^3 \text{ m becomes } 12.3 \text{ km}$$

$$0.001\ 23 \text{ } \mu\text{m becomes } 1.23 \times 10^{-3} \text{ } \mu\text{m or } 1.23 \text{ nm}$$

The prefixes shown in Table 8-4 are those that have been approved by the CGPM for use in all disciplines. For expressing a quantity by a numerical value and a unit, prefixes should preferably be chosen so that the numerical value lies between 0.1 and 1000.

An exception would occur when expressing area and volume, where the prefixes *hecto-, deka-, deci-,* and *centi-* may be used to reduce the number of digits in a quantity. For example,

$$(100 \text{ mm})^2 = 10\ 000 \text{ mm}^2 = 100 \text{ cm}^2$$
$$(100 \text{ mm})^3 = 1\ 000\ 000 \text{ mm}^3 = 1000 \text{ cm}^3 = 1 \text{ dm}^3$$

The prefixes shown with an asterisk in Table 8-4 are those that are commonly used and, with the exception of the prefix *centi-*, are those that represent powers of 1000. By reducing the variety of prefixes in a uniform way to show unit quantities for specific disciplines, the problems of misinterpretation and error are diminished. For example: the millimeter is used for all linear dimensions in mechanical engineering drawings, even when some values lie outside the range 0.1 to 1000; the centimeter is used for body measurements and clothing sizes; and the meter is used in structural and civil engineering. Compound units, which are derived units expressed in terms of two or more units and which do not have special names, shall show only one prefix when forming a multiple. The prefix is normally attached to the numerator. One exception to this permits the use of the unit kilogram in the denominator of a compound unit, since it is a base unit.

Additional precautions and guidelines for the use of prefixes as they relate to complex engineering problems can be found in the ASTM E380-76 document. (For further information on metric standards, and sponsoring bodies, see References at the end of this section.)

TABLE 8-4   SI Unit Prefixes

Multiples and submultiples	Prefix	Symbol
$1\ 000\ 000\ 000\ 000\ 000\ 000 = 10^{18}$	exa	E
$1\ 000\ 000\ 000\ 000\ 000 = 10^{15}$	peta	P
$1\ 000\ 000\ 000\ 000 = 10^{12}$	tera	T
$1\ 000\ 000\ 000 = 10^{9}$	giga	G
$1\ 000\ 000 = 10^{6}$	mega*	M
$1\ 000 = 10^{3}$	kilo*	k
$100 = 10^{2}$	hecto	h
$10 = 10^{1}$	deka	da
$0.1 = 10^{-1}$	deci	d
$0.01 = 10^{-2}$	centi*	c
$0.001 = 10^{-3}$	milli*	m
$0.000\ 001 = 10^{-6}$	micro*	μ
$0.000\ 000\ 001 = 10^{-9}$	nano	n
$0.000\ 000\ 000\ 001 = 10^{-12}$	pico	p
$0.000\ 000\ 000\ 000\ 001 = 10^{-15}$	femto	f
$0.000\ 000\ 000\ 000\ 000\ 001 = 10^{-18}$	atto	a

*Commonly used prefixes.

USE OF NON-SI UNITS WITH THE SI SYSTEM

To assist in preserving the advantage of SI as a coherent system, a minimum number of units from other systems are permitted. These units (Table 8-5) are as follows:

Time   The SI unit for time is the *second*. This unit is preferred and should be used if practical, particularly when technical calculations are involved. When time relates to life customs or calendar cycles, the minute, hour, day, and other calendar units may be necessary. For example, vehicle velocity will normally be expressed in kilometers per hour.

Plane angle   The SI unit for plane angle is the *radian*. Use of the *degree* (°) and its decimal submultiples is permissible when the radian is not a convenient unit. Use of the minute and second is permitted with the radian when measurement devices require the use of such increments.

Volume   The SI unit of volume is the *cubic meter*. This unit, or one of its regularly formed multiples, such as the cubic centimeter, is preferred for all applications. The special name *liter* has been approved for the cubic decimeter, but use of this unit is restricted to the measurement of liquids and gases. No prefix other than *milli-* should be used with liter.

Mass   The SI unit of mass is the *kilogram*. This unit, or one of the multiples formed by attaching an SI prefix to *gram*, is preferred for all applications. The *megagram* (Mg) is the appropriate unit for measuring large masses that have customarily been expressed in tons. However, the name *ton* has been given to several large mass units that are widely used in commerce and technology—the long ton of 2240 lb, the short ton of 2000 lb, and the metric ton of 1000 kg (also called the *tonne*). None of these terms is SI. The term *metric ton* should be restricted to commercial usage, and no prefixes should be used with it. Use of the term *tonne* is not recommended.

TABLE 8-5   Units Permitted for Use with SI

Quantity	Unit	Symbol	Definition
Time	minute	min	1 min = 60 s
	hour	h	1 h = 60 min = 3600 s
	day	d	1 d = 24 h = 86 400 s
	week, month, etc.	———	
Plane angle	degree	°	$1° = (\pi/180)$ rad
Volume	liter	L	$1 \text{ L} = 1 \text{ dm}^3 = 10^{-3} \text{ m}^3$
Mass	metric ton	t	$1 \text{ t} = 10^3 \text{ kg} = 1 \text{ Mg}$

Non-SI Units Permitted for Limited Use

Various disciplines still feel obligated to retain certain non-SI units that have been in popular use. These units have been allowed to remain in use for an indefinite time but may be phased out in the future. These units (see Table 8-6) are as follows:

Energy   The SI unit of energy, the *joule*, together with its multiples, is preferred for all applications. The kilowatthour (kWh) is widely used, however, as a measure of electrical energy. This unit should not be introduced into any new areas, and eventually it should be replaced by the megajoule (MJ). One kilowatthour = 3.6 megajoules.

Area   The SI unit of area is the *square meter* (m^2). Large areas are properly expressed in *square hectometers* (hm^2) or *square kilometers* (km^2). Use of the name *hectare* as an alternative to *square hectometer* is restricted to the measurement of land or water areas.

Pressure and Stress   The SI unit of pressure and stress is the *pascal* (newton per square meter, N/m^2), which with proper SI prefixes is applicable to all such measurements. Old metric gravitational units for pressure and stress, such as kilogram-force per square centimeter (kgf/cm²), shall not be used. Where the unit *torr* has been used to designate vacuum (mmHg at 0°C), its use shall be discontinued in favor of the pascal (Pa).

The *millibar* is widely used in meteorology. Its use will continue for the time being in order to permit meteorologists to communicate easily within their profession. The *kilopascal* should be used in presenting meteorological data to the public.

TABLE 8-6   Units Accepted for Limited Use

Quantity	Unit	Symbol	Definition
Energy	kilowatthour	kWh	$1 \text{ kWh} = 3.6 \text{ MJ}$
Area	hectare	ha	$1 \text{ ha} = 1 \text{ hm}^2 = 10^4 \text{ m}^2$
Pressure	bar	bar	$1 \text{ bar} = 10^5 \text{ Pa}$
Activity (for radionuclides)	curie	Ci	$1 \text{ Ci} = 3.7 \times 10^{10} \text{ Bq}$
Exposure	roentgen	R	$1 \text{ R} = 2.58 \times 10^{-4} \text{ C/kg}$
Absorbed dose	rad	rd	$1 \text{ rd} = 0.01 \text{ Gy}$

Non-SI Units and Names Not To Be Used

A great many metric units other than those of the SI system have been defined over the years. Some of these are used only in special fields; others have found broad application in countries that adopted the metric system many years ago.

Except for the special cases indicated above, non-SI units (as well as nonapproved special names for SI units) are to be avoided. A partial list of those unit names that should be avoided includes:

CGS Units   All units peculiar to the various CGS systems (measurement systems constructed using the centimeter, gram, and second as base units) are to be avoided. Among these units are the following, which are defined for mechanics, fluid mechanics, and photometry: erg, dyne, gal, poise, stokes, stilb, phot, and lambert.

Also to be avoided are the CGS units of electricity and magnetism, such as: gauss, oersted, maxwell, gilbert, biot, and franklin. This also applies to the unit names formed with the prefixes *ab-* and *stat*, for example, the abampere or the statvolt.

Unit Names to Be Avoided   Special names for multiples and submultiples of SI units are to be avoided, except for the liter, metric ton, and hectare. For example:

Units to avoid	Equivalents	
fermi (fm)	1 fm	$= 10^{-15} \text{ m}$
micron (μm)	1 μm	$= 10^{-6} \text{ m}$
millimicron (nm)	1 nm	$= 10^{-9} \text{ m}$
are	1 are	$1 \text{ dam}^2 = 100 \text{ m}^2$
gamma (magnetic flux density)	1 gamma	$= 1 \text{ nT}$
mho	1 mho	$= 1 \text{ S}$
candle	1 candle	$= 1 \text{ cd}$
candlepower	1 candlepower	$= 1 \text{ cd}$

Miscellaneous Units   Other non-SI units that are not to be used include: kilogram-force, calorie, standard atmosphere (1 atm = 101.325 kPa), technical atmosphere (1 at = 98.0665 kPa), conventional millimeter of mercury, torr, conventional centimeter of water, grade (1 grade = $\pi/200$ rad), metric carat, and metric horsepower.

CLARIFICATION OF SI MASS, FORCE, AND WEIGHT

The principal departure of SI from the gravimetric system of metric engineering units is the use of explicitly distinct units for mass and force. In SI, the name *kilogram* is restricted to the unit of mass, and *kilogram-force* (from which the word *force* was in practice often erroneously dropped) must not be used. In its place the SI unit of force, the *newton*, is used. Likewise, the newton rather than the kilogram-force is used to form derived units that include force—for example, pressure or stress (N/m² = Pa), energy (N·m = J), and power (N·m/s = W).

Considerable confusion exists as a result of the use of the term *weight* to mean *force* or *mass*. In commercial and everyday use, the term weight nearly always means mass, so that when one speaks of an object's weight, the quantity referred to is mass. It is doubtful whether this nontechnical use of the term weight can be successfully discouraged. In science and technology, the term weight of a body has usually meant the force that, if applied to the body, would give it an acceleration equal to the local acceleration of free fall (gravity). The adjective *local* has usually meant a location on the surface of the earth. In this context, the "local acceleration of free fall" (m/s², also referred to as acceleration of gravity) has the symbol g, with observed values of g differing as much as 0.5 percent at various locations on the earth's surface. The use of *force of gravity* (mass times acceleration of gravity = kg·m/s² = N) instead of the term weight, with this meaning, is recommended.

Because of the past dual use of the term weight as a quantity, this term should be avoided in technical practice except under circumstances in which its meaning is completely clear. When the term is used, it is important to know whether mass or force is intended and to use SI units properly as described above: kilograms for mass; newtons for force. Where it is necessary to convert force and/or mass from the U.S. customary units to SI units and there is some question as to which SI unit is correct, do the following:

1. Decide whether the pound unit is force or mass. Do not consider whether the term weight is involved.

2. If the pound unit is a force unit, the correct SI unit is newton (N). If the pound unit is mass, the correct SI unit is kilogram (kg).

Gravity is involved in determining the mass of an object with a balance scale or a spring scale. When a standard mass is used to determine the mass of an object, the direct effect of gravity on the two sides of the balance scale is canceled, but the indirect effect of the buoyancy of air or other fluid is generally not canceled. In using a spring scale, mass is measured indirectly, since the instrument responds to the force of gravity. Such scales may be calibrated in mass units if the variation in acceleration of gravity and buoyancy corrections are not significant in their use.

TEMPERATURE

The SI unit of temperature is the *kelvin* (K), which is to be used for expressing thermodynamic temperatures and temperature intervals. However, since the kelvin is a unit that would be quite awkward to use in everyday life, the degree Celsius (°C) has been designated for expressing temperatures and temperature intervals with SI. Kelvin will be used only for thermodynamic temperatures and temperature intervals. The Celsius scale (formerly called centigrade) is directly related to the kelvin scale as follows: The temperature interval one degree Celsius equals one kelvin, exactly.

Celsius temperature (t) is related to thermodynamic temperature (T) by the equation

$$t = T - T_0$$

where $T_0 = 273.15$ K, by definition

SPECIFIC WEIGHT, DENSITY, AND SPECIFIC GRAVITY

Many American engineering handbooks discuss specific weight and density somewhat synonymously, mainly because the differentiation between mass and force is often ignored. In other words, in the U.S. customary system, pound-force is used in all cases where the term weight is involved. However, specific weight is the weight of the unit volume of a substance, whereas density is the mass of that volume.

In the U.S. customary system, specific weight is designated in pounds-force per cubic foot (lbf/ft^3) and density is designated in $slugs/ft^3$. The name *slug* for unit mass is apparently not very popular, as many handbooks give densities of various substances in lb/ft^3. Actually, in most cases the handbook value should be indicated as lbf/ft^3 because the quantity presented is specific weight. If density is the required quantity, the handbook value must be divided by 32.174 ft/s^2 to obtain $slugs/ft^3$, which is the correct designation for use in calculations of the dynamic effects of mass.

In the metric gravitational system, the designation for specific weight is kiloponds per cubic meter (kp/m^3). A kilopond is the name given to the term *kilogram-force*, which is approximately equal to 9.81 (9.80665) $kg \cdot m/s^2$. Density has been designated in $kp \cdot s^2/m^4$.

German engineering handbooks often give specific weights of solid materials in kiloponds per cubic decimeter (kp/dm^3), which is the old standard designation times 10^{-3}. In all but the most recent handbooks, densities have not been listed, so the table values that are shown in kp/m^3 must be divided by 9.807 m/s^2 to obtain densities in $kp \cdot s^2/m^4$. It is interesting to note that since kp/m^3 is actually $(9.807 \ kg \cdot m/s^2)/m^3$, dividing by 9.807 m/s^2 for density would give kg/m^3 rather than $kp \cdot s^2/m^4$. In the SI system, density is measured in kg/m^3. There is actually no term called *specific weight*. For those disciplines that require statistical calculations with problems concerning gravity as a vertically directed mass-force of the unit volume of a material (structural and civil engineering), and where there is also a requirement in said systems for the calculation of pressure based on head and the mass-force of a liquid, the force (g) applied to the unit volume of the substance becomes N/m^3. Where handbooks show the *specific weight* of substances in N/m^3, the density of the substances can be obtained by dividing the handbook value by 9.807 m/s^2 to get kg/m^3.

Specific gravity is a term used only in the U.S. customary system to represent a ratio between the density of a substance and the density of a reference fluid, usually water.

TABLE 8-7   Average Density of Various Substances at 20°C

Substance	Density, kg/m^3	Substance	Density, kg/m^3
Liquids		Solid metals	
Water (4°C)	1000.000	Aluminum	2643.046
Sea water	1025.182	Cast iron	7080.161
Benzene	879.414	Copper	8906.265
Carbon tetrachloride	1593.837	Gold	19302.248
Ethyl alcohol	784.905	Lead	11373.109
Gasoline	680.785	Magnesium	1746.012
Kerosene	799.321	Nickel	8601.915
Lubricating oil	913.052	Silver	10508.112
Methyl alcohol	807.331	Steel	7833.028
Sulfuric acid, 100%	1830.910	Tungsten	19222.156
Turpentine	864.997	Zinc	7048.124
Mercury (0°C)	13594.870	Brass or bronze	8553.859
Nonmetallic solids		Woods	
Ice	897.034	Balsa	128.148
Concrete	2306.659	Pine, red	480.554
Glass, plate	2562.954	Maple, sugar	688.794
Granite	2643.046	Oak	720.831

Specific gravity has no metric equivalent, because it is a ratio and is therefore not restricted to any measurement system.

Examples of densities of various substances in kilograms per cubic meter are shown in Table 8-7. The densities shown are directly related to the accuracy (rounded off) of the values from which the conversions were made. No guarantee is made or implied that the values shown are infallible, since the source of the values in Table 8-7 may not correspond to the values from other sources. Where reference books show densities in pounds per cubic foot (lb/ft³ or p/ft³), as water at 4°C with a density of 62.428 lb/ft³, that value can be multiplied by 16.01846 to obtain kilogram per cubic meter (kg/m³).

THE BAROMETER AND PRESSURE GAGES

To deal with measurements of pressure requires a fundamental knowledge of atmospheric pressure. Under normal conditions, we live in an environment where the variation in atmospheric pressure may be as much as 5 percent at a given location. This change is attributed to weather conditions. Also, under normal conditions we do not concern ourselves with atmospheric pressure changes except as they relate to our personal comfort, and therefore we accept normal atmospheric pressure as the reference point from which all other pressures that we may encounter may be measured.

In the technical field there is a need to know the difference between absolute pressure and gage pressure.

- Absolute pressure is the actual pressure at a point in a fluid.
- Gage pressure is the difference between absolute pressure at a point in a fluid and the pressure of the atmosphere.

Actual (absolute) pressure is measured most frequently with a mercury barometer when an accurate measurement is desired. The common barometer (aneroid) found in the home or the office and used to approximate atmospheric pressure is calibrated against a mercury barometer in terms of inches or centimeters of mercury.

Even though the present-day barometer is graduated in inches of mercury, regardless of whether the accuracy desired is to be close or approximate, there is no need to give any thought to replacing such a device with one that is graduated in centimeters of mercury. Regardless of whether a barometer is graduated in inches or centimeters, a conversion must be made to find the actual pressure at the time and temperature of the reading where such accuracy is needed for special technology.

When weather forecasters indicate barometric pressure in centimeters of mercury and a person has retained a barometer graduated in inches of mercury, the simple conversion of 2.54 centimeters per inch, either dividing or multiplying, can be an inexpensive solution to the lack of a metric barometer.

Tables 8-8 and 8-9 are included for informational purposes.

With the SI system, pressure gages continue to be used in the same manner as with the customary system—that is, to indicate differential pressure. In this case, zero gage pressure is approximately an absolute pressure of 101.325 kPa. The gage must be adjusted to zero at the locale where it will be used.

For those areas of technology in which the transition is (or will be) taking place, it is advisable not to acquire any more pressure gages graduated in pounds per square inch (psi). The correct unit for SI pressure is the kilopascal (kPa), and the sooner the gage dial shows that SI unit, the sooner familiarity with it will be acquired. For pressure gages that have removable dials, new dials in kPa can replace the old at a nominal cost with practical marking according to Table 8-14.

Present psi tire gages can continue to be used if one divides the recommended tire pressure given in kPa by 7. For example, if metric tire pressure is recommended to be 220 kPa, dividing by 7 will give about 31.5 psi. A person who checks tire pressure and makes this conversion once or twice will soon learn that the transition is not difficult to cope with.

TABLE 8-8 Relation between Standard Barometer Reading and Absolute Pressure (Customary System)

Barometer reading $\times$ 3.3864 = absolute kPa at 0°C
Barometer reading $\times$ 3.3768 = absolute kPa at 15.6°C

Barometer, inches Hg	Pressure, kPa at 0°C	Pressure, kPa at 15.6°C
28.00	94.819	94.552
28.25	95.666	95.396
28.50	96.512	96.240
28.75	97.359	97.084
29.00	98.206	97.929
29.25	99.052	98.773
29.50	99.899	99.617
29.75	100.745	100.461
30.00	101.592	101.306
30.25	102.439	102.150
30.50	103.285	102.994
30.75	104.132	103.838
31.00	104.978	104.682
31.25	105.825	105.526
31.50	106.672	106.371

Note: Standard atmosphere = 29.92 inches of Hg at 0°C = 101.325 absolute kPa.

TABLE 8-9 Relation between Standard Barometer Reading and Absolute Pressure (Metric System)

Barometer reading $\times$ 1.3332 = absolute kPa at 0°C
Barometer reading $\times$ 1.3294 = absolute kPa at 15.6°C

Barometer, cm Hg	Pressure, kPa at 0°C	Pressure, kPa at 15.6°C
71.0	94.659	94.387
71.5	95.325	95.052
72.0	95.992	95.717
72.5	96.659	96.382
73.0	97.325	97.046
73.5	97.992	97.711
74.0	98.659	98.376
74.5	99.325	99.040
75.0	99.992	99.705
75.5	100.658	100.370
76.0	101.325	101.034
76.5	101.992	101.699
77.0	102.658	102.364
77.5	103.325	103.028
78.0	103.991	103.693
78.5	104.658	104.358
79.0	105.325	105.023
79.5	105.991	105.687
80.0	106.658	106.352

Note: Standard atmosphere = 76.0 cm of Hg at 0°C = 101.325 absolute kPa.

ENERGY AND TORQUE

The vector product of force and moment arm is widely designated by the unit *newton meter*. This unit for bending moment, or torque, results in confusion with the unit for energy, which is also newton meter. If torque is expressed as newton meters per radian, the relationship to energy is clarified, since the product of torque and angular rotation is energy:

$$(N \cdot m/rad) \cdot rad = N \cdot m$$

If the vectors were shown, the distinction between energy and torque would be obvious, since the orientation of force and length is different in the two cases. It is important to recognize this difference when indicating torque and energy. The joule must never be used as the unit for torque.

RULES FOR WRITING UNIT SYMBOLS

1. Unit symbols should be printed in roman (upright) type regardless of the type style used in the surrounding text.

2. Unit symbols are unaltered in the plural.

3. Unit symbols are not followed by a period except when used at the end of a sentence. (*Exception:* the abbreviation for inch, in, may be followed by a period in order to avoid confusion with the word *in*.)

4. Letter unit symbols are written in lowercase type unless the unit name has been derived from a proper name, in which case the first letter of the symbol is capitalized. Prefix and unit symbols retain their prescribed form regardless of the surrounding typography. (*Note:* Two exceptions to rule 4 are as follows: The international symbol for liter is the lowercase "l," which can easily be confused with the numeral. Accordingly, the symbol L is recommended for U.S. use. Also, computer type style does not allow for lowercase letters; therefore, for computer printout or other systems that have uppercase characters only, see the following section, Rules for Computer Representation of SI Symbols.)

5. In the complete expression of a quantity, a space should be left between the numerical value and the unit symbol. For example, write 35 mm, not 35mm, and 2.37 lm, not 2.37lm. *Exception:* No space is left between the numerical value and the symbol for degree, minute, and second of the plane angle, and for degrees Celsius.

6. No space is used between prefix and unit symbols.

7. Symbols, not abbreviations, should be used for units. For example, use A, not amp, for ampere.

RULES FOR COMPUTER REPRESENTATION OF SI SYMBOLS

This procedure establishes a convention for representing SI symbols in computer printouts or other systems that have uppercase characters only. Uppercase representations are not to be used in documents prepared on conventional typewriters or other systems that have both uppercase and lowercase characters available simultaneously. This procedure is based on a standard of the International Organization for Standardization (ISO 2955).

1. In narrative (free text) data, a space character should be used to separate the numerical value and the unit representation; e.g., 10 MM represents 10 millimeters. In formatted data, the use of the space character as a separator is optional, since its use or nonuse is defined in the format description.

2. To indicate multiplication of units, a period must be used between the representations of units; e.g., N.M represents newton meters, but NM represents nanometers.

3. To indicate ratios of units, the numerator and the denominator are separated by a solidus (/). Alternatively, the denominator may be expressed with a negative exponent (see rule 5); e.g., M/S or M·S—1 can be used to represent meters per second.

4. A positive exponent is indicated by a numeral (no sign) directly after the representation of the unit; e.g., M2 represents square meters (m²).

5. A negative exponent is indicated by a minus sign followed by its respective numeral directly after the representation of the unit; e.g., M—3 represents m⁻³.

6. A prefix representation is combined with a unit representation to form a decimal multiple or submultiple of the unit. This new unit representation can be raised to a positive or negative power and can be combined with other unit representations to form representations of compound units. There is no separator or space between the prefix representation and the unit representation; e.g., MS is used to represent millisecond. A prefix may not stand alone; e.g., T alone means *tesla*, not *tera-*.

RULES FOR WRITING NAMES USED WITH SI

1. Spelled-out unit names are treated as common nouns in English. Thus, the first letter of a unit name is not capitalized, except at the beginning of a sentence or in capitalized material such as a title.

2. Plurals are used when required by the rules of English grammar and are normally formed regularly; for example, *henries* for the plural of *henry*. The following irregular plurals are recommended:

Singular	Plural
lux	lux
hertz	hertz
siemens	siemens

3. No space or hyphen is used between a prefix and a unit name. There are three cases in which the final vowel in the prefix is commonly omitted: megohm, kilohm, and hectare. In all other cases in which the unit name begins with a vowel, both vowels are retained and both are pronounced.

Units Formed by Multiplication and Division

1. With unit names:
   - **Product** Use a space (preferred) or a hyphen: i.e., newton meter or newton-meter. In the case of the watt hour, the space may be omitted: watthour.
   - **Quotient** Use the work per, not the solidus: meters per second, not meters/second.
   - **Powers** Use the modifier *squared* or *cubed*, placed after the unit name: meters per second squared. In the case of area or volume, a modifier may be placed before the unit name: square millimeter, cubic meter. This exception also applies to derived units using area or volume: watts per square meter.
     Note: To avoid ambiguity in complicated expressions, symbols are preferred over words.

2. With unit symbols:
   - **Product** Use a raised dot, also known as a center point: N·m for newton meter. In the case of W·h, the dot may be omitted, thus: Wh.
     An exception to this practice is made for computer printouts, automatic-typewriter work, etc., where the raised dot is not possible; in these cases, a dot on the line may be used.
   - **Quotient** Use one of the following forms: m/s or m·s⁻¹.
     In no case should more than one solidus be used in the same expression unless

parentheses are inserted to avoid ambiguity. For example, write: J/(mol·K), J·mol^{-1}·K^{-1}, or (J/mol)/K, but not J/mol/K.

3. Symbols and unit names should not be mixed in the same expression. Write: joules per kilogram, J/kg, or J·kg^{-1}, but not joules/kilogram, joules/kg, or joules·kg^{-1}.

RULES FOR SHOWING NUMBERS

1. The recommended decimal marker is a point (period) on the line. When numbers less than one are written, a zero shall be written before the decimal marker.

2. Outside the United States, the comma is sometimes used as a decimal marker. Therefore, in some applications the common practice of using the comma to separate digits into groups of three (as in 23,479) may cause ambiguity. Where it is desired to avoid this potential source of confusion, recommended international practice calls for separating the digits into groups of three, counting from the decimal point toward the left and the right, with a small space. In numbers of four digits the space is usually not necessary, except for uniformity in tables.

    EXAMPLES: 16.018 46   27 679.90   1.355 817 9

Where this practice is followed, the space should be narrow (approximately the width of the letter "i"), and the width of the space should be constant even if, as is often the case in printing, variable-width spacing is used between words.

3. Because the word *billion* means a thousand million (prefix *giga-*) in the United States but a million million (prefix *tera-*) in some other countries, the term should be avoided in technical writing.

MODIFYING SYMBOLS

Modifying derived unit symbols as a way of qualifying the nature of the quantity for which the symbol is being used is not permitted. Thus MWe for "megawatts electrical (power)," Vac for "volts ac," and kJt for "kilojoules thermal (energy)" are not acceptable. For this reason, no attempt should be made to construct SI equivalents of the abbreviations "psia" and "psig," often used to distinguish between absolute and gage pressure. If the context leaves any doubt as to which is meant, the word *pressure* must be qualified appropriately. For example: "at a gage pressure of 220 kPa" or "at an absolute pressure of 101.325 kPa."

ENGLISH PRONUNCIATION

Recommended pronunciation in English is as indicated in the Tables 8-10 and 8-11.

RULES FOR CONVERSION AND ROUNDING

General

1. At the end of this chapter in Tables 8-12 and 8-13, are an alphabetical list of units and typical applications, which contain conversion factors either as exact values or as a seven-digit accuracy value for implementing these rules, except where the nature of the dimension makes this impractical.

2. Conversion of quantities of units between systems of measurement involves careful determination as to the number of digits to be retained after the conversion calculation has been completed. To convert "1 quart of oil" to "0.946 352 9 liters of oil" is, of course, ridiculous, because the intended accuracy of the value does not warrant the retention of so many digits.

    This following discussion provides information to be used as a guide in the conversion of any quantity from the U.S. customary system to the SI system. All con-

TABLE 8-10
Recommended Pronunciations

Prefix	Pronunciation (U.S.)
exa	ex'a (*a* as in *a*bout)
peta	pet'a (*e* as in p*e*t, *a* as in *a*bout)
tera	as in *te*rrace
giga	jig'a (*i* as in j*i*g, *a* as in *a*bout)
mega	as in *mega*phone
kilo	kill'oh
hecto	heck'toe
deka	deck'a (*a* as in *a*bout)
deci	as in *deci*mal
centi	as in *centi*pede
milli	as in *mili*tary
micro	as in *micro*phone
nano	nan'oh (*an* as in *an*t)
pico	peek'oh
femto	fem'toe (*fem* as in *fem*inine)
atto	as in an*ato*my

TABLE 8-11
Recommended Pronunciations

Selected units	Pronunciation*
candela	can-*dell*-'a
joule	rhyme with *tool*
kilometer	kill-'oh-meter
pascal	rhyme with *rascal*
siemens	same as seaman's

*The first syllable of every prefix is accented to assure that the prefix will retain its identity. Therefore, the preferred pronunciation of kilometer places the accent on the first syllable, not the second.

versions, to be logically established, must depend on the intended precision of the original quantity, implied either by a specific tolerance or by the nature of the quantity being measured. The first step, after the conversion calculation has been made, is to establish this degree of precision.

3. Proper conversion procedure is to multiply the specified quantity by the conversion factor exactly as given and then round the result to the appropriate number of significant digits to the right of the decimal point or to a realistic whole number, according to the degree of precision implied by the original quantity (see Table 8-14 and Fig. 8-1).

Precision of a Value

The intended precision of a value should relate to the number of significant digits shown. The implied precision is plus or minus one-half unit of the last significant digit in which the value is stated, because the value may be assumed to have been rounded from a greater number of digits and one-half unit of the last significant digit retained is the limit of error resulting from rounding. For example, the number 2.14 may have been rounded from any number between 2.135 and 2.145. Whether rounded or not, a quantity should always be expressed with this implication of precision in mind. For instance, 2.14 in. implies a precision of ± 0.0005 in., since the last significant digit is in units of 0.01 in.

To \ From	psi	in. H$_2$O @ 39.2°F	in. Hg @ 32°F	kgf/cm^2	bar	atmosphere (standard)	cm H$_2$O @ 4°C	mm Hg @ 0°C	kPa
psi		27.682	2.036	7.031×10^{-2}	6.895×10^{-2}	6.805×10^{-2}	70.306	51.718	6.895
in. H$_2$O @ 39.2°F	3.611×10^{-2}		7.355×10^{-2}	2.54×10^{-3}	2.491×10^{-3}	2.458×10^{-3}	2.54	1.868	0.249
in. Hg @ 32°F	0.491	13.595		3.453×10^{-2}	3.396×10^{-2}	3.342×10^{-2}	34.527	25.4	3.386
kgf/cm^2	14.223	393.71	28.959		0.981	0.968	1000	735.56	98.066
bar	14.504	401.5	29.53	1.0197		0.987	1019.7	750.1	100
atmosphere (standard)	14.696	406.82	29.921	1033.2	1.0132		1033.2	760.04	101.325
cm H$_2$O @ 4°C	1.422×10^{-2}	0.394	2.896×10^{-2}	1.00×10^{-3}	9.806×10^{-4}	9.678×10^{-4}		0.736	9.807×10^{-2}
mm Hg @ 0°C	1.934×10^{-2}	0.535	3.937×10^{-2}	1.359×10^{-3}	1.333×10^{-3}	1.316×10^{-3}	1.359		0.133
kPa	0.145	4.015	0.295	10.197×10^{-3}	0.0100	9.869	10.197	7.501	

FIG. 8-1 Pressure conversion multipliers.

Two problems interfere with this, however:

1. Quantities may be expressed in digits that are not intended to be significant. The dimension 1.1875 in. may indeed be very precise, in which case the digit in the fourth place is significant, or it may be an exact decimalization of a rough dimension given as 1³⁄₁₆ in., in which case the dimension is given with too many decimal places relative to its intended precision.

2. Quantities may be expressed omitting significant zeros. The dimension 2 in. may mean "about 2 in.," or it may mean a very precise expression, which should be written 2.000 in. In the latter case, while the added zeros are not significant in establishing the value, they are very significant in expressing the proper intended precision.

It is necessary therefore to determine an approximate implied precision before rounding. This can be done by a knowledge of the circumstances or by information on the accuracy of the measuring equipment.

If the accuracy of measurement is known, it will provide a convenient lower limit to the precision of the dimension and, in some cases, may be the only basis for establishing precision. The implied precision should never be smaller than the accuracy of measurement.

A tolerance on a dimension will give a good indication of the intended precision, although the precision will, of course, be smaller than the tolerance. A dimension of 1.635 ± 0.003 in. obviously is intended to be quite precise, and the precision implied by the number of significant digits is correct (± 0.0005 in., total 0.001 in.). A dimension of 4.625 ± 0.125 in. is obviously a different matter. The use of thousandths of an inch to express a tolerance of 0.25 in. is probably the result of decimalization of fractions, and the expression is probably better written 4.62 ± 0.12 in., with an implied precision of ± 0.005 (total implied precision 0.01 in.). The circumstances, however, should be examined and judgment applied accordingly.

A rule of thumb that is often helpful for determining implied precision of a toleranced value is to assume the precision to be one-tenth of the tolerance. Because the implied precision of the converted value should be no greater than that of the original, the total tolerance should be divided by 10 and converted, and the proper significant digits retained in both the converted value and the converted tolerance so that the total implied precision is not reduced, that is, so that the last significant digit retained is in units no larger than one-tenth the converted total tolerance.

EXAMPLE: 200 ± 15 psi. Tolerance is 30 psi; divided by 10 is 3 psi; converted it is about 20.7 kPa. The conversion of the value is 1378.9514 ± 103.421 355 kPa, which should be rounded to units of 10 kPa, since 10 kPa is the largest unit smaller than one-tenth the converted tolerance. The conversion should be 1380 ± 100 kPa.

EXAMPLE: 25 ± 0.1 oz of alcohol. Tolerance is 0.2 oz; one-tenth of the tolerance is 0.02 oz; converted it is about 0.6 cm^3. The converted value (739.34 ± 2.957 cm^3) should be rounded to units of 0.1 cm^3, which yields 739.3 ± 3 cm^3.

Conversion Procedure

In the instructions that follow, the "total implied precision" mentioned above will be referred to as TIP.

1. First determine the TIP.
2. Convert the dimension, the TIP, and the tolerance, if any, by the accurate conversion factor shown in Tables 8-12 and 8-13.
3. Choose the smallest number of decimals to retain so that the last digit retained is in units equal to or smaller than the converted TIP.
4. Round off to this number of decimals by the following rules:

   a.   Where the next digit beyond the last digit to be retained is less than 5, the last digit retained should not be changed. For example, 4.46325 rounded to three decimal places would be 4.463.

   b.   Where the digits beyond the last digit to be retained amount to more than 5, the last digit retained should be increased by one. For example, 8.37652 rounded to three places would be 8.377.

   c.   Where the next digit beyond the last digit to be retained is exactly 5, the last digit retained, if it is even, is unchanged; but if it is odd, this last digit is increased by 1. For example, 4.36500 becomes 4.36 when rounded to two places, and 4.35500 becomes 4.36 when rounded to two places.

Conversion Procedure Examples

EXAMPLE: A rod 6 in. long. Estimate of TIP (assume the intended precision to be ± ⅟₁₆ in.): ⅛ in. Converted TIP = ⅛ × 25.4 = 3.17 mm. Units to use: 1 mm. 6 in. = 152.4 mm; round to 152 mm.

EXAMPLE: 50 000 psi tensile strength. Estimate of TIP: 400 psi, from nature of use and precision of measuring equipment. Converted TIP = 2.8 MPa. Units to use: 1 MPa. 50 000 psi = 344.737 85 MPa; round to 345 MPa.

EXAMPLE: A 12.125-in. length. Estimate of TIP: 0.06 in., from nature of use. Converted TIP = 1.524 mm. Units to use: 1 mm. 12.125 in. = 307.975 mm; round to 308 mm.

SI METRIC DRAWING PRACTICE

Because the purpose of a technical drawing is to communicate technical information from the designer of a part to the producer of that part, it is imperative that the information shown on the drawing be clearly stated so that in the interpretation of the drawing there is no possibility of ambiguity. Producing a technical drawing using SI metric units for dimensions and specifications requires only minimal reorientation for an experienced drafter; however, the basic rules of clear and concise information cannot be violated.

When the decision is made within an organization or industry to begin the transition to SI metric, an early indication must be made as to what procedure should be followed in dimensional representations on technical drawings. The dimensional procedure chosen can have a direct effect on the cost of the transition in regard to whether or not existing tooling can continue to be used even though it was established with the U.S. customary system of measurement. No manufacturing facility can justify retooling expressly to make the transition to SI metric. Because of this, a dimensional procedure must take into account what tooling can continue to be used, and also whether it can be used until it must be replaced regardless of what measurement system is involved. There is only one way to adopt SI metric, and that is to let it infiltrate into the normal procedure of design and manufacture with proper planning and organized direction. If an establishment is unfamiliar with SI metric, the technical drawing policy should probably accommodate both the U.S. customary system and the SI metric system of measurement for some time, after establishing and implementing an educational-awareness program for all personnel who will be involved in designing, delineating, and processing parts and components in a manufacturing facility.

Dimensional Procedures

Two dimensional procedures have become acceptable during the early stages of the SI metric transition. It is necessary to keep one precaution in mind, however: *Discontinue either or both of these procedures as soon as it is feasible to do so.* A third procedure will probably be used during the time when metric materials and components are being prepared to supersede inch materials and components. The fourth and final dimensional procedure occurs when the transition is complete.

All procedures recommended here, except the first, recognize that SI metric will be the primary measurement, and the U.S. customary the secondary and converted value, except as noted.

Procedure 1 A drawing that is adapted to the first procedure is a soft converted drawing that combines existing inch configurations of a part with configurations that can be accepted as being of SI metric design. This type of drawing would in effect take an existing drawing and add, adjacent to an inch dimension, the metric equivalent. (For further clarification see American National Standards Institute's ANSI Y14.5-1973, subsection 7, Dual Dimensioning.) The metric equivalent could, however, be slightly modified to give the appearance of being metric-oriented if such a dimension would not have any effect on the functional interchangeability of the part. For example, if a dimension were 0.75 in. the metric counterpart could be 19 mm rather than 19.05 mm; or, if a dimension were 0.625 in., the metric counterpart could perhaps be shown as 16 mm, rather than 15.88 mm or 15.9 mm. The metric dimension, with this procedure, would be enclosed with either parentheses or brackets, depending on whether it would be acceptable to verify the finished part by using the metric dimension shown. It must be clearly understood that any information or any dimension that is enclosed in parentheses is auxiliary and cannot be used for verification purposes. When brackets are used, the information or dimension so enclosed is interpreted as having the same status as the primary information or dimension. When brackets are used, parts are permitted to be manufactured with either dimension (as on a worldwide basis), but verification of any feature by inspection gages and equipment must be accomplished in the same system of measurement that was used to produce the part.

Procedure 2 A drawing that is adapted to the second procedure can show dimensions in either of two ways, as long as the primary dimensioning is principally SI metric. In either case, the inch value that is shown on the drawing is auxiliary. In one case, the inch value is shown in parentheses adjacent to the metric dimension (see Fig. 8-2). In the other case, the inch value is shown in a conversion table that is usually placed on the left side of the drawing (see Fig. 8-3).

There are instances in which, because of high costs or nonavailability of metric-oriented tools to produce the feature, or because it is more economical to use existing components, the inch identity must continue to be shown without a metric counterpart. Such is the case of the Unified inch thread in Figs. 8-2, 8-3, and 8-4. Here it can also be seen that a metric minor thread diameter is specified. Many drawings of parts, when a thread is a design requirement, do not show the tap drill size or the minor thread

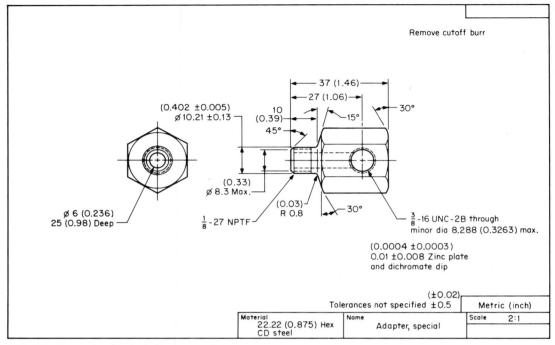

FIG. 8-2   Example of a dual-dimensioned drawing.

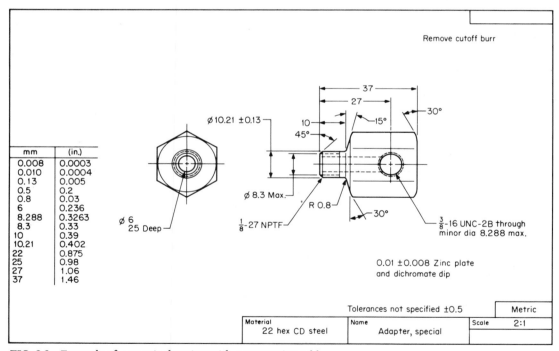

FIG. 8-3   Example of a metric drawing with a conversion table.

diameter; in a case such as this, however, when the minor thread diameter can be satisfactorily specified as metric, it is recommended that this be shown. Furthermore, it is more realistic to specify an effective thread height for a tapped hole to be 60 to 65 percent of the total that is specified in most internal thread tables. In this case, the design indicates that a reduction in the effective thread height is permitted to the size shown to reduce tool (thread-tap) breakage while retaining optimum thread strength.

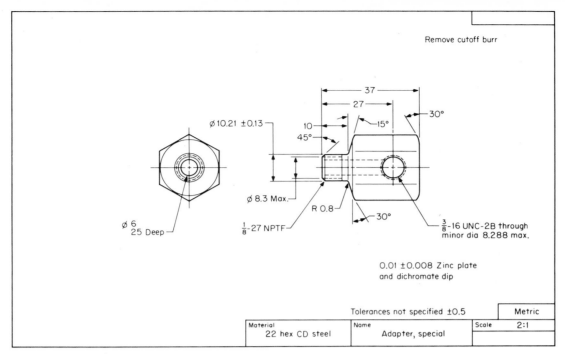

FIG. 8-4   Metric drawing.

It was earlier stated, and must be repeated because of the importance, that the design and dimensioning of a product must be shown in a manner that allows continued use of tools, jigs, and fixtures if the design does not change enough to require new tooling. Where a part is of a design that requires new tooling, regardless of what measurement system is primary, the cost of changing to SI metric can be considered to be virtually nil.

Where the design of a part can be identified with the SI metric system of measurement, care should be taken to keep from retaining the familiar inch increments. In other words, instead of specifying an increment of 0.25 inch, the increment must be 6 mm and specified on the drawing (according to this procedure) as 6(0.24); or, instead of using 1.00 inch, the increments should be 25 mm and specified on the drawing as 25(1.00) if the function of the feature is noncritical; if the function is critical, the increment would be given as 25(0.98).

Procedure 3   A drawing that is adapted to the third procedure follows the same fundamentals as the second procedure, except that no inch values appear on the drawing as a converted dimension. The use of Unified inch threads or inch-sized components may continue to control the primary design information, but otherwise the part appears to be of SI metric design (see Figs. 8-4 and 8-5).

Procedure 4   The fourth procedure, use of SI units exclusively, is followed when there is no longer any need to use inch fasteners or inch components. At this point it can be said that the transition to SI metric is complete.

Metric Measurement Application

After the decision has been made as to what type of dimensioning corresponds to the level of SI metric transition to be achieved, the following differences in drafting practice must be recognized as being indigenous to SI metric. These differences have been accepted as both the American National Standards Institute (ANSI) and the International Organization for Standardization (ISO).

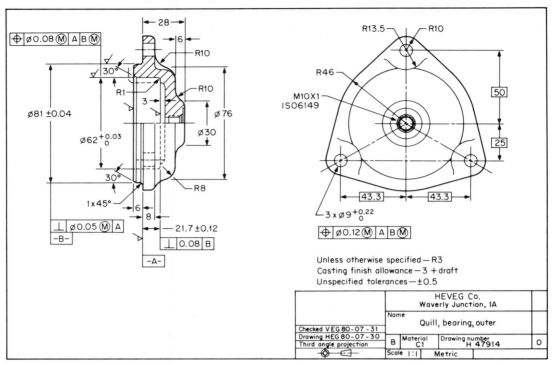

FIG. 8-5   Metric drawing.

1. *Types of dimensioning*   Dimensions are expressed as decimals, except where standardized fractional designations are necessary to identify available nominal material sizes such as for pipe or lumber.

2. *SI metric linear units*   All linear dimensions on a single drawing shall be shown in the same unit, that is, millimeter for mechanical engineering drawings, or meter for civil, structural, and architectural drawings. Units other than linear shall be shown with their symbols.

3. *Identification of linear units*   On drawings in which the majority of dimensions given are of one kind, individual identification is not required; however, a note stating that fact, such as, "Unless otherwise specified, all dimensions are in millimeters," or "Unless otherwise specified, all dimensions are in meters," shall appear on the drawing or in a document referenced on the drawing. Where the type of drawing subscribes to adjacent dual dimensioning, the title block of the drawing shall indicate the arrangement, such as "metric (inch)," "inch (metric)," or "inch [metric]," according to the implied precision of the measurements.

4. *SI metric dimensioning (mechanical engineering)*   The following shall be observed when specifying metric dimensioning on drawings:

a.   A zero precedes the decimal point for values less than one unit.

EXAMPLE: 0.15

b.   Where the dimension is a whole number, neither the decimal point nor a zero is shown.

c.   Zeros do not follow a significant digit of a value to the right of the decimal point, except where required to conform to applicable tolerancing procedures. See items *f* and *g*.

d.   Where equal plus-and-minus tolerancing is used, the metric dimension and its tolerance do not require the same number of decimal places.

EXAMPLE: 32 ± 0.008

e. Where unilateral tolerances are used and either the plus or minus tolerance is nil, a single zero is shown without a plus or minus sign.

EXAMPLE: $32 \begin{array}{c} 0 \\ -\ 0.02 \end{array}$ or $32 \begin{array}{c} +\ 0.02 \\ 0 \end{array}$

f. Where bilateral tolerancing is used, both the plus and minus tolerance have the same number of decimal places, using zeros where necessary.

EXAMPLE: $32 \begin{array}{c} +\ 0.25 \\ -\ 0.10 \end{array}$ not $32 \begin{array}{c} +\ 0.25 \\ -\ 0.1 \end{array}$

g. Where limit dimensioning is used and either the maximum or minimum value has digits following a decimal point, the value has zeros added for uniformity.

EXAMPLE: $\begin{array}{c} 25.45 \\ 25.00 \end{array}$ not $\begin{array}{c} 25.45 \\ 25 \end{array}$

h. A space will be used to separate digits into groups of three, to both the left and right of the decimal point. Commas will not be used.

EXAMPLE:
12 345	not	12,345
12 345 678	not	12,345,678
10.123 45	not	10.12345

If four digits are in a group, the space is optional, except in dimensional tables, where grouping is desirable for consistency.

EXAMPLE:
1234	optional	1 234
1234 567	optional	1 234 567
10.1234	optional	10.123 4

5. *SI metric dimensioning (civil, structural, and architectural engineering)* The instructions that apply to mechanical engineering drawings apply here as well, except for the following:

a. To differentiate between a millimeter value and a meter value, all meter numerical expressions shall be shown to three decimal places, regardless of the value's intended precision.

EXAMPLE: 1.200 or 25.000

General Drawing Principles

Except for the above differences in showing dimensions on technical drawings (inch versus millimeter or feet versus meters), all other drawing practices that have applied to the U.S. customary system of measurement can apply equally as well to an SI metric–oriented drawing. A certain number of ISO drawing practices have been studied for possible inclusion in the proposed revision of those ANSI standards for engineering drawings that are the responsibility of the American National Standards Committee (see the list below). Much of the reason for the trend to adopt ISO drawing practices is because of the need to communicate engineering design information on a worldwide basis. Language communication is difficult, but much of the difficulty can be overcome by the use of symbology rather than by words.

An example of the use of international symbology on engineering drawings is the almost total elimination of the notes that had described form and position tolerancing prior to the publication of ANSI Y14.5. The use of the diametral symbol ∅ instead of the abbreviation DIA, and the placement of that symbol or the letter R (for radius) preceding the dimensional value are examples of ISO influence. Additional symbology is being proposed as part of the next revision of ANSI Y14.5, and it is recommended that those drafting departments which desire to produce engineering drawings that can receive interpretation quality equal to that received by the organizations and facilities that are involved in helping to develop the national standards through the work of their donated personnel should acquire applicable national standards and make use of the standardized procedures contained therein.

Some of the ANSI standards for engineering drawing and related documentation practices that are available from The American Society of Mechanical Engineers (345 East 47th St., New York, NY 10017) are as follows:

Abbreviations	Y1.1-1972
Size and Format	Y14.1-1980
Line Conventions and Lettering	Y14.2m-1979
Multi and Sectional View Drawings	Y14.3-1975 (R1980)
Dimensioning and Tolerancing	Y14.5-1973
Screw Threads	Y14.6-1978
Gears and Splines	
Spur, Helical, and Rack	Y14.7.1-1971
Bevel and Hypoid	Y14.7.2-1978
Ground Vehicle Drawing Practices	
Chassis Frames	Y14.32.1-1974
Surface Texture Symbols	Y14.36-1978

International Standards

Many people have the false notion that because ISO standards are metric-oriented, the ISO standards automatically become the standard practices that will help to implement a transition to SI metric. This false notion has generated questions such as, "Do we change to first-angle projection?" or "Do we also adopt the ISO system of limits and fits?" or "Do we change our drawing paper size to the international paper sizes?"

To all these questions the answer is an emphatic *no!* None of the international practices, except perhaps ISO 1000, should be considered to be part of the SI metric transition. If an establishment feels that it is desirable to make use of certain ISO standards, that is its individual choice. In the case of the subject of limits and fits, an ANSI standard, ANSI B4.2, covers the subject for SI metric. Another standard, ANSI B4.3, covers general tolerances for metric dimensions. Both these standards were derived from ISO standards but are still not to be considered as an integral part of an SI metric transition plan.

Metric Threaded Fasteners

The improved international standard for metric threaded fasteners is still in the process of being finalized. The existing ISO standard, however (ISO 965, parts I, II, and III), satisfies the requirements needed to specify metric threads on an engineering drawing. Regardless of the outcome of the standards activity, the identification of a thread for a bolt, capscrew, or nut should remain the same.

A metric thread is first indicated with the capital letter M, followed by the nominal size and the pitch in millimeters. For example, M12 $\times$ 1.5 designates a 12-mm-diameter thread with a coarse 1.5-mm pitch (the pitch is the distance between adjacent threads).

In addition to the above size designation, the fit of a thread, in relation to its mating thread, is designated by grade numbers and letters. Smaller numbers carry closer tolerances, with numbers 4 through 8 applying to internal threads and numbers 3 through 9 applying to external threads. Grade 6 is preferred for both internal and external general-purpose threads. Letters designate the position of the thread tolerance relative to the basic diameters. The ISO standard indicated above provides an excellent reference and is obtained from ANSI.

DEFINITIONS

Base Units

meter   The length equal to 1 650 763.73 wavelengths, in vacuum, of the orange-red line of the spectrum of the krypton-86 atom.

kilogram   The standard for the unit of mass is a cylinder of platinum-iridium alloy kept by the International Bureau of Weights and Measures at Paris. A duplicate, in the custody of the National Bureau of Standards, serves as the mass standard for the United States. This is the only base unit defined by an artifact.

second   The duration of 9 192 631 770 periods of the radiation corresponding to the transition between two hyperfine levels of the ground state of the cesium 133 atom.

ampere   That constant current which, if maintained in two straight parallel conductors of infinite length and of negligible circular cross section placed 1 meter apart in a vacuum, would produce between these conductors a force equal to 2×10^{-7} newtons per meter of length.

kelvin   The fraction 1/273.16 of the thermodynamic temperature of the triple point of water. The temperature 0 (zero) K is called *absolute zero*.

candela   The luminous intensity, in the perpendicular direction, of a surface of 1/600 000 square meter of a blackbody at the temperature of freezing platinum (2045 K) under a pressure of 101 325 newtons per square meter.

mole   The amount of substance of a system that contains as many elementary entities as there are atoms in 0.012 kilogram of carbon 12.

Supplementary Units

radian   The plane angle between two radii of a circle which cut off on the circumference an arc equal in length to the radius.

steradian   The solid angle which, having its vertex in the center of a sphere, cuts off an area of the surface of the sphere equal to that of a square with sides of length equal to the radius of the sphere.

Derived Units with Special Names

hertz   The frequency of a periodic phenomenon of which the period is one second.

newton   That force which, when applied to a body having a mass of one kilogram, gives it an acceleration of one meter per second squared.

pascal   The pressure or stress of one newton per square meter.

joule   The work done when the point of application of a force of one newton is displaced a distance of one meter in the direction of the force.

watt   The power that gives rise to the production of energy at the rate of one joule per second.

coulomb   The quantity of electricity transported in one second by a current of one ampere.

volt   The difference of electric potential between two points of a conductor carrying a constant current of one ampere, when the power dissipated between these points is equal to one watt.

farad   The capacitance of a capacitor between the plates of which there appears a difference of potential of one volt when it is charged by a quantity of electricity equal to one coulomb.

ohm   The electric resistance between two points of a conductor when a constant difference of potential of one volt applied between these two points produces in this conductor a current of one ampere, this conductor not being the source of any electromotive force.

siemens   The electric conductance of a conductor in which a current of one ampere is produced by an electric potential difference of one volt.

weber   The magnetic flux which, linking a circuit of one turn, produces in it an electromotive force of one volt as it is reduced to zero at a uniform rate in one second.

tesla   The magnetic flux density given by a magnetic flux of one weber per square meter.

henry   The inductance of a closed circuit in which an electromotive force of one volt is produced when the electric current in the circuit varies uniformly at a rate of one ampere per second.

lumen   The luminous flux emitted in a solid angle of one steradian by a point source having a uniform intensity of one candela.

lux   The illuminance produced by a luminous flux of one lumen uniformly distributed over a surface of one square meter.

gray The energy imparted by ionizing radiation to a mass of matter corresponding to one joule per kilogram.

becquerel The activity of a radionuclide having one spontaneous nuclear transition per second.

degree Celsius The units of kelvin and Celsius temperature intervals are identical: A temperature expressed in degrees Celsius is equal to the temperature expressed in kelvin minus 273.15.

CONVERSION FACTORS WITH AN ALPHABETICAL LIST OF UNITS

The following fairly commonplace units, listed in alphabetical order in Table 8-12, are intended to serve two purposes:

1. To express the various units of measure as exact numerical multiples of coherent SI metric units. Relationships that are exact in terms of the base unit are followed by an asterisk. Relationships that are not followed by an asterisk are either the results of physical measurements or only approximate.

2. To provide multiplying factors for converting expressions of measurements of the U.S. customary system of measurement to corresponding SI metric units.

Conversion factors are presented for ready adaptation to computer readout and electronic data transmission. The factors are written as a number greater than one and less than ten with six or fewer decimal places. The letter following the number, E for exponent, a plus or minus symbol, and two digits, which indicate the power of 10 by which the number must be multiplied to obtain the correct value, are interpreted as follows:

$$3.523\ 907\ E - 02 \quad \text{is} \quad 3.523\ 907 \times 10^{-2}$$

or $\qquad\qquad\qquad 0.035\ 239\ 07$

$$3.386\ 389\ E + 03 \quad \text{is} \quad 3.386\ 389 \times 10^{3}$$

or $\qquad\qquad\qquad 3\ 386.389$

APPLICATION OF SI UNITS

Table 8-13 illustrates recommended SI metric use for application in the mechanical industries, particularly those related to automotive and agricultural industries. The particular recommendations are not mandatory but have been found to be quite acceptable and preferred.

Arrangement The unit applications are arranged in alphabetical order of quantities, by principal nouns. Thus, to find SI use for *surface tension*, look under *tension, surface;* or for *specific energy*, look under *energy, specific.*

Metric units Some unit expressions continue to be the same as at present, but the designation has been changed to remove a possible communication ambiguity. Such is the case, for instance, where revolutions per minute is now designated as rpm, whereas the future designation to be used in conjunction with the SI metric system is r/min. In other words, the use of "p" for "per" in a unit designation is no longer permitted. Instead, the solidus (/) is correct usage.

Rates and other derived quantities It is not practical to list all possible applications, but others, such as rates, can be derived. For example, if guidance is desired for *heat energy per unit volume*, looking up *energy and volume* will show the recommendation to be kJ/m³.

Conversion factors Conversion factors are shown from old units to SI metric units to seven significant digits (see Table 8-14).

TABLE 8-12 Alphabetical List of Units

(Symbols of SI units given in parentheses)

To convert from	to	Multiply by
abampere	ampere (A)	1.000 000*E+01
abcoulomb	coulomb (C)	1.000 000*E+01
abfarad	farad (F)	1.000 000*E+09
abhenry	henry (H)	1.000 000*E+09
abmho	siemens (S)	1.000 000*E+09
abohm	ohm (Ω)	1.000 000*E−09
abvolt	volt (V)	1.000 000*E−08
acre foot (U.S. survey)[a]	meter³ (m³)	1.233 489 E+03
acre (U.S. survey)	meter² (m²)	4.046 873 E+03
ampere hour	coulomb (C)	3.600 000*E+03
are	meter² (m²)	1.000 000*E+02
angstrom	meter (m)	1.000 000*E−10
astronomical unit	meter (m)	1.495 979 E+11
atmosphere (standard)	pascal (Pa)	1.013 250*E+05
atmosphere (technical = 1 kgf/cm²)	pascal (Pa)	9.806 650*E+04
bar	pascal (Pa)	1.000 000*E+05
barn	meter² (m²)	1.000 000*E−28
barrel (for petroleum, 42 gal)	meter³ (m³)	1.589 873 E−01
board foot	meter³ (m³)	2.359 737 E−03
British thermal unit (International Table)[b]	joule (J)	1.055 056 E+03
Btu (mean)	joule (J)	1.055 87 E+03
Btu (thermochemical)	joule (J)	1.054 350 E+03
Btu (39°F)	joule (J)	1.059 67 E+03
Btu (59°F)	joule (J)	1.054 80 E+03
Btu (60°F)	joule (J)	1.054 68 E+03
Btu (IT)·ft/h·ft²·°F (k, thermal conductivity)	watts per meter kelvin (W/m·K)	1.730 735 E+00
Btu (thermochemical)·ft/h·ft²·°F (k, thermal conductivity)	watts per meter kelvin (W/m·K)	1.729 577 E+00
Btu (IT)·in/h·ft²·°F (k, thermal conductivity)	watts per meter kelvin (W/m·K)	1.442 279 E−01
Btu (thermochemical)·in/h·ft²·°F (k, thermal conductivity)	watts per meter kelvin (W/m·K)	1.441 314 E−01
Btu (IT)·in/s·ft²·°F (k, thermal conductivity)	watts per meter kelvin (W/m·K)	5.192 204 E+02
Btu (thermochemical)·in/s·ft²·°F (k, thermal conductivity)	watts per meter kelvin (W/m·K)	5.188 732 E+02
Btu (IT)/h	watt (W)	2.930 711 E−01
Btu (thermochemical)/h	watt (W)	2.928 751 E−01
Btu (thermochemical)/min	watt (W)	1.757 250 E+01
Btu (thermochemical)/s	watt (W)	1.054 350 E+03
Btu (IT)/ft²	joule per meter² (J/m²)	1.135 653 E+04
Btu (thermochemical)/ft²	joule per meter² (J/m²)	1.134 893 E+04
Btu (thermochemical)/ft²·h	watt per meter² (W/m²)	3.152 481 E+00
Btu (thermochemical)/ft²·min	watt per meter² (W/m²)	1.891 489 E+02
Btu (thermochemical)/ft²·s	watt per meter² (W/m²)	1.134 893 E+04
Btu (thermochemical)/in²·s	watt per meter² (W/m²)	1.634 246 E+06
Btu (IT)/h·ft²·°F (C, thermal conductance)	watt per meter² kelvin (W/m²·K)	5.678 263 E+00
Btu (thermochemical)/h·ft²·°F (C, thermal conductance)	watt per meter² kelvin (W/m²·K)	5.674 466 E+00
Btu (IT)/s·ft²·°F	watt per meter² kelvin (W/m²·K)	2.044 175 E+04
Btu (thermochemical)/s·ft²·°F	watt per meter² kelvin (W/m²·K)	2.042 808 E+04
Btu (IT)/lb	joule per kilogram (J/kg)	2.326 000*E+03
Btu (thermochemical)/lb	joule per kilogram (J/kg)	2.324 444 E+03

TABLE 8-12 Alphabetical List of Units (*Continued*)

To convert from	to	Multiply by
Btu (IT)/lb·°F (c, heat capacity)	joule per kilogram kelvin (J/kg · K)	4.186 800*E+03
Btu (thermochemical)/lb·°F (c, heat capacity)	joule per kilogram kelvin (J/kg · K)	4.184 000 E+03
bushel (US)	meter³ (m³)	3.523 907 E−02
caliber (inch)	meter (m)	2.540 000*E−02
calorie (IT)	joule (J)	4.186 800*E+00
calorie (mean)	joule (J)	4.190 02 E+00
calorie (thermochemical)	joule (J)	4.184 000*E+00
calorie (15°C)	joule (J)	4.185 80 E+00
calorie (20°C)	joule (J)	4.181 90 E+00
calorie (kilogram, IT)	joule (J)	4.186 800*E+03
calorie (kilogram, mean)	joule (J)	4.190 02 E+03
calorie (kilogram, thermochemical)	joule (J)	4.184 000*E+03
cal (thermochemical)/min	watt (W)	6.973 333 E−02
cal (thermochemical)/s	watt (W)	4.184 000*E+00
cal (thermochemical)/cm²·min	watt per meter² (W/m²)	6.973 333 E+02
cal (thermochemical)/cm²·s	watt per meter² (W/m²)	4.184 000*E+04
cal (thermochemical)/cm·s·°C	watt per meter kelvin (W/m·K)	4.184 000*E+02
cal (thermochemical)/cm²	joule per meter² (J/m²)	4.184 000*E+04
cal (IT)/g	joule per kilogram (J/kg)	4.186 800*E+03
cal (thermochemical)/g	joule per kilogram (J/kg)	4.184 000*E+03
cal (IT)/g·°C	joule per kilogram kelvin (J/kg · K)	4.186 800*E+03
cal (thermochemical)/g·°C	joule per kilogram kelvin (J/kg · K)	4.184 000*E+03
carat (metric)	kilogram (kg)	2.000 000*E−04
centimeter of mercury (0°C)	pascal (Pa)	1.333 22 E+03
centimeter of water (4°C)	pascal (Pa)	9.806 38 E+01
centipoise	pascal second (Pa·s)	1.000 000*E−03
centistokes	meter² per second (m²/s)	1.000 000*E−06
circular mil	meter² (m²)	5.067 075 E−10
clo	kelvin meter² per watt (K·m²/W)	2.003 712 E−01
cup	meter³ (m³)	2.365 882 E−04
curie	becquerel (Bq)	3.700 000*E+10
day (mean solar)	second (s)	8.640 000 E+04
day (sidereal)	second (s)	8.616 409 E+04
degree (angle)	radian (rad)	1.745 329 E−02
degree Celsius	kelvin (K)	$t_K = t_{°C} + 273.15$
degree centigrade	same as degree Celsius	
degree Fahrenheit	degree Celsius (°C)	$t_{°C} = (t_{°F} - 32)/1.8$
degree Fahrenheit	kelvin (K)	$t_K = (t_{°F} + 459.67)/1.8$
degree Rankine	kelvin (K)	$t_K = t_{°R}/1.8$
°F·h·ft²/Btu (IT) (R, thermal resistance)	kelvin meter² per watt (K·m²/W)	1.761 102 E−01
°F·h·ft²/Btu (thermochemical) (R, thermal resistance)	kelvin meter² per watt (K·m²/W)	1.762 280 E−01
denier	kilogram per meter (kg/m)	1.111 111 E−07
dyne	newton (N)	1.000 000*E−05
dyne cm	newton meter (N·m)	1.000 000*E−07
dyne/cm²	pascal (Pa)	1.000 000*E−01
electronvolt	joule (J)	1.602 19 E−19
EMU of capacitance	farad (F)	1.000 000*E+09
EMU of current	ampere (A)	1.000 000*E+01
EMU of electric potential	volt (V)	1.000 000*E−08

To convert from	to	Multiply by
EMU of inductance	henry (H)	1.000 000*E−09
EMU of resistance	ohm (Ω)	1.000 000*E−09
ESU of capacitance	farad (F)	1.112 650 E−12
ESU of current	ampere (A)	3.335 6 E−10
ESU of electric potential	volt (V)	2.997 9 E+02
ESU of inductance	henry (H)	8.987 554 E+11
ESU of resistance	ohm (Ω)	8.978 554 E+11
erg	joule (J)	1.000 000*E−07
erg/cm²·sec	watt per meter² (W/m²)	1.000 000*E−03
erg/sec	watt (W)	1.000 000*E−07
faraday (based on carbon-12)	coulomb (C)	9.648 70 E+04
faraday (chemical)	coulomb (C)	9.649 57 E+04
faraday (physical)	coulomb (C)	9.652 19 E+04
fathom	meter (m)	1.828 8 E+00
fermi (femtometer)	meter (m)	1.000 000*E−15
fluid ounce (U.S.)	meter³ (m³)	2.957 353 E−05
foot (U.S. survey)	meter (m)	3.048 006 E−01
foot of water (39.2°F)	pascal (Pa)	2.988 98 E+03
foot	meter (m)	3.048 000*E−01
ft²	meter² (m²)	9.290 304*E−02
ft²/h (thermal diffusivity)	meter² per second (m²/s)	2.580 640*E−05
ft²/sec	meter² per second (m²/s)	9.290 304*E−02
ft³ (volume; section modulus)	meter³ (m³)	2.831 685 E−02
ft³/min	meter³ per second (m³/s)	4.719 474 E−04
ft³/sec	meter³ per second (m³/s)	2.831 685 E−02
ft⁴ (moment of section)c	meter⁴ (m⁴)	8.630 975 E−03
ft/h	meter per second (m/s)	8.466 667 E−05
ft/min	meter per second (m/s)	5.080 000*E−03
ft/sec	meter per second (m/s)	3.048 000*E−01
ft/sec²	meter per second squared (m/s²)	3.048 000*E−01
footcandle	lux (lx)	1.076 391 E+01
footlambert	candela per meter squared (cd/m²)	3.426 259 E+00
ft·lbf	joule (J)	1.355 818 E+00
ft·lbf/hr	watt (W)	3.766 161 E−04
ft·lbf/min	watt (W)	2.259 697 E−02
ft·lbf/sec	watt (W)	1.355 818 E+00
ft·poundal	joule (J)	4.214 011 E−02
free fall, standard (g)	meter per second squared (m/s²)	9.806 650*E+00
gal	meter per second squared (m/s²)	1.000 000*E−02
gallon (Canadian liquid)	meter cubed (m³)	4.546 090 E−03
gallon (U.K. liquid)	meter cubed (m³)	4.546 092 E−03
gallon (U.S. liquid)	meter cubed (m³)	3.785 412 E−03
gallon (U.S. dry)	meter cubed (m³)	4.404 884 E−03
gallon (U.S. liquid)/day	meter³ per second (m³/s)	4.381 264 E−08
gallon (U.S. liquid)/min	meter³ per second (m³/s)	6.309 020 E−05
gallon (U.S. liquid)/hp·hr (SFC, specific fuel consumption)	meter³ per joule (m³/J)	1.410 089 E−09
gamma	tesla (T)	1.000 000*E−09
gauss	tesla (T)	1.000 000*E−04
gilbert	ampere (A)	7.957 747 E−01
gill (U.K.)	meter cubed (m³)	1.420 654 E−04
gill (U.S.)	meter cubed (m³)	1.182 941 E−04
grad	degree (angular)	9.000 000*E−01
grad	radian (rad)	1.570 796 E−02
grain (1/7000 lb avoirdupois)	kilogram (kg)	6.479 891*E−05
grain (lb avoirdupois/7000/gallon) (U.S. liquid)	kilogram per meter³ (kg/m³)	1.711 806 E−02

TABLE 8-12 Alphabetical List of Units *(Continued)*

To convert from	to	Multiply by
gram	kilogram (kg)	1.000 000*E−03
g/cm³	kilogram per meter³ (kg/m³)	1.000 000*E+03
gram-force/cm²	pascal (Pa)	9.806 650*E+01
hectare	meter squared (m²)	1.000 000*E+04
horsepower (550 ft·lbf/s)	watt (W)	7.456 999 E+02
horsepower (boiler)	watt (W)	9.809 50 E+03
horsepower (electric)	watt (W)	7.460 000*E+02
horsepower (metric)	watt (W)	7.354 99 E+02
horsepower (water)	watt (W)	7.460 43 E+02
horsepower (UK)	watt (W)	7.457 0 E+02
hour (mean solar)	second (s)	3.600 000 E+03
hour (sidereal)	second (s)	3.590 170 E+03
hundredweight (long)	kilogram (kg)	5.080 235 E+01
hundredweight (short)	kilogram (kg)	4.535 924 E+01
inch	meter (m)	2.540 000*E−02
in²	meter squared (m²)	6.451 600*E−04
in³ (volume; section modulus)^d	meter cubed (m³)	1.638 706 E−05
inch of mercury (32°F)	pascal (Pa)	3.386 38 E+03
inch of mercury (60°F)	pascal (Pa)	3.376 85 E+03
inch of water (39.2°F)	pascal (Pa)	2.490 82 E+02
inch of water (60°F)	pascal (Pa)	2.488 4 E+02
in³/min	meter³ per second (m³/s)	2.731 177 E−07
in⁴ (moment of section)^e	meter⁴ (m⁴)	4.162 314 E−07
in/sec	meter per second (m/s)	2.540 000*E−02
in/sec²	meter per second squared (m/s²)	2.540 000*E−02
kayser	1 per meter (1/m)	1.000 000*E+02
kelvin	degree Celsius	$t_{°C} = t_K − 273.15$
kilocalorie (IT)	joule (J)	4.186 800*E+03
kilocalorie (mean)	joule (J)	4.190 02 E+03
kilocalorie (thermochemical)	joule (J)	4.184 000*E+03
kilocalorie (thermochemical)/min	watt (W)	6.973 333 E+01
kilocalorie (thermochemical)/sec	watt (W)	4.184 000*E+03
kilogram-force (kgf)	newton (N)	9.806 650*E+00
kgf·m	newton meter (N·m)	9.806 650*E+00
kgf·s²/m (mass)	kilogram (kg)	9.806 650*E+00
kgf/cm²	pascal (Pa)	9.806 650*E+04
kgf/m²	pascal (Pa)	9.806 650*E+00
kgf/mm²	pascal (Pa)	9.806 650*E+06
km/h	meter per second (m/s)	2.777 778 E−01
kilopond	newton (N)	9.806 650*E+00
kW h	joule (J)	3.600 000*E+06
kip (1000 lbf)	newton (N)	4.448 222 E+03
kip/in² (ksi)	pascal (Pa)	6.894 757 E+06
knot (international)	meter per second (m/s)	5.144 444 E−01
lambert	candela per meter² (cd/m²)	$1/\pi$ *E+04
lambert	candela per meter² (cd/m²)	3.183 099 E+03
langley	joule per meter² (J/m²)	4.184 000*E+04
league	meter (m)	4.828 042 E+03
light year	meter (m)	9.460 55 E+15
liter^f	meter (m)	1.000 000*E−03
maxwell	weber (Wb)	1.000 000*E−08
mho	siemens (S)	1.000 000*E+00
microinch	meter (m)	2.540 000*E−08
micron	meter (m)	1.000 000*E−06
mil	meter (m)	2.540 000*E−05
mile (international)	meter (m)	1.609 344*E+03
mile (statute)	meter (m)	1.609 3 E+03
mile (U.S. survey)	meter (m)	1.609 347 E+03

To convert from	to	Multiply by
mile (international nautical)	meter (m)	1.852 000*E+03
mile (U.K. nautical)	meter (m)	1.853 184*E+03
mile (U.S. nautical)	meter (m)	1.852 000*E+03
mile² (international)	meter² (m²)	2.589 988 E+06
mile² (U.S. survey)	meter² (m²)	2.589 998 E+06
mi/h (international)	meter per second (m/s)	4.470 400*E−01
mi/h (international)	kilometer per hour (km/h)	1.609 344*E+00
mi/min (international)	meter per second (m/s)	2.682 240*E+01
mi/sec (international)	meter per second (m/s)	1.609 344*E+03
millibar	pascal (Pa)	1.000 000*E+02
millimeter of mercury (0°C)	pascal (Pa)	1.333 22 E+02
minute (angle)	radian (rad)	2.908 882 E−04
minute (mean solar)	second (s)	6.000 000 E+01
minute (sidereal)	second (s)	5.983 617 E+01
month (mean calendar)	second (s)	2.628 000 E+06
oersted	ampere per meter (A/m)	7.957 747 E+01
ohm centimeter	ohm meter (Ω·m)	1.000 000*E−02
ohm circular-mil per ft	ohm millimeter² per meter (Ω·mm²/m)	1.662 426 E−03
ounce (avoirdupois)	kilogram (kg)	2.834 952 E−02
ounce (troy or apothecary)	kilogram (kg)	3.110 348 E−02
ounce (U.K. fluid)	meter cubed (m³)	2.841 307 E−05
ounce (U.S. fluid)	meter cubed (m³)	2.957 353 E−05
ounce-force	newton (N)	2.780 139 E−01
ozf·in	newton meter (N·m)	7.061 552 E−03
oz(avoirdupois)/gal (U.K. liquid)	kilogram per meter³ (kg/m³)	6.236 021 E+00
oz(avoirdupois)/gal (U.S. liquid)	kilogram per meter³ (kg/m³)	7.489 152 E+00
oz(avoirdupois)/in³	kilogram per meter³ (kg/m³)	1.729 994 E+03
oz(avoirdupois)/ft²	kilogram per meter² (kg/m²)	3.051 517 E−01
oz(avoirdupois)yd²	kilogram per meter² (kg/m²)	3.390 575 E−02
parsec	meter (m)	3.085 678 E+16
peck (U.S.)	meter cubed (m³)	8.809 768 E−03
pennyweight	kilogram (kg)	1.555 174 E−03
perm (0°C)	kilogram per pascal second meter² (kg/Pa·s·m²)	5.721 35 E−11
perm (23°C)	kilogram per pascal second meter² (kg/Pa·s·m²)	5.745 25 E−11
perm·in (0°C)	kilogram per pascal second meter (kg/Pa·s·m)	1.453 22 E−12
perm·in (23°C)	kilogram per pascal second meter (kg/Pa·s·m)	1.459 29 E−12
phot	lumen per meter² (lm/m²)	1.000 000*E+04
pica (printer's)	meter (m)	4.217 518 E−03
pint (U.S. dry)	meter cubed (m³)	5.506 105 E−04
pint (U.S. liquid)	meter cubed (m³)	4.731 765 E−04
point (printer's)	meter (m)	3.514 598*E−04
poise (absolute viscosity)	pascal second (Pa·s)	1.000 000*E−01
pound (lb avoirdupois)ᵍ	kilogram (kg)	4.535 924 E−01
pound (troy or apothecary)	kilogram (kg)	3.732 417 E−01
lb·ft² (moment of inertia)	kilogram meter² (kg·m²)	4.214 011 E−02
lb·in² (moment of inertia)	kilogram meter² (kg·m²)	2.926 397 E−04
lb/ft·hr	pascal second (Pa·s)	4.133 789 E−04
lb/ft·sec	pascal second (Pa·s)	1.488 164 E+00
lb/ft²	kilogram per meter² (kg/m²)	4.882 428 E+00
lb/ft³	kilogram per meter³ (kg/m³)	1.601 846 E+01
lb/gal (U.K. liquid)	kilogram per meter³ (kg/m³)	9.977 633 E+01
lb/gal (U.S. liquid)	kilogram per meter³ (kg/m³)	1.198 264 E+02
lb/hr	kilogram per second (kg/s)	1.259 979 E−04
lb/hp·hr (SFC, specific fuel consumption)	kilogram per joule (kg/J)	1.689 659 E−07

TABLE 8-12 Alphabetical List of Units *(Continued)*

To convert from	to	Multiply by
lb/in³	kilogram per meter³ (kg/m³)	2.767 990 E+04
lb/min	kilogram per second (kg/s)	7.559 873 E−03
lb/sec	kilogram per second (kg/s)	4.535 924 E−01
lb/yd³	kilogram per meter³ (kg/m³)	5.932 764 E−01
poundal	newton (N)	1.382 550 E−01
poundal/ft²	pascal (Pa)	1.488 164 E+00
poundal s/ft²	pascal second (Pa·s)	1.488 164 E+00
pound-force (lbf)[h]	newton (N)	4.448 222 E+00
lbf·ft	newton meter (N·m)	1.355 818 E+00
lbf·ft/in	newton meter per meter (N·m/m)	5.337 866 E+01
lbf·in	newton meter (N·m)	1.129 848 E−01
lbf·in/in	newton meter per meter (N·m/m)	4.448 222 E+00
lbf·s/ft²	pascal second (Pa·s)	4.788 026 E+01
lbf/ft	newton per meter (N/m)	1.459 390 E+01
lbf/ft²	pascal (Pa)	4.788 026 E+01
lbf/in	newton per meter (N/m)	1.751 268 E+02
lbf/in² (psi)	pascal (Pa)	6.894 757 E+03
lbf/lb (thrust/weight [mass] ratio)	newton per kilogram (N/kg)	9.806 650 E+00
quart (U.S. dry)	meter cubed (m³)	1.101 221 E−03
quart (U.S. liquid)	meter cubed (m³)	9.463 529 E−04
rad (radiation dose absorbed)	gray (Gy)	1.000 000*E−02
rhe	1 per pascal second (1/Pa·s)	1.000 000*E+01
rod	meter (m)	5.029 210 E+00
roentgen	coulomb per kilogram (C/kg)	2.58 E−04
second (angle)	radian (rad)	4.848 137 E−06
second (sidereal)	second (s)	9.972 696 E−01
section	meter² (m²)	2.589 998 E+06
shake	second (s)	1.000 000*E−08
slug	kilogram (kg)	1.459 390 E+01
slug/ft·s	pascal second (Pa·s)	4.788 026 E+01
slug/ft³	kilogram per meter³ (kg/m³)	5.153 788 E+02
statampere	ampere (A)	3.335 640 E−10
statcoulomb	coulomb (C)	3.335 640 E−10
statfarad	farad (F)	1.112 650 E−12
stathenry	henry (H)	8.987 554 E+11
statmho	siemens (S)	1.112 650 E−12
statohm	ohm (Ω)	8.987 554 E+11
statvolt	volt (V)	2.997 925 E+02
stere	meter cubed (m³)	1.000 000*E+00
stilb	candela per meter² (cd/m²)	1.000 000*E+04
stokes (kinematic viscosity)	meter² per second (m²/s)	1.000 000*E−04
tablespoon	meter cubed (m³)	1.478 676 E−05
teaspoon	meter cubed (m³)	4.928 922 E−06
tex	kilogram per meter (kg/m)	1.000 000*E−06
therm	joule (J)	1.055 056 E+08
ton (assay)	kilogram (kg)	2.916 667 E−02
ton (long, 2240 lb)	kilogram (kg)	1.016 047 E+03
ton (metric)	kilogram (kg)	1.000 000*E+03
ton (nuclear equivalent of TNT)[i]	joule (J)	4.184 E+09
ton (refrigeration)	watt (W)	3.516 800 E+03
ton (register)	meter cubed (m³)	2.831 685 E+00
ton (short, 2000 lb)	kilogram (kg)	9.071 847 E+02
ton (long)/yd³	kilogram per meter³ (kg/m³)	1.328 939 E+03
ton (short)/hr	kilogram per second (kg/s)	2.519 958 E−01
ton-force (2000 lbf)	newton (N)	8.896 444 E+03
tonne	kilogram (kg)	1.000 000*E+03

To convert from	to	Multiply by
torr (mm Hg, 0°C)	pascal (Pa)	1.333 22 E+02
township	meter squared (m²)	9.323 994 E+07
unit pole	weber (Wb)	1.256 637 E−07
W·hr	joule (J)	3.600 000*E+03
W·sec	joule (J)	1.000 000*E+00
W/cm²	watt per meter² (W/m²)	1.000 000*E+04
W/in²	watt per meter² (W/m²)	1.550 003 E+03
yard	meter (m)	9.144 000*E−01
yd²	meter squared (m²)	8.361 274 E−01
yd³	meter cubed (m³)	7.645 549 E−01
yd³/min	meter³ per second (m³/s)	1.274 258 E−02
year (calendar)	second (s)	3.153 600 E+07
year (sidereal)	second (s)	3.155 815 E+07
year (tropical)	second (s)	3.155 693 E+07

Note: An asterisk after the sixth decimal place indicates that the conversion factor is exact.
[a]For an explanation of the U.S. basis of length measurement, see ASTM E380-76.
[b]This value was adopted in 1956. Some of the earlier International Tables use the value 1.055 04 E+03. The exact conversion factor is 1.055 055 852 62*E+03.
[c]This is sometimes called moment of inertia of a plane section about a specified axis.
[d]The exact conversion is 1.638 706 4*E−05.
[e]This is sometimes called moment of inertia of a plane section about a specified axis.
[f]In 1964 the General Conference on Weights and Measures adopted the name *liter* as a special name for the cubic decimeter. Prior to this decision the liter differed slightly (previous value, 1.000 028 dm³), and in expression of precision volume measurement this fact must be kept in mind.
[g]The exact conversion factor is 4.535 923 7*E−01.
[h]The exact conversion factor is 4.448 221 615 260 5*E+00.
[i]Defined (not measured) value.

TABLE 8-13 Application of SI Units

Quantity	Typical application	From old units (U.S. customary and CGS)	To SI metric units	Multiply by
Acceleration, angular	General	rad/s²	rad/s²	no change
Acceleration, linear	Vehicle	(mile/h)/s	(km/h)/s	1.609 344*[a]
	General (includes acceleration of gravity)[b]	ft/s²	m/s²	0.304 8*
Angle, plane	Rotational calcuations	r (revolution)	r (revolution)	no change
		rad	rad	no change
	Geometric and general	° (deg)	° (deg)	no change
		′ (min)	decimalize	1/60*
		″ (sec)	decimalize	1/3600*
Angle, solid	Illumination calculations	sr	sr	no change
Area	Cargo platforms, frontal areas, fabrics, general	in²	m²	0.000 645 16*
		ft²	m²	0.092 903 04*
	Small areas, orifices	in²	mm²	645.16*
	Brake and clutch contact area, glass, radiators	in²	cm²	6.451 6*
	Land area	acre	ha[c]	0.404 687 3
Bending moment	(See *Moment of force*)			
Capacitance, electric	Capacitors	μF	μF	no change
Capacity, electric	Battery rating	A·h	A·h	no change
Capacity, heat	General	Btu/°F	kJ/K[d]	1.899 101
Capacity, heat specific	General	Btu/(lb·°F)	kJ/(kg·K)[d]	4.186 8*
Capacity, volume	(See *Volume*)			

TABLE 8-13 Application of SI Units *(Continued)*

Quantity	Typical application	From old units (U.S. customary and CGS)	To SI metric units	Multiply by
Coefficient of heat transfer	General	Btu/(h·ft²·°F)	W/(m²·K)[d]	5.678 263
Coefficient of linear expansion	Shrink fit, general	°F⁻¹, (1/°F)	K⁻¹, (1/K)[d]	1.8*
Conductance, electric	General	mho	S	1*
Conductance, thermal	(See *Coefficient of heat transfer*)			
Conductivity, electric	Material property	mho/ft	S/m	3.280 840
Conductivity, thermal	General	Btu·ft/(h·ft²·°F)	W/(m·K)[d]	1.730 735
Consumption, fuel	(See *Efficiency, fuel*)			
Consumption, oil	Vehicle performance testing	qt/1000 miles	L/1000 km	0.588 036 4
Consumption, specific fuel	(See *Efficiency, fuel*)			
Consumption, specific oil	Engine testing	lb/(hp·h)	g/(kW·h)	608.277 4
		lb/(hp·h)	g/MJ	168.965 9
Current, electric	General	A	A	no change
Density, current	General	A/in²	kA/m²	1.550 003
		A/ft²	A/m²	10.763 91
Density, magnetic flux	General	kilogauss	T	0.1*
Density (mass)	Solid	lb/yd³	kg/m³	0.593 276 3
		lb/in³	kg/m³	27 679.90
		lb/ft³	kg/m³	16.018 46
		ton (short)/yd³	t/m³	1.186 553
		ton (long)/yd³	t/m³	1.328 939
	Liquid	lb/gal	kg/L	0.119 826 4
	Gas	lb/ft³	kg/m³	16.018 46
Density of heat flow rate	Irradiance, general	Btu/(h·ft²)	W/m²	3.154 591
Drag	(see *Force*)			
Economy, fuel	(See *Efficiency, fuel*)			
Efficiency, fuel	Highway vehicles			
	Economy	mile/gal	km/L	0.425 143 7
	Consumption	mile/gal	L/100 km	e
	Off-highway vehicles	hp·h/gal	kW·h/L	0.196 993 1
		lb/(hp·h)	g/(kW·h)	608.277 4
		lb/(hp·h)	kg/MJ	0.168 965 9
Energy, work, enthalpy, quantity of heat	Impact strength	ft·lbf	J	1.355 818
	Heat	Btu	kJ	1.055 056
	Energy usage	kW·h	kW·h	no change
		kW·h	MJ	3.6*
	Mechanical, general	ft·lbf	J	1.355 818
		ft·pdl	J	0.042 140 11
		hp·h	MJ	2.684 520
Energy, specific	General	cal/g[f]	J/g	4.186 8*
		Btu/lb	kJ/kg	2.326*
Enthalpy	(see *Energy*)			
Entropy	(See *Capacity, heat*)			
Entropy, specific	(See *Capacity, heat specific*)			

Quantity	Typical application	From old units (U.S. customary and CGS)	To SI metric units	Multiply by
Floor loading	(See *Mass per area*)			
Flow, heat, rate	(See *Power*)			
Flow, mass, rate	General	lb/min	kg/min	0.453 592 4
		lb/sec	kg/s	0.453 592 4
Flow, volume	Air, gas, general	ft³/sec	m³/s	0.028 316 85
		ft³/sec	m³/min	1.699 011
	Liquid flow, pump capacity	gal/sec	L/s	3.785 412
		gal/sec	m³/s	0.003 785 412
		gal/min	L/min	3.785 412
	Seal and packing leakage	oz/sec	mL/s	29.573 53
		oz/min	mL/min	29.573 53
Flux, luminous	Light bulbs	lm	lm	no change
Flux, magnetic	Coil rating	maxwell	Wb	0.000 000 01*
Force, thrust, drag	Pedal, spring, belt, hand lever, general	lbf	N	4.448 222
		ozf	N	0.278 013 9
		pdl	N	0.138 255
		kgf	N	9.806 650
	Drawbar, breakout	lbf	kN	0.004 448 222
Frequency	System, sound and electrical	Mc/sec	MHz	1*
		kc/sec	kHz	1*
		Hz, c/s	Hz	no change
	Mechanical events, rotational	r/sec (rps)	s⁻¹, r/s	no change
		r/min (rpm)	min⁻¹, r/min	no change
Hardness	Conventional hardness numbers, BHN, R, etc., not affected by change to SI			
Heat	(See *Energy*)			
Heat capacity	(See *Capacity, heat*)			
Heat capacity, specific	(See *Capacity, heat specific*)			
Heat flow rate	(See *Power*)			
Heat flow, density of	(See *Density of heat flow*)			
Heat, specific	General	cal/g^g	kJ/kg	4.186 8*
		Btu/lb	kJ/kg	2.326*
Heat transfer coefficient	(See *Coefficient of heat transfer*)			
Illuminance, illumination	General	fc	lx	10.763 91
Impact strength	(See *Strength, impact*)			
Inductance, electric	Filters and chokes, permeance	H	H	no change
Intensity, luminous	Light bulbs	candlepower	cd	1*
Intensity, radiant	General	W/sr	W/sr	no change
Leakage	(See *Flow, volume*)			
Length	Land distance, maps, odometers	mile	km^h	1.609 344*
	Field size, turning circle, braking distance, cargo platforms	rod	m^h	5.029 210 1*
		yd	m	0.914 4*
		ft	m	0.304 8*

TABLE 8-13 Application of SI Units *(Continued)*

Quantity	Typical application	From old units (U.S. customary and CGS)	To SI metric units	Multiply by
	Engineering drawings, engineering part specifications, motor vehicle dimensions	in	mm	25.4*
	Coating thickness, filter rating	mil	μm	25.4*
		μin	μm	0.025 4*
		micron	μm	1*
	Surface texture			
	Roughness, average	μin	μm	0.025 4*
	Roughness sampling length, waviness height, and spacing	in	mm	25.4*
	Radiation wavelengths, optical measurements (interference)	μin	nm	25.4*
Load	(See *Mass*)			
Luminance	Brightness	foot lambert	cd/m^2	3.426 259
Magnetization	Coil field strength	A/in	A/m	39.370 08
Mass	Vehicle mass (weight) axle rating, rated load, tire load, lifting capacity, tipping load, load, general	ton (long)	Mg, t	1.016 047
		ton (short)	Mg, t	0.907 184 7
		lb	kg	0.453 592 4
		slug	kg	14.593 90
	Small mass	oz	g	28.349 52
Mass per area	Fabric, surface coatings	oz/yd^2	g/m^2	33.905 75
		lb/ft^2	kg/m^2	4.882 428
		oz/ft^2	g/m^2	305.151 7
	Floor loading	lb/ft^2	kg/m^2	4.882 428
Mass per length	General	lb/ft	kg/m	1.488 164
		lb/yd	kg/m	0.496 054 7
Modulus of elasticity	General	lbf/in^2	MPa	0.006 894 757
Modulus of rigidity	(See *Modulus of elasticity*)			
Modulus, section	General	in^3	mm^3	16 387.06
		in^3	cm^3	16.387 06
Moment, bending	(See *Moment of force*)			
Moment of area, second	General	in^4	mm^4	416 231.4
		in^4	cm^4	41.623 14
Moment of force, torque, bending moment	General, engine torque, fasteners	lbf·in	N·m	0.112 984 8
		lbf·ft	N·m	1.355 818
		kgf·cm	N·m	0.098 066 5*
	Locks, light torque	ozf·in	mN·m	7.061 552
Moment of inertia	Flywheel, general	lb·ft^2	kg·m^2	0.042 140 11
Moment of mass	Unbalance	oz·in	kg·mm	0.720 077 8
Moment of momentum	(See *Momentum, angular*)			
Moment of section	(See *Moment of area, second*)			
Momentum	General	lb·ft/sec	kg·m/s	0.138 255
Momentum, angular	Torsional vibration	lb·ft^2/sec	kg·m^2/s	0.042 140 11
Permeability	Magnetic core properties	H/ft	H/m	3.280 840

Quantity	Typical application	From old units (U.S. customary and CGS)	To SI metric units	Multiply by
Permeance	(See *Inductance*)			
Potential, electric	General	V	V	no change
Power	General, light bulbs	W	W	no change
	Air conditioning, heating	Btu/min	W	17.584 27
		Btu/h	W	0.293 071 1
	Engine, alternator, drawbar, power takeoff, general	hp (550 ft·lbf/sec)	kW	0.745 699 9
Pressure	All pressures except very small	lbf/in²	kPa	6.894 757
		inHg (60°F)	kPa	3.376 85
		inH₂O (60°F)	kPa	0.248 84
		mmHg (0°C)	kPa	0.133 322
		kgf/cm²	kPa	98.066 5*
		bar	kPa	100*
	Very small pressures (high vacuum)	lbf/in²	Pa	6 894.757
Pressure, sound, level	Accoustical measurement	dB	dB	no change
Quantity of electricity	General	C	C	no change
Radiant intensity	(See *Intensity, radiant*)			
Resistance, electric	General	Ω	Ω	no change
Resistivity, electric	General	Ω·ft	Ω·m	0.304 8*
		Ω·ft	Ω·cm	30.48*
Sound pressure level	(See *Pressure, sound level*)			
Speed	(See *Velocity*)			
Spring rate, linear	General spring properties	lbf/in	N/mm	0.175 126 8
Spring, rate torsional	General	lbf·ft/deg	N·m/deg	1.355 818
Strength, field, electric	General	V/ft	V/m	3.280 840
Strength, field, magnetic	General	oersted	A/m	79.577 47
Strength, impact	Materials testing	ft·lbf	J	1.355 818
Stress	General	lbf/in²	MPa	0.006 894 757
Surface tension	(See *Tension, surface*)			
Temperature	General use	°F	°C	$t_{°C} = (t_{°F} - 32)/1.8*$
	Absolute temperature, thermodynamics, gas cycles	°R	K	$T_K = T_{°R}/1.8*$
Temperature interval	General use	°F	K^i	1 K = 1°C = 1.8°F*
Tension, surface	General	lbf/in	mN/m	175 126.8
		dyne/cm	mN/m	1*
Thrust	(See *Force*)			
Time	General	s	s	no change
		h	h	no change
		min	min	no change
Torque	(See *Moment of force*)			
Toughness, fracture	Metal properties	ksi $\sqrt{in}$	MPa·m$^{1/2}$	1.098 843
Vacuum	(See *Pressure*)			
Velocity, angular	(See *Velocity, rotational*)			

TABLE 8-13   Application of SI Units   (*Continued*)

Quantity	Typical application	From old units (U.S. customary and CGS)	To SI metric units	Multiply by
Velocity, linear	Vehicle	mile/h	km/h	1.609 344*
	General	ft/s	m/s	0.304 8*
		ft/min	m/min	0.304 8*
		in/s	mm/s	25.4*
Velocity, rotational	General	rad/s	rad/s	no change
		r/sec	r/s	1*
		r/min	r/min	no change
Viscosity, dynamic	General liquids	centipoise	mPa·s	1*
Viscosity, kinematic	General liquids	centistokes	mm²/s[j]	1*
Volume	Truck body, shipping or freight, general	yd³	m³	0.764 554 9
		ft³	m³	0.028 316 85
		bushel	m³	0.035 239 07
	Automobile luggage capacity	ft³	L	28.316 85
	Gas pump displacement, air compressor, small gaseous, air reservoir	in³	cm³	16.387 06
	Engine displacement			
	Large engines	in³	L	0.016 387 06
	Small engines	in³	cm³	16.387 06
	Liquid—fuel, lubricant, etc.	gal	L	3.785 412
		qt	L	0.946 352 9
		pt	L	0.473 176 5
	Small liquid	oz	mL	29.573 53
	Irrigation, reservoir	acre·ft	ha·m[c]	0.123 348 9
Weight	May mean either mass or force—see text			
Work	(See *Energy*)			
Young's modulus	(See *Modulus of elasticity*)			

[a]The asterisk indicates that conversation factor is exact.

[b]Standard acceleration of gravity = 9.806 650 m/s², exactly.

[c]Official use in surveys and cartography involves the U.S. survey mile (frequently called the U.S. statute mile), based on the U.S. survey foot, which is longer than the international foot by two parts per million. The factors used here for acre, acre foot, and rod are based on the U.S. survey foot. Factors for all other old length-based units are based on the international foot. For further details on this matter, see ASTM E380-76.

[d]In these expressions K indicates temperature interval. Therefore K may be replaced with °C if desired without changing the value or affecting the conversion factor. kJ/(kg·K) = kJ/(kg·°C).

[e]Convenient conversion: 235.215 ÷ (mile/gal) = L/100 km.

[f]Not to be confused with kcal/g. kcal is often called calorie.

[g]Standard acceleration of gravity = 9.806 650 m/s², exactly.

[h]Official use in surveys and cartography involves the U.S. survey mile (frequently called the U.S. statute mile), based on the U.S. survey foot, which is longer than the international foot by two parts per million. The factors used here for acre, acre foot, and rod are based on the U.S. survey foot. Factors for all other old length-based units are based on the international foot. For further details of this matter, see ASTM E380-76 or IEEE Std 268-1976.

[i]In this expression K indicates temperature interval. Therefore K may be replaced with °C if desired without changing the value or affecting the conversion factor. kJ/(kg·K) = kJ/(kg·°C).

[j]Viscosity is frequently expressed in SUS (Saybolt Universal Seconds). SUS is the time in seconds for 60 mL of fluid to flow through a standard orifice at a specified temperature. Conversion between kinematic viscosity, mm²/s (centistokes) and SUS can be made by referring to ANSI Z11.129-1975, American National Standard Method for Conversion of Kinematic Viscosity to Saybolt Universal Viscosity or to Saybolt Furol Viscosity.

TABLE 8-14 Pressure Conversions

(1 psi = 6.894 757 kPa)

psi	kPa (conversion)	kPa (practical)	psi	kPa (conversion)	kPa (practical)
1	6.894 757	7	56	386.106 392	385
2	13.789 514	14	57	393.001 149	395
3	20.684 271	21	58	399.895 906	400
4	27.579 028	28	59	406.790 663	405
5	34.473 785	35	60	413.685 420	415
6	41.368 542	40	61	420.580 177	420
7	48.263 299	50	62	427.474 934	430
8	55.158 056	55	63	434.369 691	435
9	62.052 813	60	64	441.264 448	440
10	68.947 57	70	65	448.159 205	450
11	75.842 327	75	66	455.053 962	455
12	82.737 084	85	67	461.948 719	460
13	89.631 841	90	68	468.843 476	470
14	96.526 598	95	69	475.738 233	475
15	103.421 355	105	70	482.632 990	480
16	110.316 112	110	71	489.527 747	490
17	117.210 869	115	72	496.422 504	495
18	124.105 626	125	73	503.317 261	505
19	131.000 383	130	74	510.212 018	510
20	137.895 14	140	75	517.106 775	515
21	144.789 897	145	76	524.001 532	525
22	151.684 654	150	77	530.896 289	530
23	158.579 411	160	78	537.791 046	540
24	165.474 168	165	79	544.685 803	545
25	172.368 925	170	80	551.580 560	550
26	179.263 682	180	81	558.475 317	560
27	186.158 439	185	82	565.370 074	565
28	193.053 196	195	83	572.264 831	570
29	199.947 953	200	84	579.159 588	580
30	206.842 71	205	85	586.054 345	585
31	213.737 467	215	86	592.949 102	595
32	220.632 224	220	87	599.843 859	600
33	227.526 981	230	88	606.738 616	605
34	234.421 738	235	89	613.633 373	615
35	241.316 495	240	90	620.528 130	620
36	248.211 252	250	91	627.422 887	625
37	255.106 009	255	92	634.317 644	635
38	262.000 766	260	93	641.212 401	640
39	268.895 523	270	94	648.107 158	650
40	275.790 28	275	95	655.001 915	655
41	282.685 037	280	96	661.896 672	660
42	289.579 794	290	97	668.791 429	670
43	296.474 551	295	98	675.686 186	675
44	303.369 308	305	99	682.580 943	685
45	310.264 065	310	100	689.475 7	690
46	317.158 822	320	125	861.844 625	860
47	324.053 579	325	150	1 034.213 550	1 035
48	330.948 336	330	175	1 206.582 475	1 205
49	337.843 093	340	200	1 378.951 400	1 380
50	344.737 85	345	225	1 551.320 325	1 550
51	351.632 607	350	250	1 723.689 250	1 725
52	358.527 364	360	275	1 896.058 175	1 900
53	365.422 121	365	300	2 068.427 100	2 070
54	372.316 878	370	320	2 206.322 240	2 205
55	379.211 635	380	325	2 240.796 025	2 240

TABLE 8-14   Pressure Conversions (*Continued*)

psi	kPa (conversion)	kPa (practical)	psi	kPa (conversion)	kPa (practical)
335	2 309.743 595	2 310	1 800	12 410.562 6	12 410
350	2 413.164 950	2 415	1 950	13 444.776 15	13 445
375	2 585.533 875	2 585	2 000	13 789.514	13 790
400	2 757.902 800	2 760	2 100	14 478.989 7	14 480
425	2 930.271 725	2 930	2 150	14 823.727 55	14 825
450	3 102.640 650	3 100	2 200	15 168.465 4	15 170
475	3 275.009 575	3 275	2 250	15 513.203 25	15 515
500	3 447.378 500	3 450	2 300	15 857.941 1	15 860
525	3 619.747 425	3 620	2 350	16 202.678 95	16 200
550	3 792.116 350	3 790	2 400	16 547.416 8	16 550
575	3 964.485 275	3 965	2 450	16 892.154 65	16 890
600	4 136.854 200	4 135	2 500	17 236.892 5	17 240
625	4 309.223 125	4 310	2 700	18 615.843 9	18 615
650	4 481.592 050	4 480	3 000	20 684.271	20 685
675	4 653.960 975	4 655	3 500	24 131.649 5	24 130
700	4 826.329 900	4 825	3 600	24 821.125 2	24 820
725	4 998.698 825	5 000	3 900	26 889.552 3	26 890
750	5 171.067 750	5 170	4 000	27 579.028	27 580
775	5 343.436 675	5 345	4 500	31 026.406 5	31 025
800	5 515.805 600	5 515	4 800	33 094.833 6	33 095
825	5 688.174 525	5 690	5 000	34 473.785	34 475
850	5 860.543 450	5 860	5 500	37 921.163 5	37 920
875	6 032.912 375	6 035	6 000	41 368.542	41 370
900	6 205.281 300	6 205	6 500	44 815.920 5	44 815
925	6 377.650 225	6 380	7 000	48 263.299	48 265
950	6 550.019 150	6 550	7 500	51 710.677 5	51 710
975	6 722.388 075	6 720	8 000	55 158.056	55 160
1 000	6 894.757 000	6 895	8 500	58 605.434 5	58 605
1 200	8 273.708 400	8 275	9 000	62 052.813	62 050
1 250	8 618.446 250	8 620	9 500	65 500.191 5	65 500
1 500	10 342.135 500	10 340	10 000	68 947.57	68 950
1 725	11 893.455 820	11 895			

REFERENCES

The following documents are recommended to all who propose to make the transition to, or to become acquainted with, SI metric.

The International System of Units (SI), NBS Special Publication 330 (order by SD Catalog no. C13.10:330/4).

Metric Practice, ASTM E380.

Recommended Practice for the Use of Metric (SI) Units in Building Design and Construction, NBS Technical Note 938 (order by SD Catalog no. C13.46:938).

SI Units and Recommendations for the Use of Their Multiples and of Certain Other Units, ISO 1000 (available from ANSI).

Identifications and addresses of sponsoring bodies are listed below.

ANSI—American National Standards Institute, Inc., 1430 Broadway, New York, NY 10018.

ASME—The American Society of Mechanical Engineers, 345 East 47th Street, New York, NY 10017.

ASTM—American Society for Testing and Materials, 1916 Race Street, Philadelphia, PA 19103.

NBS—National Bureau of Standards. These documents can be ordered from the Superintendent of Documents, U.S. Government Printing Office, Washington, DC 20402.

It is advisable to carefully review additional material in establishing an SI reference library. The most appropriate source of advice concerning metric information is the American National Metric Council, 1625 Massachusetts Avenue, N.W., Washington, DC 20036.

Interactive Graphics

Robert L. Myers

Interactive graphics (IAG) is a technology that brings the power of computers to design and drafting. Combining interactive computing techniques with graphic input/output allows the user to communicate with the computer using pictures (drawings), i.e., the drafting language, not the obscure combination of coded numbers that is the computer's language. Interactive computing means that the computer interacts with the user—receiving inputs, reacting to them instantaneously, and outputting the results of the computations to the user within seconds. When input devices automatically convert sketches and drawings into the required internal computer format and the output devices produce drawings, a computer can help relieve the user from much of the repetitious, error-prone tasks a drafter faces.

Interactive graphics is also referred to as computer-aided design (CAD) or computer-aided design and drafting (CADD). As the name *computer-aided* implies, an IAG system helps an engineer, designer, or drafter do the job, but it does not replace human intellect, intuition, or experience. By performing such routine tasks as tracing, dimensioning, sectioning, and extracting material takeoffs, an IAG system increases the productivity of the users; it does not not replace them.

The range of equipment configurations that can be called an IAG system is quite large. A terminal to help a mechanical designer create an axle, analyze it using stress-analysis programs, and then produce a paper tape to machine it using numerical controlled (NC) machine tools will be quite different from a digitizing system used to input a printed-circuit-board (PCB) layout and output a magnetic tape to generate photo-plotted artwork. The techniques and equipment described in this chapter can be combined in a variety of configurations, depending on the application of the system.

I. GLOSSARY

In addition to the terms defined within the body of this section, there are certain terms that require clarification. There are also synonyms used by various suppliers that can cause confusion. These synonyms are listed here, along with the definitions.

alignment   A function used to line up drawing elements horizontally or vertically.

alphanumeric display   A cathode-ray tube (CRT) at a work station used to display menus, prompting, and other alphanumeric information, or a CRT that is used as a control console for an entire system.

application units   The units of measure for the drawing database.

attentioned; selected   An item currently active and highlighted by a white block background, by blinking on and off, or with special marking.

auxiliary menu   Operating-level functions for top-level software; a subset.

calibrate   A function through which drawing, menu, and tablet areas may be located and drawing scale set.

center point   A point that is an equal distance from all points on the circumference of an arc.

construction point   A temporary position used to construct drawing elements.

coordinate   The number value of a position located in a system using an X axis and a Y axis.

copy   To make a duplicate of a drawing element.

cursor   A plus sign (+) or other symbol on the graphics display screen, used in pointing and selecting.

deactivate; store   The ability to save a drawing after completion.

delete; erase   To remove a drawing element.

digitizer   A large graphics tablet used to convert raw graphics data into a digital format.

drawing dump   A function used to print the contents of the drawing database.

drawing elements   The object geometry, text, and symbols of which a drawing consists.

drawing file   A drawing in digital format.

drawing file library   An area in the drawing file used to store symbols for that drawing.

drawing status   A function through which drawing levels, grids, and intensities can be created and displayed.

edit   To add to, delete from, and/or rearrange.

end point   A construction point that indicates the termination of a particular line, arc, or curve segment.

font   The style of text characters.

free pick   To place the construction points of a drawing element anywhere on the display screen.

full view   A function used to display graphically the entire contents of the drawing database.

gap   An invisible line segment that connects two visible line segments on the same line.

graphics display screen   The CRT at a work station that displays the drawing.

graphics tablet   An electronic surface with which, by using a stylus, the operator controls the screen cursor.

grid   A network of visible or invisible horizontal and vertical lines or tick marks used to aid drawing construction.

grid pick   A function through which the construction points of a drawing element are connected to the closest grid intersection point.

group; symbol   Two or more drawing elements that are treated as one item.

intensity   The degree of brightness of a displayed item.

job accounting   The summary of operation data for specific drawing files such as drawing time.

joystick   A picture controller that allows panning and zooming and/or controls the cursor.

key-in   To signal the IAG system via the keyboard.

kind   A miscellaneous field in which a qualifying parameter can be specified.

level; layer   An independent layer on a drawing.

major grid   A grid in which the grid line spacing is a multiple of either a reference or a working grid.

master library   A storage facility for symbols that may be accessed by multiple operators at multiple work stations.

master menu   A list of top-level functions within a single application.

menu   A set of drawing functions.

midpoint   The middle point of a displayed line or arc.

mirror   To reverse the image of an object.

move   To change the position of a display drawing element.

move to point   A function that allows a gap to be created on a line when multisegmented lines are created.

multisegmented lines   A function that creates a series of connected line segments.

nonorthogonal   Lines in which the construction points may be anywhere on the display screen.

object geometry; primitives   Special lines, lines, arcs, and points.

object snap   A function that connects the construction points of a drawing element to the drawing element closest to it.

offset grid   A function that allows all grids to be moved a specified distance in the X or Y axis.

operator's console   An alphanumeric terminal that provides an independent communication path to the central processing unit (CPU).

orthogonal lines   Lines in which the construction points are aligned horizontally or vertically.

orthogonal snap   A function which forces drawing elements to be aligned horizontally or vertically.

pan   To move a drawing across the display screen to view all parts of it.

pen number   The pen-station number of a plotter.

peripherals   Various input/output devices.

picture controller   A hardware box with a joystick that allows panning and zooming.

picture processor   A graphics or drawing processor.

plot   To produce hard-copy graphics output.

plot queue   A series of drawings in an active file awaiting plotting.

pointing   Aligning the cursor with a drawing element.

predefined   Automatically selected by default.

preexisting point   A position that has been previously established and recorded.

prompt   Instructions on the operator action required to implement a function.

properties; attributes   The nongraphics characteristics of a drawing element.

puck   A device with a reticle that is used on a digitizer for pointing, selecting drawing elements, and digitizing graphics data.

reference grid   A display network of vertical and horizontal lines or tick marks used to aid drawing element placement.

reject   A function used to cancel the last work station operation.

rotate   To swivel a drawing element in a clockwise or counterclockwise direction.

scale   To change the size of a drawing element.

selecting; picking   Signaling that a drawing element should be recorded and processed.

step and repeat   A function used to copy an object a specified number of times at a specified distance.

stylus   An electric pencillike device used with a graphics tablet for positioning the cursor on the graphics display.

symbols; groups   Multiple geometry and text treated as one unit.

system processing facility   The central processing unit (CPU), operator's console, and mass storage devices.

text   Alphabetic, numeric, and puncutuation characters.

title block   The area on a drawing that contains the drawing number, revision, date, etc.

vertex   The point at which two lines intersect.

window; viewport   A rectangular area inside a drawing.

working grid   A network of lines or tick marks to which the construction points of drawing elements are automatically attached.

workstation   The device at which the operator can display and manipulate graphics data.

X axis   On a drawing, the horizontal axis.

Y axis   On a drawing, the vertical axis.

Z axis   On a drawing, the axis in and out of the screen.

zooming   Magnifying an area of a drawing on the graphics display screen.

II. THEORY OF OPERATION

2.0 Introduction

Computer graphics consists of devices that bidirectionally change graphic information to digital data with the power of a computer. Interactive graphics uses interactive computing techniques, which allow almost instantaneous feedback. Once a geometric shape has been entered into the IAG system, the operator can quickly and accurately manipulate the shape while viewing the results graphically. Sophisticated software is used to perform repetitive analytical and/or manipulative operations at the operator's discretion.

Since its advent, IAG has become widely accepted as a cost-effective tool, and in some instances requirements for highly precise graphics have dictated its use.

2.1 Database Characteristics

When a drawing is transformed into a digital representation internal to the computer, the collection of digital numbers that describes the graphics and any related information is referred to as a *database*.

2.1.1 Primitives or Drawing Elements

Every shape, no matter how complex, is composed of a few basic geometric components. The following sections describe the basic geometric building blocks of an IAG system.

2.1.1.1 Points   Points (see Fig. 9-1) are used to establish any uniquely definable location. Points have no length, width, or depth, merely a location.

2.1.1.2 Polygons   A polygon (see Fig. 9-2) is defined as a simple straight line defined by its two end points (see 2.1.1.1) or as a connected series of many straight line segments. At every instance of a polygon change of direction (the point where straight line segments meet), the point of directional change must be defined by a vertex.

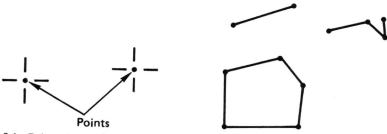

Points

FIG. 9-1   Points. Vertical and horizontal lines are for locating points.

FIG. 9-2   Polygons.

2.1.1.3 Arcs and Circles   Arcs and circles (see Fig. 9-3) are defined by their radius, center location, and start and stop points. Variations of these parameters are usually possible.

2.1.1.4 Polyarcs   Polyarcs (see Fig. 9-4) are the combination of straight line segments and arc segments to create a unique graphic entity.

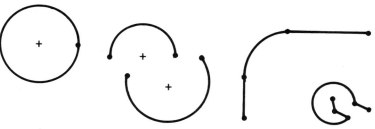

FIG. 9-3   Arcs and circles.

FIG. 9-4   Polyarcs.

2.1.1.5 Paths   Paths (see Fig. 9-5) are polygons (see 2.1.1.2) with a defined width. The polygon describes their centerline, or a specific side of the path.

2.1.1.6 Conic Sections   Conical shapes (see Fig. 9-6) can be defined by their type, location or focus, reference plane(s), and start and stop points.

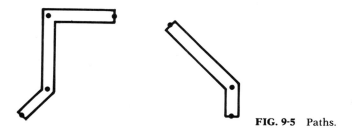

FIG. 9-5   Paths.

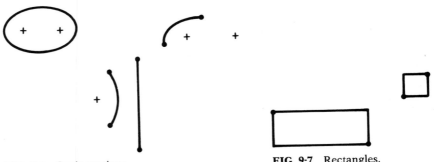

FIG. 9-6   Conic sections.   **FIG. 9-7**   Rectangles.

2.1.1.7 Rectangles   Rectangles (see Fig. 9-7) consist of four orthogonal straight line segments. A rectangle may be specified by the location of diagonally opposite corners. The IAG system ensures the orthogonality of the sides.

2.1.1.8 Curves   Curves require the use of polynomial spline techniques or other fitting techniques for data to represent graphical objects.

2.1.2 Symbols or Groups

Symbols are combinations of primitives defined as a unique entity. A symbol can be simple or complex. Once the symbol is defined, the primitives lose their individual identity and are treated as part of an entity. Figure 9-8 is a symbol comprising a polygon and a circle.

2.1.2.1 Nesting of Symbols   In the same manner that primitives are combined to create symbols (see 2.1.2), symbols may be combined, or nested, to form more complex symbols. Figure 9-9 shows a symbol created by nesting the four symbols described in Fig. 9-8.

2.1.3 Library

A library is a collection of all symbols and primitives associated with a drawing file (see 2.1.4.1).

2.1.4 Drawing File

A drawing file is a coordinated collection of data needed to define a unique drawing.

2.1.4.1 Library References   Drawing files are created by specifying geometric entities from a library and indicating their location and orientation. If an entity is repeated several times on a drawing, it is defined only once in the library, and the required number of references are made.

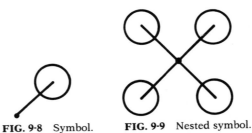

FIG. 9-8   Symbol.   **FIG. 9-9**   Nested symbol.

2.1.5 Layers or Levels

All but the most elementary databases are layered. Layers may be thought of as sheets of very fine vellum that are overlaid with exact registration and no stretch. Information may be added to, deleted from, modified, and displayed on any layer individually or in any combination simultaneously. Symbols and primitives (see 2.1.1 and 2.1.2) can be defined on multiple layers (see Fig. 9-10).

2.1.6 Grids

The computer in an IAG system is designed to manipulate numbers and equations, not geometric shapes. Grids are a means of describing points in space by numbers and are a necessary component of IAG. Every point (see 2.1.1.1), every start, vertex, and stop locations of polygons, every center, start, and stop points of arcs, etc., must coincide with a grid point.

2.1.6.1 Cartesian Coordinate System With reference to Fig. 9-11, the horizontal line represents the X axis ($y = 0$), and the vertical line the Y axis ($x = 0$). The location of a point in the XY plane is specified by the orthogonal distance to the X and Y axes, the X coordinate being the distance to the Y axis and the Y coordinate the distance to the X axis. Defining a point in three dimensions requires a third axis, called the Z axis. It is perpendicular to the X and Y axes.

2.1.6.2 Polar Coordinate System Any point in a plane can be defined by its distance from an arbitrary point called the *origin* and its angle in a given direction (see Fig. 9-12*a*). Three-dimensional polar coordinates require a distance and the angles from two reference directions (see Fig. 9-12*b*).

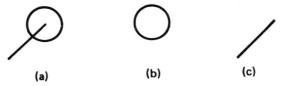

(a)   (b)   (c)

FIG. 9-10 The symbol, *a*, is a multilayered symbol. The circle, *b* is on layer 1, and the polygon, *c*, is on layer 5.

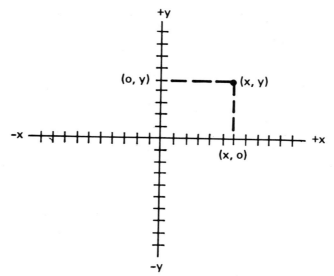

FIG. 9-11 Cartesian coordinates.

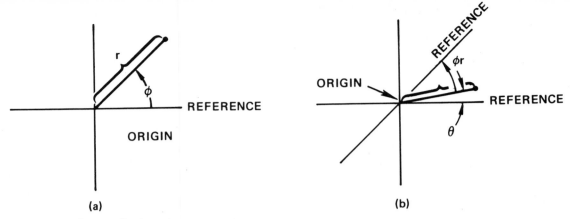

FIG. 9-12   Polar coordinates.

2.1.6.3 Basic Grid, Resolution, or Working Grid   The specified size of the basic grid increments determines the accuracy to which an object is defined and its maximum size. If, for example, the grid system contains 10,000 divisions in each axis, and each division is defined as 0.001 in., a graphic entity can be described to an accuracy of ± 0.001 in. The maximum size of the entity or combination of entities in any direction for this grid relationship is 10 in.

2.1.6.4 Reference Grid   A reference grid is a grid whose size is a multiple of the basic grid. This grid is easily changed to facilitate use.

2.1.7 Text

Although text is also made up of primitives, it is not treated in the same manner as graphic data are. If the shape "E" in Fig. 9-13 were entered as the letter E, even though E is a symbol composed of four polygons (see 2.1.2 and 2.1.1.2), it would be treated as the letter E.

2.1.7.1 Fonts   Any text font can be drawn by an IAG system; however, since arcs take longer to draw than polygons (see 2.3.2.4) do, most IAG systems use characters made from straight line segments. Fonts can usually be created by the operator.

2.1.7.2 Size   Text size is easily specified and/or changed.

2.1.7.3 Orientation   Text can be in a straight line, at any angle, or associated with a graphic entity such as a polyarc (see 2.1.1.4), in which case the text parallels the associated object (see Fig. 9-14).

2.1.7.4 Text Strings   More than one character of text can be manipulated by the IAG system as a single entity. An arbitrary number of characters can be a text string and attached to one text node (see 2.1.7.5).

2.1.7.5 Nodes   Text strings in a graphic system database (see 2.1) are located by a reference point, referred to as a *text node*. If a text node is part of a symbol definition (see 2.1.2), the text experiences the same graphic manipulations as the symbol does.

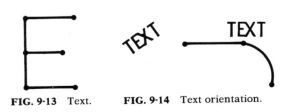

FIG. 9-13   Text.   **FIG. 9-14**   Text orientation.

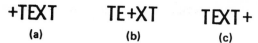

FIG. 9-15 Text justification: *a*, left; *b*, center; and *c*, right.

2.1.7.6 Justification Text strings can be associated with a location, either left-, center-, or right-justified (see Fig. 9-15).

2.1.7.7 Off-Line Text Entry Entering text does not require graphic resources, but it does prevent the operator from using the graphic capabilities of the IAG system when involved with text. Various methods are available to create text strings on other devices and merge them with the graphic data in a drawing file, minimizing the system time required for text entry.

2.2 Creating a Database

Creating a database is the conversion of graphic, text, and numeric information into digital data and organizing this information into logical groupings.

2.2.1 Manual Digitizing

A general definition of digitizing is "converting to numbers," which includes filling out computer forms for punched cards and many other methods. In IAG terminology, manual digitizing implies using a digitizer (see 3.1).

2.2.1.1 Calibration Graphic information input to an IAG system must be associated with its real-world size in order to correlate the scale of the material to be digitized with the basic grid of the system. If at least three points (see 2.1.1.1) on a plane are indicated and the coordinates of each are defined, the actual size of the object to be digitized is equated to a given number of grid units. Referring to Fig. 9-16, the coordinates 0,0 are entered, then point 1 is physically located by the digitizer (see 3.1), and points 2 and 3 are entered in the same manner. Any other point on the plane can now be located proportionately from these three points. Point 4 will have an x value equal to $(a/b) \times 100$ and a y value of $(c/d) \times 100$.

2.2.1.1.1 Scale When a digitizer is calibrated (see 2.2.1.1), the magnitude of actual space between the three points is related to size as defined by the number of basic grid units in the database. If 3 in. on the digitizer surface were equated to 0.003 in. in the database, the geometric figure being digitized would be interpreted to be represented at a scale of 1000$\times$, a thousand times larger.

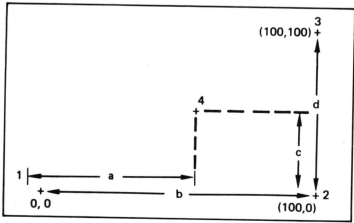

FIG. 9-16 Calibrating a digitizer.

2.2.1.2 Using Grids while Digitizing   When a graphic entity is defined to a computer, all descriptive points such as the start and stop points of arcs (see 2.1.1), the vertexes of polygons and polyarcs, and the corners of rectangles, must be coincident with a basic grid point. Digitizing the shapes in Figs. 9-1 through 9-8 would require locating each of the indicated points on grid.

2.2.1.2.1 Grid Lock   Referring to Fig. 9-17, the dots numbered 1 through 9 represent the grid, and points *a*, *b*, and *c* are points (see 2.1.1.1) to be digitized. When a point indicated is not on the grid, it is snapped to the nearest grid. Point *b* would be snapped to grid point 3. Point *c*, being equidistant from four grid points, could be snapped to any one of the four.

2.2.1.2.2 Dead Band   The finer the grid (see 2.1.6), the more likely a point to be digitized will be located so that it could be snapped to more than one grid point (see 2.2.1.2.1). To help reduce the ambiguous area between grid points, a dead band area is established (see Fig. 9-18). Any attempt to digitize a point in a dead band area results in the point being refused by the IAG system. Point *b* is accepted as being on grid point 3, but point *c* is not accepted. When an attempt to digitize a point is refused, the IAG system indicates that a dead band error has occurred (see 4.1.5).

2.2.1.2.3 Attainable Accuracy   Accuracy attainable with a digitizer cursor (see 3.1.5.2) is limited by reticle cross hair width, accuracy of the graphic material being digitized, and the parallax associated with the human eye. A reasonable limit to the minimum positioning of the digitizer is 30 divisions per inch. If more accurate digitizing is required a larger scale input must be used (see 2.2.1.1.1).

2.2.1.2.4 Speed   The coarser a digitizing grid (see 2.2.1.2), the larger will be the area around a given grid point that snaps the digitized point (see 2.1.1.1) to the grid point. Conversely, a fine grid requires more effort to locate the point to be digitized; therefore the operator must digitize at a slower rate.

2.2.1.3 Object Definition, or Geometry   The different types of primitives (see 2.1.1) are described by varying numbers of points. Each unique method of describing a geometric entity requires a different approach to digitizing.

2.2.1.3.1 Points   Digitizing a point (see 2.1.1.1), requires locating the point using the digitizer cursor (see 3.1.5.2) and notifying the IAG system when the cursor is in the proper position.

2.2.1.3.2 Polygons   Digitizing polygons (see 2.1.1.2) requires informing the system that a polygon is to be entered; digitizing each vertex as a point (see 2.2.1.3.1), starting at one end and continuing sequentially to the other end; and informing the IAG system when the last vertex has been entered (see Fig. 9-19).

2.2.1.3.3 Arcs and Circles   Arcs and circles are digitized by entering the start and stop points of the arc (these points are coincident for a circle) and the center as points

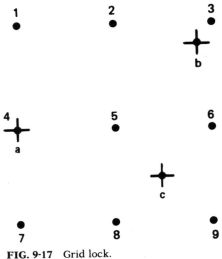

FIG. 9-17   Grid lock.

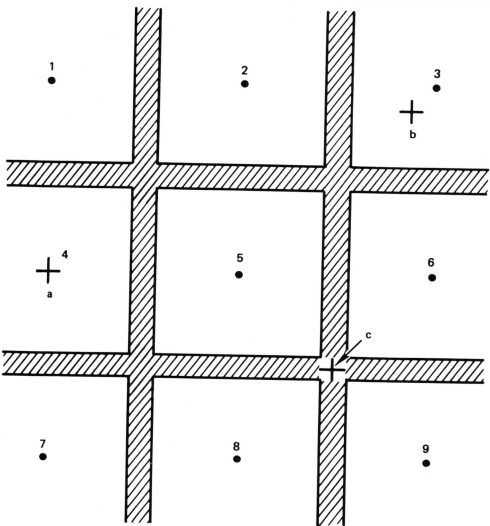

FIG. 9-18 Dead band.

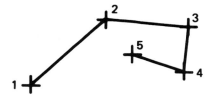

FIG. 9-19 Polygon definition.

(see 2.2.1.3.1). Another method is to digitize as points the start and stop points of the arc and another point on the arc (see Fig. 9-20).

2.2.1.3.4 Polyarcs Polyarcs are digitized in the same manner as polygons (see 2.2.1.3.2), except that the IAG system must be told whether the next points to be input are to define an arc segment (see 2.2.1.3.3) or a polygon segment (see Fig. 9-21).

2.2.1.3.5 Paths Digitizing a path (see Fig. 9-22) follows the same procedure used for digitizing a polygon (see 2.2.1.3.2).

2.2.1.3.6 Conic Sections Conic sections are digitized by entering the start, stop points, and focuses of the curve as points (see 2.2.1.3.1) and any reference lines as polygons (see 2.2.1.3.2 and Fig. 9-23).

2.2.1.3.7 Rectangles Rectangles are digitized by entering two opposing diagonal corners as points (see 2.2.1.3.1 and Fig. 9-24).

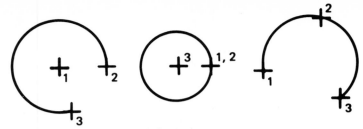

FIG. 9-20   Arc and circle definition.

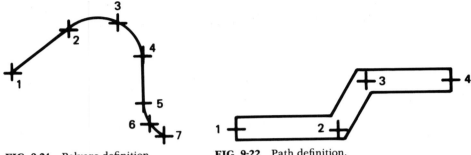

FIG. 9-21   Polyarc definition.   **FIG. 9-22**   Path definition.

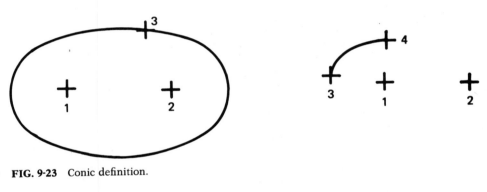

FIG. 9-23   Conic definition.

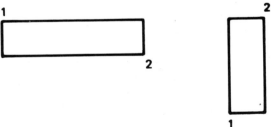

FIG. 9-24   Rectangle definition.

2.2.1.3.8 Symbols   Digitizing a symbol (see 2.1.2) is identical to digitizing a point (see 2.2.1.3.1), with the reference point of the symbol located as a point (see Fig. 9-25).

2.2.1.3.9 Curves   Curves are digitized by entering points along their path. These points are then used as defining points for spline, polynomial, or other fitting techniques.

2.2.1.4 Menu   A menu is a shorthand method of communicating with an IAG system. Instead of typing the name of the library component to be entered, selecting the appropriate area of a menu matrix with drawn library components communicates the choice to the IAG system. Referring to Fig. 9-26, positioning the cursor as if digitizing a point (see 2.2.1.3.1) in the upper-right-hand square tells the IAG system that the next digitizer inputs are to locate polyarcs. Graphic commands (see 2.3.3) can also be invoked using a menu.

2.2.1.5 Step and Repeat   Specifying the number of repetitions and the spacing of a step-and-repeat–generated pattern (see 2.3.3.7) can be performed using a digitizer. Referring to Fig. 9-27, after the component has been selected from the library (see 2.1.4.1), three points (see 2.2.1.3.1) are required for each axis. Point 1 indicates the position of one end of the pattern in the x direction; point 2 indicates the spacing, and point 3 the position of the last element in the matrix. Points 4, 5, and 6 perform the same functions in the Y axis.

2.2.2 Automatic Digitizing

By optical or electronic methods, graphic data can be scanned and input to a computer. This, however, is only the first step in automatically creating a database (see 2.1). Additional processing of the graphic data is needed to recognize and group primitives into symbols and/or text. The vast majority of benefits an IAG system can provide depend on the database being made up of primitives, symbols, and text. Without this type of database, not much more than reproduction of the input is possible.

2.2.3 Digitizing Solid Objects

Three-dimensional-coordinate measuring machines that determine the x, y, z coordinates of points on a solid object are available. They are basically used for quality control inspection of finished manufactured parts, but they can also be used to input three-dimensional geometry. One technique is to hold one axis constant and trace a two-dimensional curve over the surface of the object. After one pass, the axis being held constant is moved, and another pass over the object creates a parallel section.

REFERENCE POINT   **FIG. 9-25**   Symbol digitizing.

POINT	POLYGON	ARC	POLYARC
PATH	CONIC	RECTANGE	SYMBOL 1
SYMBOL 2	PICK	COPY	DRAW

FIG. 9-26   Menu.

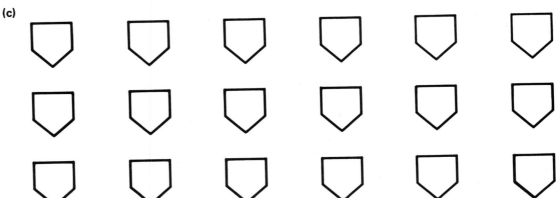

(a)

(b)

(c)

FIG. 9-27 Step and repeat. The symbol, *a*, was picked and the spacings in *X* and *Y* defined by points 1 through 6 in *b*. The IGS will determine how many symbols are needed and then fill in the matrix, *c*.

After the entire object is scanned, the parallel sections are processed by the IAG computer to regenerate the scanned surface. This processing is quite complex, and the recreation of the surface is an approximation of varying accuracy.

2.2.4 Analysis Programs

Many engineering analysis and design programs exist that use regular computer output devices to produce numerical results that are best visualized by graphics. These programs can be modified so their data is suitable for entry into an IAG system. Once the data are in the IAG system, the capabilities of the system are used to display and interact with the results of the programs. The system can also be used as a preprocessor of data for input into the analysis programs.

2.2.5 Drawing Storage

The total memory capacity of the computer in an IAG system is made up of several devices with varying storage capabilities and speed of data access and transfer. The combined capacity is segmented into two areas for drawing files (see 2.1.4): an active area for any drawing file that is being manipulated, and an inactive area for storing drawings for future recall. Storing a drawing consists of moving the drawing from the active area to the inactive area.

2.3 Interacting with a Database

2.3.1 Retrieving a Drawing

Retrieving a drawing is the reverse of storing a drawing (see 2.2.5), and consists of moving a file from the inactive area to the active area of the IAG computer memory.

2.3.2 Display

The display, with its ability to communicate, is the heart of the IAG system. Creating a drawing using the power of a computer to help with geometric construction and manipulation while viewing the results within seconds is a much more productive method than using pencil and paper. Clarity and speed of the display, coupled with the extensive drawing aids, are the keys to the effectiveness of an IAG system.

2.3.2.1 Scale Changes, View Windows, and Viewports   With an IAG system and its finite screen size, viewing a drawing at different scale factors is important. The IAG system has the capability to fill its screen with an arbitrarily chosen rectangular area of the drawing file (see 2.1.4). The effect of displaying a small area (see Fig. 9-28), over the entire screen is a close-range (magnified) view of the selected area. Similarly, if the entire drawing is displayed, the effect is one of standing back from the drawing.

2.3.2.2 Multiple Views and Viewports   The viewing area of the display can be separated and used for different views of the same drawing file (see 2.1.4). In a two-dimensional system, varying scales of different areas of the drawing can be displayed simultaneously. In a three-dimensional system, multiple projections of the same shape can be simultaneously displayed.

2.3.2.3 Layers or Levels   Layers of a drawing file (see 2.1.5 and 2.1.4) can be viewed in any combination (see Fig. 9-29). For display purposes, any layer in the drawing file can be viewed and modified, viewed and not modified, or not viewed. The display status of layers is easily changed; this capability is very useful when the operator is interested in only one or a few layers of a database that may contain many layers. When a symbol (see 2.1.2) that is defined to be on multiple layers is moved, the information on all the layers moves, maintaining the same layer-to-layer registration.

2.3.2.4 Circular Interpolation   A circle is treated by its definition in the IAG database (see 2.1.1.3). Displays can operate only on vectors, or on straight line segments. When displaying an arc or a circle, the IAG computer will approximate the curve by a number of polygons (see 2.1.1.2). The higher the number of straight line segments, the closer the circular approximation, and the longer the time required to display the circle or arc (see Fig. 9-30).

2.3.3 Drawing Modification

Adding to, changing, or deleting from a drawing file (see 2.1.4) is accomplished using graphic commands. These commands can vary in name from one IAG system to another, but they perform the same function. Menus can also be used to issue these graphic commands (see 2.2.1.4).

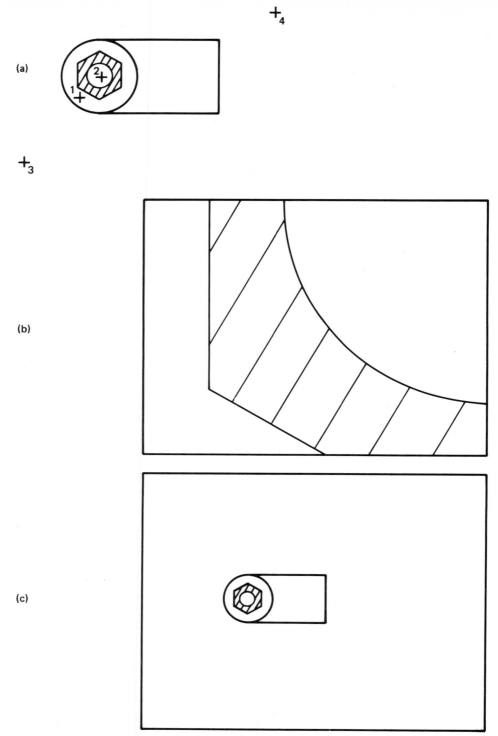

FIG. 9-28 The drawing file: *a* can be viewed at close range by specifying points 1 and 2; *b* is the screen display as a result of specifying points 1 and 2; *c* is the screen display as a result of specifying points 3 and 4.

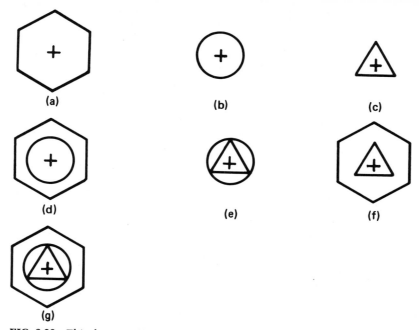

FIG. 9-29 This drawing file has *a* on layer 1, *b* on layer 2, and *c* on layer 3; *d* is the display of layers 1 and 2, *e* layers 2 and 3, *f* layers 1 and 3, and *g* all three layers.

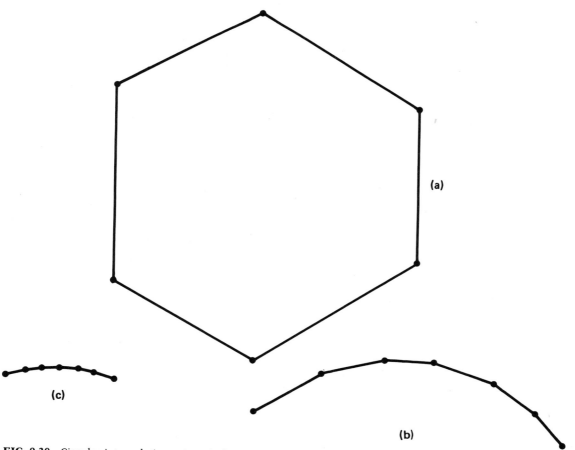

FIG. 9-30 Circular interpolation: *a* is a circle approximated by six polygons; *b* is a finer approximation with more, shorter polygons; and *c* is still finer.

2.3.3.1 DRAW   Adding a graphic entity to a drawing file (see 2.1.4) is accomplished by using a DRAW command. When the DRAW command is given, the appropriate number of coordinate points must be specified to define the primitive or symbol (see 2.1).

2.3.3.2 PICK or SELECT   Before a command can be executed on a graphic entity already in the drawing file (see 2.1.4), the entities to be affected must be identified. The entities chosen may be blinking or have some form of graphic notation added for visual identification by the operator (see Fig. 9-31). All PICKed entities are acted upon by the same graphic command.

2.3.3.3 ERASE   PICKed (see 2.3.3.2) entities can be deleted from a drawing file (see 2.1.4) using an ERASE command.

2.3.3.4 MOVE   PICKed entities can be relocated using a MOVE command (see Fig. 9-32). This command does not affect orientation of the moved entities.

2.3.3.5 ROTATE   Orientation of PICKed (see 2.3.3.2) entities can be changed using a ROTATE command (see Fig. 9-33).

2.3.3.6 COPY   Any PICKed (see 2.3.3.2) graphic entity or combination of entities can be duplicated using the COPY command.

2.3.3.7 STEP AND REPEAT   A matrix of graphic entities can be created using the STEP AND REPEAT command. The basic graphic element of the matrix is PICKed (see 2.3.3.2). The number of repetitions in X, the spacing between the repetitions in X, the number of repetitions in the Y axis, the spacing between these repetitions, the number of repetitions in the Z axis (for a three-dimensional array), and the spacing between the Z repetitions are input to the IAG system. The system then creates the matrix by duplicating the basic entity and spacing the graphics as directed (see Fig. 9-27).

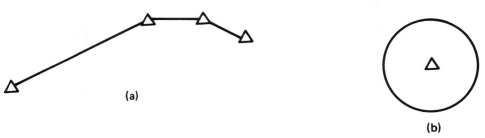

(a)

(b)

FIG. 9-31   PICKed entity identification. The polygon in *a* has a triangle on each vertex to indicate its PICKed status. A circle being PICKed is signified by a triangle on the circle's center.

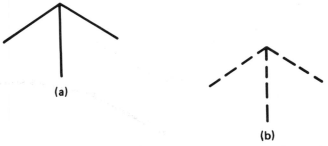

(a)

(b)

FIG. 9-32   MOVE command. The dashed figure, *b*, is the result of a MOVE (down and to the right) command executed on the solid figure. Orientation is not affected.

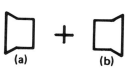

FIG. 9-33 ROTATE command. The dashed symbol represents the results of a ROTATE command on the solid symbol.

FIG. 9-34 MIRROR command. Shape *b* was created by MIRRORing shape *a* in the *X* axis about the indicated point.

2.3.3.8 MIRROR   A mirror image of a graphic entity can be created using a MIRROR command (see Fig. 9-34). The entity or group of entities is first PICKed (see 2.3.2.2), and then the axis to be mirrored about is defined.

2.3.3.9 Nested Commands or Macros   Whenever two or more graphic commands are frequently executed in the same sequence, they can be combined, or nested, into another command. If COPY (see 2.3.2.6) and MOVE (see 2.3.2.4) were frequently used together, they could be nested and the new command called COMO. When the IAG system received the command COMO, it would COPY all PICKed (see 2.3.2.2) entities and then MOVE them (see Fig. 9-35).

2.4 On-Line Design

In many applications of IAG systems, the database (see 2.1) is not created by digitizing as described in 2.2.1. The database is generated by an operator using the graphic commands listed in 2.3.3. The power of the IAG system is used to provide considerable geometric construction assistance to the operator. Application software (see 4.2) can also increase operator productivity.

2.5 Output

Once a database (see 2.1) is created, there are many different methods of producing useful documentation.

2.5.1 Hard Copy

2.5.1.1 Graphic   Any portion or all of a drawing file (see 2.1.4) can be outputted using a pen plotter (see 3.3.1) to create ink drawings. A portion of the drawing file can be selected to be outputted, and a scale factor can also be specified. Varying line widths or colors can be associated with predetermined data groups in the drawing file to distinguish different information.

2.5.1.2 Text   Information in the database (see 2.1) that is best described by alphanumeric text can be extracted and output in the appropriate format. A bill of materials is an example.

FIG. 9-35 Nested commands. Executing a MORSO command consisting of a MOVE followed by a ROTATE produced *b* from *a*.

2.5.2 Drawing Files

Drawing files (see 2.1.4) can be treated like any other digital data by computers. Drawing files can be transferred from one IAG system to another if their data formats are compatible. A drawing file can be output to storage media, such as magnetic tape (see 3.4.4), to be saved for future use.

2.5.3. Coordinate Data for Analysis Programs

Many computer programs, such as finite-element stress analysis, require numerous coordinate inputs to perform their function. An IAG system provides an easy method of extracting coordinate information from a geometric shape. Once a shape is defined and a database (see 2.1) created, the desired points (see 2.1.1.1) are indicated, and the system extracts the coordinates, formats them, and prepares them for input to the analysis program.

2.5.4 Numerical-Control (NC) Programming

The technique of NC parts programming requires studying a drawing of a part to be machined, manually coding each position and movement of the cutting tool in the proper sequence, and inputting this code into a computer for processing into the actual series of instructions to the machine tool controller. If the part to be machined is already in the database of an IAG system (see 2.1), the system can accept feed rates, cutter sizes, and other pertinent information, calculate the coordinates of the path the cutting tool should follow, and generate a centerline file. This centerline can be displayed for approval and, if approved, formatted for transfer to the computer that performs the final processing.

2.5.5 Precision Graphics

2.5.5.1 Photoplotting   High-quality precision artwork can be generated by photoplotters (see 3.3.4) from an IAG drawing file (see 2.1.4). Available photoplotter reticle sizes are used by the IAG system to transform the drawing file into a series of commands called a *photoplotter program.* The photoplotter program operates the photoplotter and reproduces the drawing file on film layer by layer (see 2.1.5) or in any combination of layers.

2.5.5.2 Pattern Generation   Extremely precise artwork primarily for use in fabricating integrated or hybrid circuits is usually generated by a pattern generator (see 3.3.5) from an IAG drawing file (see 2.1.4). Aperture information is used by the IAG system to transform the drawing file into the appropriate series of instructions to the pattern generator to reproduce the drawing file at the desired scale.

2.5.6 Microfilm

Computer output microfilm (COM) devices generate aperture cards directly from computer output. An IAG system can generate the appropriate series of commands to program a COM device, putting the graphic information contained in a drawing file (see 2.1.4) directly on an aperture card, thereby eliminating the need for photographing a drawing.

III.   HARDWARE

The electronic and electromechanical components of an IAG system are called *hardware.* As a general rule, hardware has capability but no intelligence and must be directed by a detailed series of simple commands called *software* (see IV).

3.1 Digitizers

Digitizers provide the fastest means for entering predefined graphic data. They resemble regular drafting tables with special electromechanical devices added. The information to be digitized is placed on the table surface and secured with masking tape in the usual drafting manner.

3.1.1 Free Cursor

A free-cursor digitizer derives its name from the minimum amount of restraint imposed on its use (see 3.1.5.2, and Fig. 9-36). The cursor assembly is attached to the digitizer table only by a thin multiconductor cable and can be electronically or sonically detected. Free-cursor digitizers are either absolute or relative devices. In an absolute electronic digitizer, a fine orthogonal grid of conductors, either wire or a two-sided printed-circuit board, is located directly beneath the table surface, creating a two-dimensional cartesian coordinate system (see 2.1.6.1). Each conductor in the grid radiates at a different frequency, and the cursor acts as an antenna. Cursor position on the digitizer table surface is established by the strongest frequencies received.

The conductors beneath the surface of relative electronic free-cursor digitizers are not necessarily in a two-dimensional orthogonal array. Operation depends on the cursor being placed in an initializing position that is fixed on a corner of the digitizer table surface and notifying the IAG system that the cursor is in the initializing position; the digitizer then keeps track of the relative movement of the cursor from the initialization position. The electronics that sense the relative movement sometimes dictate that the cursor be kept either in contact with or very close to the table surface. If the proximity tolerance of the cursor is exceeded by lifting it away from the table surface, the initialization procedure must be repeated. This is called *calibrating the digitizer* and is not to be confused with calibrating a digitizer for entering a drawing (see 2.2.1.1).

In a sonic digitizer, the cursor emits upon command a spark that is detected by microphones spaced along an orthogonal grid. The time delay for the spark to reach

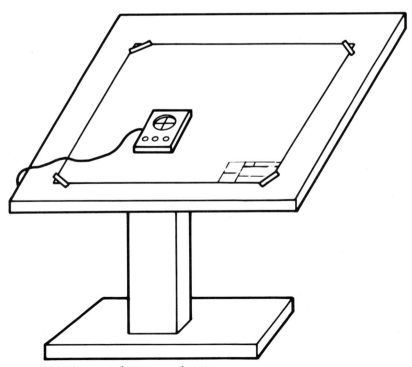

FIG. 9-36   Electronic free-cursor digitizer.

the differently positioned microphones provides cursor position. This technique is easily expanded to three dimensions (see Fig. 9-37).

3.1.2 Fixed Arm

In a fixed-arm or gantry-type digitizer, a vertical beam is mounted on the table. It can be moved horizontally across the table surface (see Fig. 9-38). The cursor assembly is mounted on the beam and can be moved vertically. Coordinate positions are read by electronic and/or optical sensors on the beam and its supports.

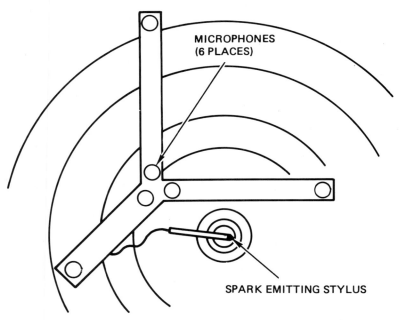

FIG. 9-37 Three-dimensional sonic digitizer.

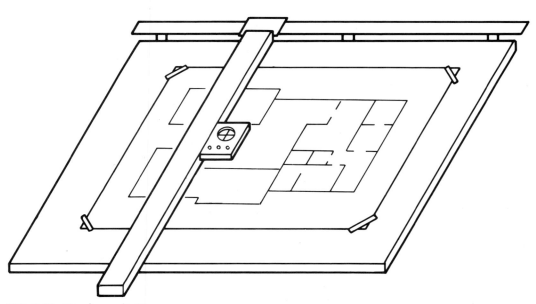

FIG. 9-38 Fixed-arm digitizer.

3.1.3 Digitizer-Plotter

A digitizer-plotter combines the capability of a digitizer and a flatbed plotter (see 3.3.1.1) in one device (see Fig. 9-39). The cursor and pen-block assemblies (see Fig. 9-49) are mounted on a horizontal beam. The beam can move vertically across the table surface, and the cursor and pen block can move horizontally on the beam. The cursor is positioned by means of a joystick (see 3.2.3.2) mounted on the beam and attached to servo motors that move the beam and cursor–pen-block assembly. Position coordinates are determined by resolvers on the servo motors. When operating as a plotter, the digitizer-plotter performs as a flatbed plotter.

3.1.4 Stereo Plotter

Stereo plotters use pictures taken by aerial survey cameras as their input. Working with two pictures simultaneously, an operator can determine and plot constant-elevation contours from the photographs by using special glasses to view the colored or polarized light. Stereo plotters can be connected to an IAG system, with the coordinate pairs being recorded and automatically input to the system.

3.1.5 Common Features

3.1.5.1 Table and Base   The digitizing surface can be raised, lowered, and tilted in the same manner as most modern drafting tables (see Fig. 9-36). Operator posture when digitizing is governed by the same considerations that apply to a regular drafting machine.

3.1.5.2 Cursor/Puck Assembly   The cursor, also referred to as a bombsight or a reticle, consists of a pair of orthogonal cross hairs and is used to indicate position to the digitizer. This cursor is not to be confused with the cursor indicator of a display (see 3.2.1). When the operator has the cursor positioned over the proper point, the coordinates of the point (see 2.1.1.1) are read by the digitizer electronics for transfer to the IAG computer. Pushing the appropriate button causes the coordinates to be entered (see 3.1.5.3). Magnifying lenses are available to increase the accuracy of positioning the cursor (see Fig. 9-40).

3.1.5.3 Programmable Buttons   Push buttons located on the cursor assembly are used to communicate with the IAG computer (see Fig. 9-40). More than one button is

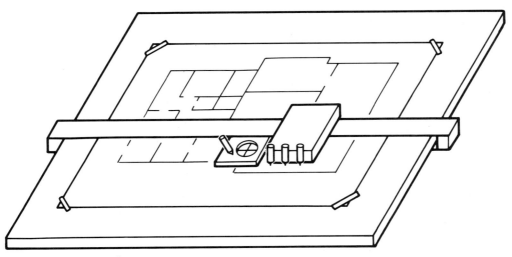

FIG. 9-39   Digitizer/plotter.

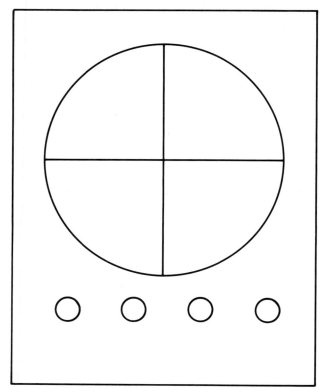

FIG. 9-40   Cursor assembly.

required to identify which of the different type points (see 2.2.1.3) is being entered when graphic objects are digitized.

3.1.5.4 Free-Cursor Puck and Stylus   A free-cursor digitizer constrains its cursor assembly by only a thin multiconductor cable. This allows another hardware mechanization to perform the cursor's function of locating the coordinates of points (see Fig. 9-41). A stylus resembling a ballpoint pen may be used in lieu of a cursor assembly to determine point locations. It contains either a microswitch inside its body or an accompanying push button to be held in the operator's free hand. Closing the microswitch in the stylus by exerting greater pressure when touching the tip to the table surface, or pushing the accompanying button a specified number of times, communicates to the IAG system what type of point is being entered (see 3.1.5.3).

3.1.5.5 Backlighting   Digitizers are available with a backlit surface similar to a light table.

3.1.5.6 Rear Projecting   Rear-projecting digitizers can project a photographic negative or similar source material onto a digitizer surface from the rear. Digitizing is then performed in the normal manner.

3.1.5.7 Program-Function Keyboards   The menu (see 2.2.1.4) technique of communicating with an IAG system can significantly improve digitizing throughput. A program-function keyboard is the hardware mechanization of a menu consisting of a matrix of push buttons (see Fig. 9-42). These push buttons, like menus, can be assigned specific functions; pressing the selected button eliminates the need to move the cursor over the appropriate area of a menu.

3.1.5.8 Cursor Position Indicators   Cursor position can be continually monitored by using the cursor position indicators, which are digital readouts identifying the present cursor position (see Fig. 9-43). They are updated frequently, and they change as the cursor is moved.

3.1.5.9 Keyboard A keyboard unit that contains a standard typewriter keyboard and possibly special buttons, keys, and/or lights is used to communicate with the system.

3.2 Interactive Console

The interactive console is used for on-line design (see 2.4) or to modify an existing drawing file (see 2.1.4).

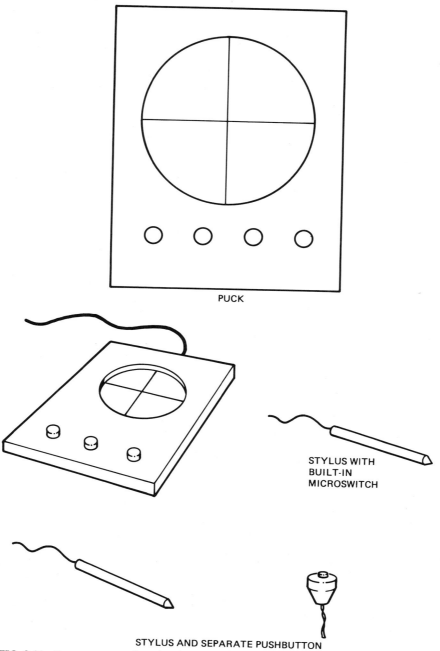

PUCK

STYLUS WITH
BUILT-IN
MICROSWITCH

STYLUS AND SEPARATE PUSHBUTTON

FIG. 9-41 Free-cursor puck and stylus.

FIG. 9-42   Program function keyboard.

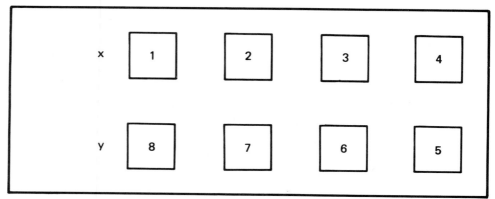

FIG. 9-43   Cursor position indicators.

3.2.1 Graphic Display

Graphic displays can display both graphic and alphanumeric data, and they make possible the IAG capabilities listed in 2.3.2. Instantaneous display of the results of creating or modifying drawings is a key to an IAG system. Cursor (see 3.1.5.2) position is indicated on the display, making it possible to sketch on an input device (see 2.3.3) and view the drawing file (see 2.1.4) as it is being created or changed. Cursor position display and scale changes (see 2.3.2.1) provide drafting accuracy and range unattainable with manual methods.

3.2.1.1 Storage Tube   The direct-view-storage tube (DVST) displays graphic data in the form of vectors or straight line segments at any angle (see 2.3.2.4). Vectors are "painted" on the screen by a writing pen whose stream of emitted electrons is deflected in the horizontal and vertical directions to impact at predetermined spots on the inside surface of the tube. Where the electron stream hits the inside surface of the tube, a phosphorus coating is electrically excited and is visible on the outside screen face of the tube. The phosphorus is kept excited by a flood gun. Once data are painted on the screen, they are not selectively erasable. For a graphic entity to be erased from the display, it must first be removed from the drawing file (see 2.1.4), the entire screen erased, and the updated version of the drawing file repainted on the screen. Storage-tube light output is limited. The screen is best viewed in low ambient lighting.

3.2.1.2 Vector Refresh   Vector-refresh displays are also known as stroke writers or directed beam-oriented CRTs. They employ a CRT with a single cathode gun that has its emitted beam of electrons deflected in the horizontal and vertical directions to draw vectors and straight line segments (see 2.3.2.4) in the phosphorus coating on the inside of the viewing screen. The vectors are visible from outside the screen. The persistence of the phosphor is short, requiring the vectors to be repainted, or refreshed, a minimum of 30 times per second. When a graphic entity has been deleted from the drawing file (see 2.1.4) being displayed, it will disappear from the screen or be selectively erased. Vector-refresh displays are easily viewed in normal ambient lighting.

3.2.1.3 Raster Refresh   Raster-refresh displays operate on the same principles as the CRT in a television set. An electron beam emitted from the tube cathode is scanned across the inside of the screen in a raster pattern (see Fig. 9-44) and can be turned on or off while it is being scanned. When the beam is turned on, electrons strike the phosphorus coating on the inside of the tube, and the resulting dot can be seen from the outside of the screen, creating a dense two-dimensional array of discrete points. Vectors are converted by the IAG computer into series of contiguous points for display. Depending on the orientation of the vector and the viewing scale (see 2.3.2.1), the display can appear to be granular. Raster-refresh displays can be easily viewed in normal ambient lighting.

3.2.2 System Communication

3.2.2.1 Alphanumeric Cathode-Ray Tube (CRT)   Many communications between an operator and an IAG system are nongraphic. A CRT terminal that can display letters and numbers (alphanumerics) can display messages from the system to the operator, leaving the graphics CRT (see 3.2.1) free for graphic information. A keyboard (see 3.1.5.9) is usually associated with the CRT to transmit the operator's commands to the IAG computer.

3.2.2.2 Teletypewriter   Teletypewriters and teletypewriterlike devices combine a keyboard (see 3.1.5.9) and a typewriterlike printing mechanism for two-way communication between the operator and the IAG system.

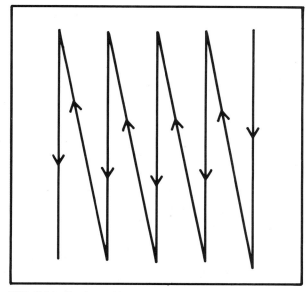

FIG. 9-44   Raster pattern.

3.2.3 Analog Input

Analog input is a means of controlling the display cursor (see 3.2.1) position manually. The graphics display reacts to the analog input immediately and continually displays cursor position.

3.2.3.1 Tablet   A tablet is a free-cursor digitizer (see 3.1.1) with a smaller digitizing surface area. Cursor control (see 3.2.1) is accomplished using a stylus with a built-in microswitch (see 3.1.5.4 and Fig. 9-45).

3.2.3.2 Joystick   A joystick is a short shaft protruding from a base that translates tilting of the shaft in the X and Y directions into voltages to drive the display cursor (see 3.2.1) in the direction the shaft is tilted (see Fig. 9-46) and/or controls panning and zooming.

3.2.3.3 Thumbwheel Switches   Thumbwheel switches are potentiometers, conveniently turned using one's thumb, that are used to generate voltages to position the display cursor (see 3.2.1). Thumbwheel switches independently move the cursor in their associated X, or Y axes.

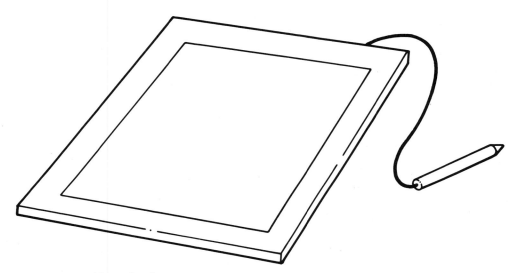

FIG. 9-45   Tablet and stylus.

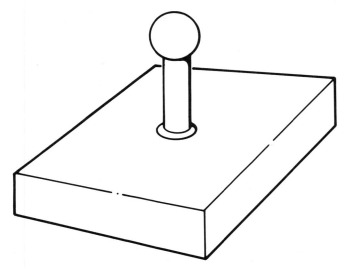

FIG. 9-46   Joystick.

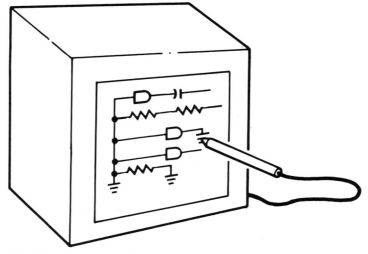

FIG. 9-47   Light pen and CRT display.

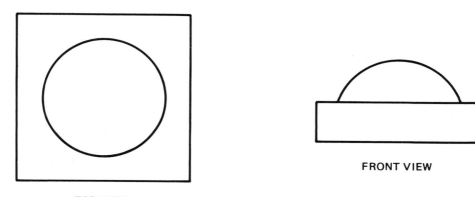

TOP VIEW

FRONT VIEW

FIG. 9-48   Trackball.

3.2.3.4 Light Pen   A light pen performs the same function as the stylus associated with a tablet (see 3.2.3.1). It works with refresh displays only and senses light incident on its tip. When the light pen is activated, the tip held against the display screen and light is sensed (see Fig. 9-47); the IAG system computes the cursor position in the drawing file (see 2.1.4).

3.2.3.5 Trackballs   A trackball consists of a spherical ball mounted in a housing that contains electronics to move the cursor (see 3.2.1) as the trackball is rotated (see Fig. 9-48).

3.2.4 Digital Input

3.2.4.1 Keyboard   See 3.1.5.9.

3.2.4.2 Program-Function Keyboard   See 3.1.5.7.

3.2.4.3 Function Buttons   Function buttons are push buttons mounted in a convenient location so that when pressed they instruct the IAG system to perform a preassigned function.

3.3 Plotters

Devices that produce graphic hard-copy output from an IAG system are called plotters. Plotters may be on-line or off-line. On-line plotters are connected electrically to the IAG computer, and the drawing file (see 2.1.4) can be plotted directly by the computer. Off-line plotters are not directly connected and require the graphic information to be transported to the plotter using magnetic tape (see 3.4.4) or a similar data-transfer medium.

3.3.1 Pen Plotters

Pen plotters produce drawings using ballpoint, felt-tip, or liquid-ink pens on paper, Mylar, or equivalent material. The pens are held and controlled by a pen block (see Fig. 9-49). The pen block can be moved in two axes across the surface of the paper. A PEN UP command from the IAG computer causes the pen to not contact the paper. A PEN DOWN command pushes the pen tip into contact with the paper. The IAG computer plots a drawing file (see 2.1.4) by combining PEN UP, PEN DOWN, and movement commands to the plotter.

3.3.1.1 Flatbed A flatbed plotter consists of a flat surface that holds a piece of paper in a fixed position using vacuum or electrostatic means, a pen block (see 3.3.1), and incremental or linear motors to position the pen block in two orthogonal axes (see Fig. 9-50).

3.3.1.2 Drum The drum plotter's characteristic element is a drum. The paper to be plotted upon is held to the drum by vacuum columns and take-up reels on both sides of the drum. A reel of paper that runs the width of the drum and has sprocket holes close to its edges must be used. The sprocket holes are needed to move the paper with the drum as the drum rotates. The pen block (see 3.3.1) moves along the logitudinal axis of the drum to provide, with drum rotation, two axes of motion (see Fig. 9-51).

3.3.2 Electrostatic Plotter

Electrostatic plotters produce drawings by using a closely spaced row of small electrodes. Spacing between electrodes can be as narrow as 0.005 in. Paper is drawn across the electrodes perpendicular to their row alignment, and as the paper crosses the row the electrodes can be energized. When an electrode is energized, it creates a small electrically charged spot that turns dark after passing through a toner (see Fig. 9-52). The resulting plot can be noticeably granular depending on the graphic arrangement being plotted. Electrostatic plotters can also be used as computer line printers (see 3.4.6).

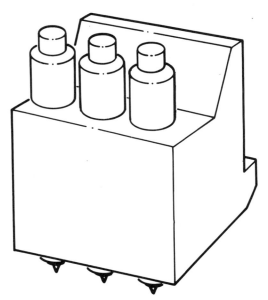

FIG. 9-49 Pen block.

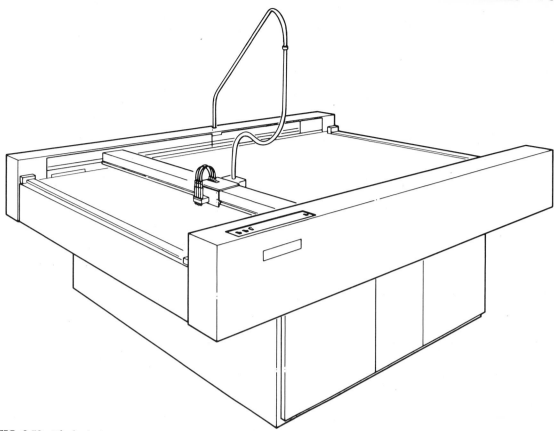

FIG. 9-50 Flatbed plotter.

FIG. 9-51 Drum plotter.

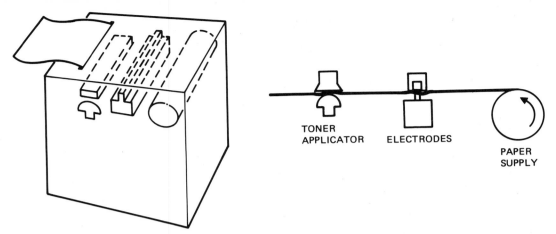

FIG. 9-52 Electrostatic plotter.

3.3.3 Hard-Copy Unit

A hard-copy unit reproduces the information displayed on a storage tube on specially coated paper. The information is taken directly from the storage tube electronically and is reproduced exactly as it appears on the screen.

3.3.4 Photoplotter

Photoplotters contain a light source, a shutter, a lens system, precise apertures called reticles, a vacuum hold-down table for film, and drive motors with associated gearing to move the film or light source with respect to the other in two dimensions. With the shutter open, light is projected through a reticle onto film being held in place on the vacuum table. By opening and closing the shutter, selecting various reticles, and moving the light source with respect to the film in two dimensions, precise graphics can be plotted on the film.

3.3.5 Pattern Generator

With a xenon strobe as a light source, a focusing lens assembly, and rectangular apertures of different sizes, or one variable aperture, pattern generators can create extremely accurate artwork. The rectangular reticles are positioned in two dimensions, rotated to any angle, and light is flashed repeatedly through the lens assembly to create the desired geometries.

3.4 Computer Equipment

3.4.1 Central Processing Unit (CPU)

The CPU is that area of the computer in which computing and logical choices take place. The CPU interprets stored programs and software and controls the overall IAG system operation.

3.4.2 Memory

The total storage capability of an IAG system is provided by a random-access memory that closely supports the CPU (see 3.3.1) disk drives (see 3.3.3) and magnetic-tape drives (see 3.3.4). The term *memory*, when used alone, usually refers to the random-access memory logically associated with the CPU.

3.4.3 Disk Drive

Disk memory is generally slower and cheaper, and has much more capacity than the random-access memory (see 3.4.2) associated with a CPU. It is necessary for system operation and can also store drawings (see 2.2.5).

3.4.4 Magnetic Tape

Magnetic tape is currently the cheapest medium available to reasonably store and retrieve large amounts of digital data. Magnetic tape is used in an IAG environment for archival drawing storage and as a means of transporting data from one IAG system to another, from an IAG system to another computer, or to an off-line device (see 3.3).

3.4.5 Paper-Tape Reader and Punch

Paper tape is a slow, cheap method of transferring digital data. Prior to the advent of magnetic tape (see 3.3.4), it was the most widely used data-transfer medium. Many NC devices still rely on paper tape, since the large magnetic fields usually present in a factory environment could alter the data on a magnetic tape. A paper-tape reader can read data from the holes in a punched paper tape, format it, and input it to a computer. A paper-tape punch can take data from a computer, format it, and punch it into paper tape.

3.4.6 Line Printer

When large amounts of alphanumeric information are to be outputted from a computer, a line printer is usually used. The data can be formatted as desired and outputted at a high rate.

3.4.7 Teletypewriter and Control Console

Teletypewriters and teletypewriterlike devices (see 3.2.2.2) are often used to communicate with a computer. The design of the computer may dictate the use of a teletypewriter in lieu of an operator's control panel.

3.5 Power Requirements

IAG systems, like most other computer-based systems, require a stable, transient-free power source. Spikes and momentary depressions in the primary power supplied can cause intermittent erratic system behavior and/or loss of data.

3.6 Environmental Requirements

Error-free operation of an IAG system requires that ambient temperature and humidity be kept within the manufacturer's specifications. When the environment is outside the stated specifications, system performance becomes unreliable and tends to degrade gradually. The presence of dust or static electricity can also preclude reliable system operation.

IV. SOFTWARE

Any computer or computer-based system must be programmed before its hardware (see III) can be useful. *Software* refers to programming, the key to IAG or any other type of computer application.

4.1 System Software

Software needed to make the system perform its basic functions is called *system software*, sometimes referred to as the *operating system*. An example of a basic function would be calibration of a digitizer (see 2.2.1.1).

4.1.1 Drawing Management

Drawing-management software allows an operator to create library components (see 2.1.3) and drawing files (see 2.1.4), store and retrieve drawing files, and manage the database (see 2.1).

4.1.1.1 Active and Storage Areas   Referring to Fig. 9-53, the total storage capacity of a disk is separated into areas for software, active drawing files (see 2.1.4) and inactive drawing files. When an operator is interacting with a drawing file, the file must be located in the active area. Storing the drawing (see 2.2.5) involves moving the drawing file from the active to the storage area. Recalling a drawing is the reverse.

4.1.1.2 Logging On and Off   The first step for an operator in using an IAG system is usually to log onto the system. The system asks for the operator's name and possibly a department name or other identification. As the operator types in the information, it is displayed on the screen. The system also asks for a password, which is not displayed on the screen as it is entered. This feature allows an operator to restrict access to his or her drawing files (see 2.1.4), which have been associated with the predetermined password. When finished using the system, an operator logs off by indicating that he or she is through with the system. At this point, the IAG system computes how long the operator has used the system. All system usage is compiled and tabulated for a given period.

4.1.1.3 File Protection   Data files, which include drawing files (see 2.1.4), can be protected from unauthorized reading, modification, and deletion by the system software (see 4.1).

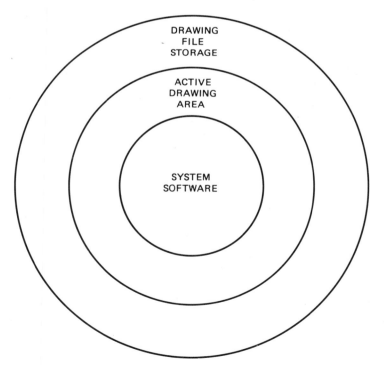

FIG. 9-53   Disk data areas.

4.1.2 Initialization

Starting an IAG system from a "power-off" condition requires an initialization procedure. The random-access memory associated with the CPU (see 3.3.2) is usually volatile; that is, when the power is off, the data and programming stored in the memory are lost. Initialization requires transferring a short "bootstrap" program into the CPU from a nonvolatile storage device and transferring control to the program to start the system operating.

4.1.3 Crash Recovery

Crash is the computer term for a system failure. If the computer "gets lost" or a power failure occurs, drawing files (see 2.1.4) and other data may be lost or destroyed. If the system is not restarted properly, an additional risk of losing information is incurred. A crash recovery procedure is intended to restart the system with a minimum loss of data.

4.1.4 Prompting

With the quick two-way communication capabilities of IAG devices, the computer can provide assistance to an operator, especially a new user. Messages can be displayed to guide the user through the proper sequence of commands needed to perform complicated operations.

4.1.5 Error Messages

A misspelled command, illegal sequence of steps, or attempts to misuse the system will result in the display of an error message by the IAG system. The message is intended to inform the operator that an error condition exists and to give some insight into the problem.

4.1.6 Multiple Users

In order to minimize the cost of an IAG system by maximizing the usage of its components, a time-sharing operating system (see 4.1) allows several operators to use one computer.

4.1.7 Input/Output (I/O) Handlers

Any device that transfers data into or out of a computer needs a *software routine*—a term used to describe a short program segment—to control the device and the data transfer. The routines that control attached devices are called I/O handlers.

4.1.8 Foreground and Background

With its tremendous computing speed, quite often the computer in an IAG system is idle and waiting for another user command. To utilize this available computer time, sophisticated operating systems (see 4.1) have a foreground/background capability. The functions in which response time is critical or high priority are designated as *foreground tasks*. Other tasks are designated as *background tasks*. Low-priority functions are assigned background status. Any foreground function will be performed first, and when there is no foreground computing to be done, background programs are executed. When a foreground request is made of the computer, background computing is suspended until all foreground processing demands have been satisfied.

4.1.9 Links to the Database

Provisions are made in most IAG systems to allow a programmer access to the database (see 2.1). If a user has a programming capability and wants to improve the system for his or her application, that user can write his or her own programs and thus access the database and all the resources of the IAG system.

4.2 Application Software

IAG hardware (see III), coupled with system software (see 4.1), provides a powerful general-purpose design and drafting tool that increases overall operator productivity. The IAG computer can have special features programmed for a particular design or drafting discipline; this programming is called *application software*. The scope of capabilities which are *not* turnkey ranges from the drafting of two-dimensional drawings to the sophisticated modeling and machining of complex parts. The capabilities required to permit economical use of a system are dependent upon each application.

IAG systems are tools that must be fine-tuned or specifically chosen relative to their characteristics, which in turn are closely related to their application. Turnkey or various applications cannot be reasonably or effectively addressed by any one of the tools available. It is therefore necessary to closely scrutinize the application and then focus on a specific portion that can produce substantial economical results. The final result over a period of years may be the use of several different systems, each being maximized relative to its capabilities. Application can be categorized by the type of data manipulation required to produce an end result such as a drawing. The categories are two-dimensional, two-and-a-half dimensional, and three-dimensional.

4.2.1 Two-Dimensional (2D) Drawings

Drawings, which are two-dimensional (2D), are still the principal means of communication. Examples of 2D drawings are electrical schematics, instrumentation drawings, and control schematics. Dimensioned drawings (details, plans and elevations, assemblies, etc.) are 2D in presentation but often represent various views which really describe the 3D views and dimensions of a product, assembly, or plant. The principal area for improvement is in 2D drawings, because of the amount of manual labor expended on them. An IAG system that effectively improves the completion of 2D drawings can stand on its own without any further sophistication.

4.2.2 Two-and-a-Half Dimensional (2½ D) Drawings

Drawings completed in an IAG system can include additional information that is associated with the 2D drawing description. This information can include:

- Nongraphical data for bill-of-material generation
- Z axis description

The Z axis data can be used to:

- Calculate pseudo 3D representations for display.
- Calculate NC travel for simple machining operations such as drilling and simple pocketing and contouring.

4.2.3 Three-Dimensional (3D) Drawings

A true 3D database theoretically provides a reference for extraction of specific data as they relate to drawing creation, sophisticated design modeling, and NC machining.

In reality, design modeling and NC machining is currently optimized, but drawing creation with automatic updates based on changes to the 3D database is neither a reality nor practical. There is no single modeling technique that offers a total solution. The 3D modeling and NC machining capabilities, however, are noteworthy, because certain designs can now be easily handled and in many instances for the first time really described without building an actual model.

4.2.4 2D versus 3D

Each category has its advantages and disadvantages, which must be evaluated relative to the specific requirements mix of a company. Some typical scenarios may be:

- 2D system for drafting only
- 3D system for modeling only
- 3D system for modeling and NC machining only
- 2D and 3D system for drafting, modeling, and NC machining

Preparation of Patent Drawings[1]

Gilbert A. Thomas

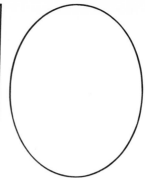

CORRECT DRAWINGS REQUIRED FOR ACCEPTABLE PATENT APPLICATION

Patent law is a highly structured body of statutes precisely describing the requirements of an acceptable patent application, including the drawings. The drafter who fails to prepare proper drawings when required will find that the entire application will be

[1]Figures 10-5 through 10-14 are taken from the *Guide for Patent Draftsmen*, published by the U.S. Department of Commerce.

placed in an informal status. Drawings will be required to be submitted within two months before the application can be formally accepted. Failure to provide satisfactory drawings could result in a rejection of the entire application.

The drafter who provides drawings that are incorrect or improperly prepared will find that the entire application will be rejected. New or corrected drawings will be required to overcome the rejection. The additional cost to the patent applicant in time and money can be considerable.

This section will enable the designer or drafter to prepare patent drawings acceptable to the U.S. Patent Office.

STATUTORY BASIS FOR PATENT DRAWINGS

The basic federal statute relating to patent drawings is found in 35 U.S.C. 113, quoted here in full:

> Drawings—The applicant shall furnish a drawing where necessary for the understanding of the subject matter sought to be patented. When the nature of such subject matter admits of illustration by a drawing and the applicant has not furnished such a drawing, the Commissioner may require its submission within a time period of not less than two months from the sending of a notice thereof. Drawings submitted after the filing date of the application may not be used (i) to overcome any insufficiency of the specification due to lack of an enabling disclosure or otherwise inadequate disclosure therein, or (ii) to supplement the original disclosure thereof for the purpose of interpretation of the scope of any claim.

In addition to the basic statute, Rules 1.51 and 1.81 of "Rules of Practice in Patent Cases" state that drawings when necessary are required for a complete application. (Found in 37 C.F.R. [*Code of Federal Regulations*] 1977.)

A complete application, sufficient to entitle an applicant to a serial number and filing date from the U.S. Patent Office, must include the drawings when necessary. The acceptance of applications is a responsibility of the Applications Division of the Patent Office. The personnel of that division decide whether or not a drawing is required for issuance of a serial number and transfer of the examination to the Examining Division.

SUBJECT MATTER REQUIRING DRAWINGS

Utility Patents

The U.S. district courts and the Patent Appeals Board have determined in various legal cases the subject matter that requires the submission of drawings with an application.

Article of Manufacture

A drawing must accompany an application for an article of manufacture. For example, see Fig. 10-1*a*, a nut lock, and *b*, a stud. This requirement applies to all manufactured items in which the utility of the item lies in the function or use of the article and not solely in its composition.

Apparatus

In all cases in which the application describes a machine, motor, or apparatus, a drawing is required to complete the application. Figure 10-2 is a patent drawing of a fluid jet nozzle.

Process

The Commissioner of Patents has decided that drawings need not accompany an application describing a process or method. Good practice, however, requires that the drafter provide a block or flow diagram to illustrate the various aspects of the method or process claimed, as in Fig. 10-4*a* which illustrates a process flowchart. (By comparison, Fig. 10-4*b* depicts a device and so needs a drawing.)

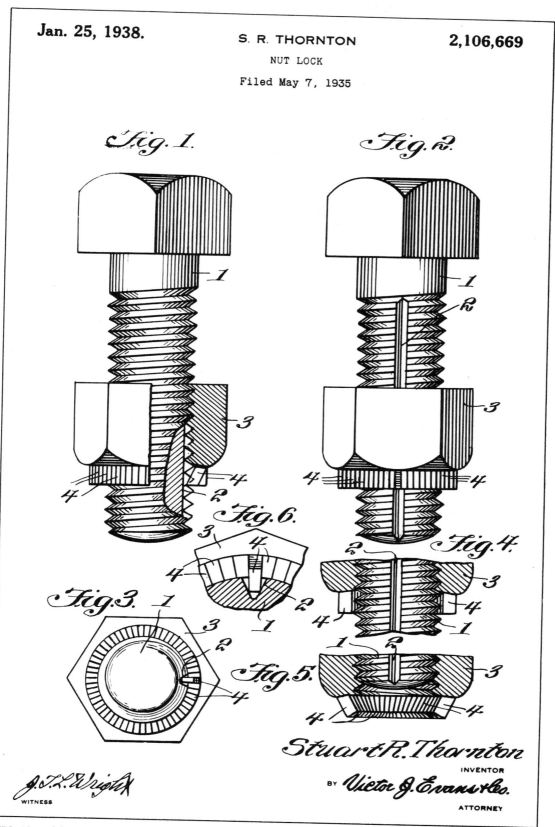

Fig.1.

Fig.2.

Fig.6.

Fig.3.

Fig.4.

Fig.5.

Stuart R. Thornton

INVENTOR

BY *Victor J. Evans & Co.*

ATTORNEY

J. T. L. Wright

WITNESS

FIG. 10-1   (*a*) Article of manufacture: nut lock.

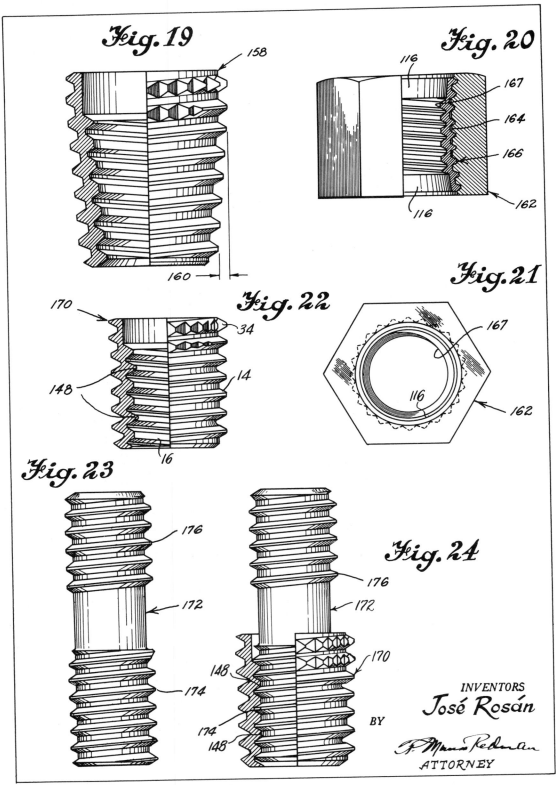

Fig. 19

Fig. 20

Fig. 22

Fig. 21

Fig. 23

Fig. 24

INVENTORS
José Rosán

BY

F. Mario Redman
ATTORNEY

FIG. 10-1 (*b*) Article of manufacture: stud.

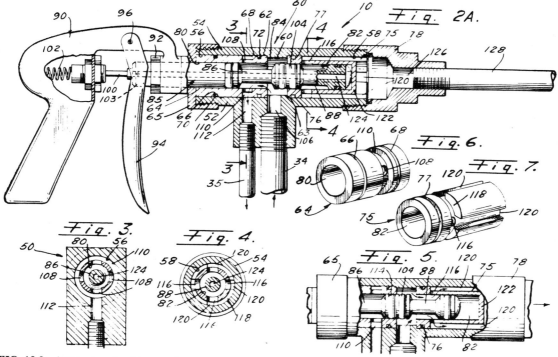

FIG. 10-2 Apparatus: fluid jet nozzle.

Composition of Matter

Patent applications directed to compositions of matter generally will be accepted without drawings. This classification includes chemical compounds defined by formula, medical compositions, good or beverage products, etc.

Other No-Drawing Situations

The *Manual of Patent Examining Procedures*, section 608.02, has described certain other situations in which drawings need not be submitted.

Coated Articles or Products

Drawings are not required when the invention resides solely in coating or impregnating a conventional sheet, e.g., paper or cloth, or an article of known and conventional character with a particular composition; the application should contain claim to the coated or impregnated sheet or article, unless significant details of structure or arrangement are involved in the article claims.

Articles Made from a Particular Material or Composition

Drawings need not be submitted when the invention consists of making an article of a particular material or composition, unless significant details of structure or arrangement are involved in the article claims.

Laminated Structures

Drawings are not required when the claimed invention involves only laminations of sheets (and coatings) of specified material, unless significant details of structure or arrangement (other than the mere order of the layers) are involved in the article claims.

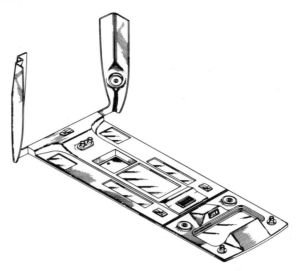

FIG. 10-3   Design patent drawing: van console.

Articles, Apparatus, or Systems in Which the Sole Distinguishing Feature Is the Presence of a Particular Material

When the invention resides solely in the use of a particular material in an otherwise old article, apparatus, or system that is recited broadly in the claims, drawings need not be supplied. For example:

1. For a hydraulic system distinguished solely by the use therein of a particular hydraulic fluid
2. For packaged structures wherein the structure and arrangement of the package are conventional and the only distinguishing feature is the use of a particular material contained therein

Illustration Subsequently Required

The acceptance of an application without a drawing does not preclude the examiner from requiring an illustration in the form of a drawing under the second sentence of Rule 81, 37 C.F.R. (1977). In requiring such a drawing, the examiner will clearly indicate that the requirement is made under the second sentence of Rule 81, and will be careful not to state that to do so is "because the nature of the case admits" of a drawing, as that statement might give rise to an erroneous impression as to the completeness of the application when filed. Examiners making such requirements are to require specifically, as a part of the applicant's next response, at least an ink sketch or permanent print of any drawing proposed in response to the requirement, even though no allowable subject matter has yet been indicated. This will afford the examiner an early opportunity to determine the sufficiency of the illustration and the absence of new matter. The description should be amended to contain references to the new illustration.

Special No-Drawing Categories

The Patent Office is willing to accept black-and-white photographs or photomicrographs (not photolithographs or other reproductions of photographs) printed on sensitized paper in lieu of india-ink drawings, to illustrate inventions that are incapable of being accurately or adequately depicted by india-ink drawings, restricted to the following categories:

- Crystalline structures
- Metallurgical microstructures

- Textile fabrics
- Grain structures
- Ornamental effects

The photographs or photomicrographs must show the invention more clearly than can be done by india-ink drawings and must otherwise comply with the rules concerning drawings.

To be acceptable, such photographs must be made on photographic paper having the following characteristics that are generally recognized in the photographic trade: double-weight paper with a surface described as smooth, and a white tint.

Design Patent

All design-patent applications require drawings, because by definition the subject of a design patent is the ornamental *design* of an article of manufacture (see Fig. 10-3).

Content of Drawing

Rule 1.83 (Rule of Procedure for Patent Cases)

The drawing must show every feature of the invention specified in the claims, but these need not be manufacturing claims. The District Court of Kansas has said in a case concerning the amount of detail required in drawings: "Disclosure of claimed invention is performed not only by claims made, but by patent specifications and accompanying drawings, which elucidate claims; test of disclosure of claimed invention is only that one skilled in art must be able to ascertain invention without undue experimentation; there is no requirement that drawings accompanying patent be so detailed as to be production specification."

In fact, conventional features disclosed in the description and claims, when their detailed illustration is not essential for a proper understanding of the invention, should be illustrated in the drawing in the form of a graphical drawing symbol or a labeled representation (e.g., a labeled rectangular box). In Fig. 10-4a this is illustrated by the

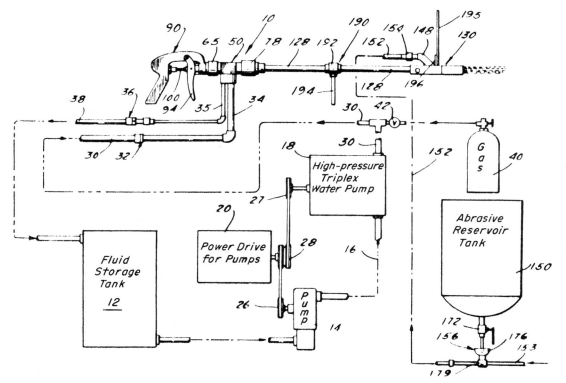

FIG. 10-4 (*a*) Fluid jet abrasive system.

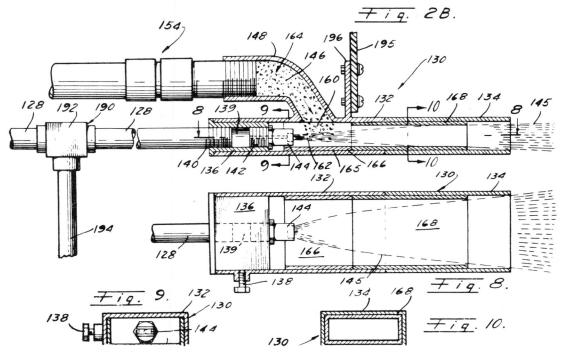

FIG. 10-4   (*b*) Fluid jet abrasive device.

high-pressure triplex water pump, power drive for pumps, and so forth. The Patent Office Board of Appeals has stated that clarity, and not rigid adherence to detail, is what is required for patent drawings. "Functional-type block diagrams and accompanying description that serve with rest of specification to enable person skilled in art to practice claimed invention with only reasonable degree of routine operation or experimentation is preferable in complex systems disclosure in which claimed invention would be obscured by conventional drawings and circuit descriptions."

Improvements to Existing Devices

When the invention consists of an improvement on an old machine, the drawing must when possible exhibit, in one or more views, the improved portion itself, disconnected from the structure, and also, in another view, only as much of the old structure as will suffice to show the connection of the invention therewith. In Fig. 10-4*b*, number 154 identifies an old coupling structure, 128 a known conduit, and 194 a previously known holding arm or leg.

PRODUCING THE DRAWINGS

Function

The function of patent drawings is to explain the principle of the invention. They are not required to be working drawings from which the invention can be produced or to be patterns for the model maker. The various portions of the invention shown in the drawings need not be to scale, show dimensions, or be proportionate.

Technical Requirements for Drawings

Paper and Ink

Drawings must be made upon pure white paper of a thickness corresponding to two-ply or three-ply bristol board. The surface of the paper must be calendered and smooth and of a quality that will permit erasure and correction with india ink. India ink, or its equivalent in quality, must be used for pen drawings to secure perfectly black solid lines. *Important:* The use of white pigment to cover lines is not acceptable.

Size of Sheet and Margins

The size of a sheet on which a drawing is made must be exactly 8½ by 14 in. (21.6 by 35.6 cm). One of the short sides of the sheet is regarded as its top. The drawing must include a top margin of 2 in. (5.1 cm), and bottom and side margins of ¼ in. (6.4 mm) from the edges, thereby leaving a "sight" precisely 8 by 11¾ in. (20.3 by 29.8 cm). Margin border lines are not permitted. All work must be included within the "sight." The sheets may be provided with two ¼-inch (6.4-mm) diameter holes having their centerlines spaced ¹¹⁄₁₆ in. (17.5 mm) below the top edge and 2¾ in. (7.0 cm) apart, said holes being equally spaced from the respective side edges.

Character of Lines

All drawings must be made with drafting instruments or by a process that will give them satisfactory reproduction characteristics. Every line and letter must be absolutely black and permanent; the weight of all lines and letters must be heavy enough to permit adequate reproduction. This direction applies to all lines, however fine; to shading; and to lines representing cut surfaces in sectional views. All lines must be clean, sharp, and solid, and fine or crowded lines should be avoided. Solid black should not be used for sectional or surface shading. Freehand work should be avoided wherever it is possible to do so. It is absolutely essential that drafting instruments be used. Freehand drawing or sketching is unacceptable unless the drafter can demonstrate to the Patent Office that the figure cannot be produced by the use of drafting instruments. Only black permanent ink is acceptable. Pencil drawings, since they lack permanency, are not permissible. Colored inks are not acceptable.

Hatching and Shading

1. Hatching should be made by oblique parallel lines, which may be not less than about ¹⁄₂₀ in. (1.3 mm) apart.

2. Heavy lines on the shaded side of objects should be used except where they tend to thicken the work and obsure reference characters. The light should come from the upper left-hand corner at an angle of 45°. Surface delineations should be shown by proper shading, which should be open. The *Guide for Patent Draftsmen* states that any style of lettering may be used, and several acceptable styles are shown in Fig. 10-5, as well as correct shading to create the proper 45° light approach angle.

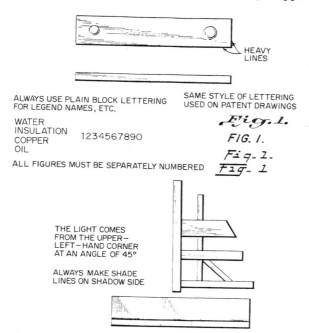

FIG. 10-5 Lettering and light shading.

Cylindrical and Spherical Objects

Figure 10-6a indicates correct shading for various size pipes, shafts, and spherical objects. Surface delineations should be shown by proper shading. Whereas a single heavy line on the shadow side is sufficient for small pipes, rods, shafts, or spherical objects, multiple shade lines are required for larger objects. The shade lines are blended from the second line, the outer line being the brightest.

Shading for Perspective and Depth

Figure 10-6c indicates the correct technique for perspective shading and mirror indication. In these perspective views heavy shade lines are placed on the edge closest to

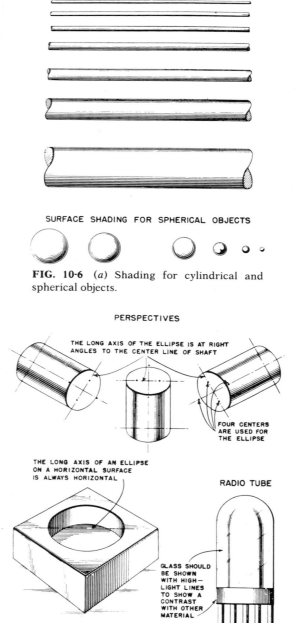

SURFACE SHADING FOR
VARIOUS SIZES OF PIPES AND SHAFTS

SURFACE SHADING FOR SPHERICAL OBJECTS

FIG. 10-6 (a) Shading for cylindrical and spherical objects.

PERSPECTIVES

THE LONG AXIS OF THE ELLIPSE IS AT RIGHT
ANGLES TO THE CENTER LINE OF SHAFT

FOUR CENTERS
ARE USED FOR
THE ELLIPSE

THE LONG AXIS OF AN ELLIPSE
ON A HORIZONTAL SURFACE
IS ALWAYS HORIZONTAL

RADIO TUBE

GLASS SHOULD
BE SHOWN
WITH HIGH—
LIGHT LINES
TO SHOW A
CONTRAST
WITH OTHER
MATERIAL

FIG. 10-6 (b) Shading for perspective and depth.

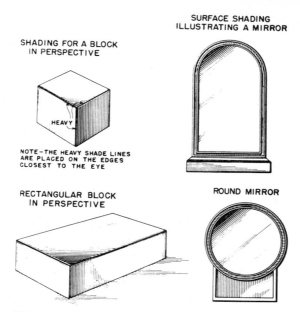

FIG. 10-6 (c) Shading for perspective and depth.

the eye. The rule of light approach from the 45° left-hand corner does not apply to perspective views. Horizontal ground lines reinforce the perspective view appearance. Advanced perspective techniques are illustrated in Fig. 10-6b. Centerlines, not to be shown on submitted drawings, clearly indicate the fundamental rules for determining the long axis of the ellipse. As in Fig. 10-6c, glass is shown by light lines to contrast with other materials.

Beveled-Edge and Irregular-Surface Shading

Inclined and irregular surfaces are distinguished from flat surfaces by the use of shading as shown in Fig. 10-6d. Note that the outer line is always a light line and that the surface intensity increases until a heavy line used to outline or edge the surface is reached.

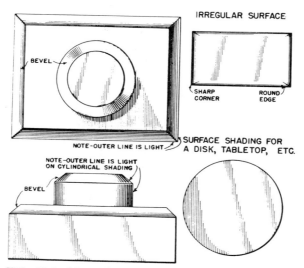

FIG. 10-6 (d) Surface shading for beveled edges and irregular surfaces.

Threads—Conventional Method

Figure 10-7 illustrates several methods of depicting conventional threads in bolts, studs, and openings. The designer should notice that convex and concave surface shading is related to the incidence of light from the 45° angle.

Gears and Bearings

Figure 10-8a and b depicts the conventional method for illustrating the most common forms of gears and bearings. The drafter must take particular care to show the correct spacing between gear teeth and must be careful concerning the weight of the shading lines used.

Special Figures and Forms

Figure 10-9a and b contains illustrations of special figures and forms most commonly encountered in patent drawings.

1. Link chain, lighted objects, and cones follow the conventional methods previously described. Freehand shading not otherwise permitted is acceptable for illustrating fabric, abrasive materials, sand, gravel, adhesive coatings, and other materials that are not linear or are highly irregular or amorphous in form.

2. Certain acceptable graphic representations of materials and colors have been established by the Patent Office. These are shown in Fig. 10-9b. These graphic representations must be used to depict the indicated colors or materials in patent drawings.

Scale

The scale to which a drawing is made ought to be large enough to show the mechanism without crowding when the drawing is reduced in reproduction, and views of portions of the mechanism on a larger scale should be used when necessary to show details clearly; two or more sheets should be used if one does not give sufficient room to accomplish this end, but the number of sheets should not be more than is necessary.

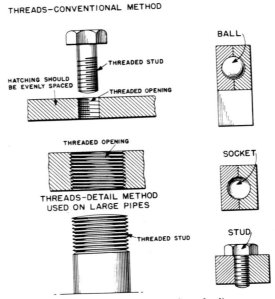

FIG. 10-7   Threads (conventional method).

BEVEL GEARS

BALL BEARING

NOTE—TEETH OF
EACH GEAR
HAVE THE
SAME
SLANT

INNER
RACE

OUTER
RACE

NOTE—ALL TEETH CONVERGE IN A CENTRAL
POINT. BROKEN LINES ARE FOR INSTRUCTION
PURPOSES AND ARE NOT TO BE PLACED ON DRAWINGS

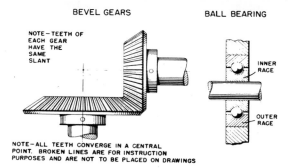

TOP PLAN VIEW

ROLLER BEARING

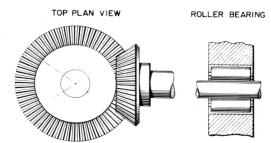

FIG. 10-8 (*a*) Gears and bearings.

SPUR GEAR

HELICAL GEAR

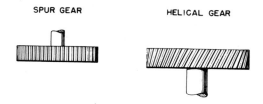

TWO SPUR GEARS IN MESH

WORM

GEAR

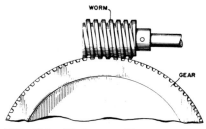

FIG. 10-8 (*b*) Gears and bearings.

Reference Characters

The different views should be consecutively numbered figures. Reference numerals (and letters, but numerals are preferred) must be plain, legible, and carefully formed, and they should not be encircled. They should, if possible, measure at least ⅛ in. (3.2 mm) in height so that they may bear reduction to ¹⁄₂₄ in. (1.1 mm), and they may be slightly larger when there is sufficient room. They must not be so placed in the close and complex parts of the drawing as to interfere with a thorough comprehension of the same, and therefore they should rarely cross or mingle with the lines. When necessarily grouped around a certain part, they should be placed a short distance away, at the closest point where there is available space, and connected by lines to the parts

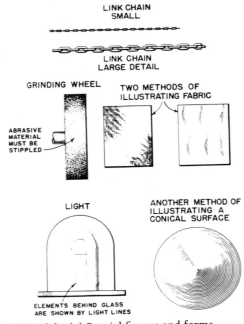

FIG. 10-9 (*a*) Special figures and forms.

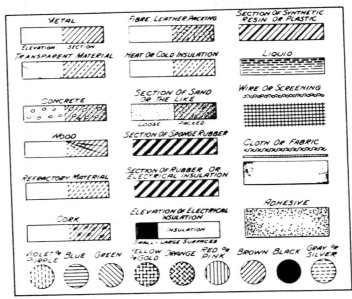

FIG. 10-9 (*b*) Special figures and forms.

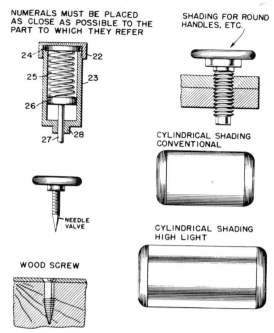

FIG. 10-10 Reference characters: advanced shading techniques (wood).

to which they refer. They should not be placed upon hatched or shaded surfaces, but when necessary a blank space may be left in the hatching or shading where the character occurs so that it shall appear perfectly distinct and separate from the work. The same part of an invention appearing in more than one view of the drawing must always be designated by the same character, and the same character must never be used to designate different parts (see Fig. 10-10).

Reference Characters—Alternative Shading Systems

Reference characters should be placed a short distance from the parts to which they refer (see Fig. 10-10). When necessary, blank spaces must be left in hatched or shaded areas to permit the insertion of reference characters. The character should be connected to the part by the shortest possible line.

Descriptive matter, trademarks, and commercial names are not permitted on patent drawings. Legends if necessary should be added in plain block lettering.

The views should always be properly planned so that one figure is not placed upon another.

Symbols and Legends

Graphical drawing symbols and other labeled representations may be used for conventional elements when appropriate, subject to approval by the Office. Figure 10-11a and b illustrates electrical symbols; Fig. 10-12a and b shows mechanical symbols.

The Patent Office has generally accepted for use in patent drawings the symbols listed in the following publications:

ANSI Y32.2-1975 (IEEE 315-1975): Graphic Symbols for Electrical and Electronics Diagrams
ANSI Y32.10-1967(R1974): Graphic Symbols for Fluid Power Diagrams
ANSI Y32.11-1961: Graphic Symbols for Process Flow Diagrams in the Petroleum and Chemical Industries
ANSI Y32.14-1973 (IEEE 91-1973): Graphic Symbols for Logic Diagrams
ANSI Z32.2.3-1949(R1953): Graphic Symbols for Pipe Fittings, Valves, and Piping
ANSI Z32.2.4-1949(R1953): Graphic Symbols for Heating, Ventilating, and Air Conditioning
ANSI Z32.2.6-1950(R1956): Graphic Symbols for Heat-Power Apparatus

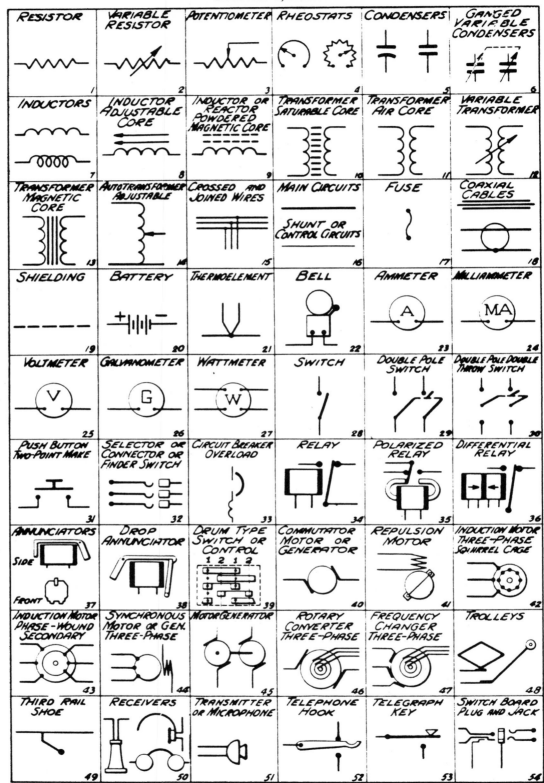

FIG. 10-11 (*a*) Electrical symbols.

Electrical Symbols — continued

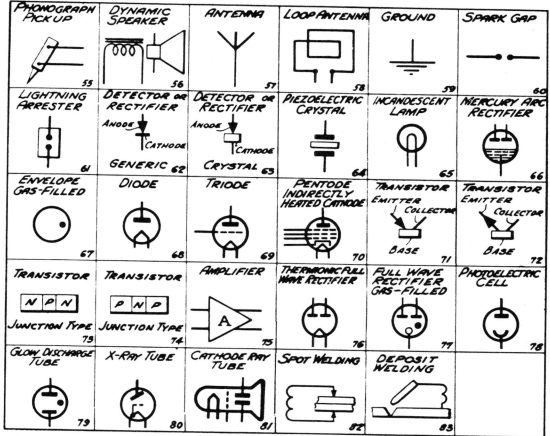

FIG. 10-11 (*b*) Electrical symbols.

Mechanical Symbols

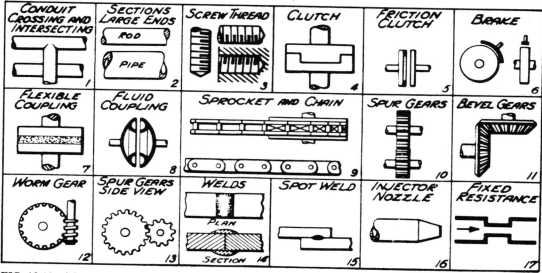

FIG. 10-12 (*a*) Mechanical symbols.

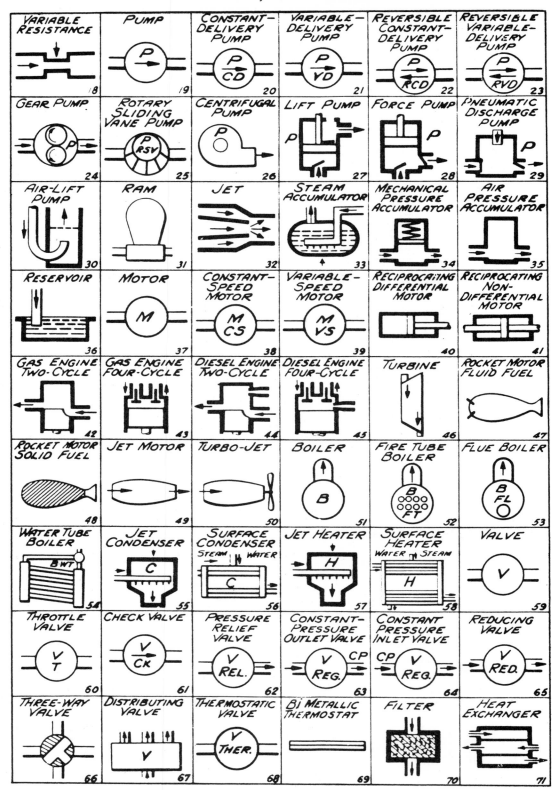

FIG. 10-12 (*b*) Mechanical symbols.

These publications can be obtained from the American National Standards Institute, 1430 Broadway, New York, NY 10018. The elements for which such symbols and labeled representations are used must be adequately identified in the specification. Although descriptive matter on drawings is not permitted, suitable legends may be used, or may be required, in proper cases, as in diagrammatic views and flowsheets, to show materials, or where labeled representations are employed to illustrate conventional elements. Arrows may be required in proper cases to show direction of movement. The lettering should be as large as or larger than the reference characters.

Views

The drawing must contain as many figures as may be necessary to show the invention, and the figures should be consecutively numbered if possible in the order in which they appear. The figures may be plan, elevation, section, or perspective views, and detail views of portions or elements on a larger scale, if necessary, may also be used. Exploded views, such as in Fig. 10-13, which shows the separated parts of the same figure embraced by a bracket, may be used to show the relationship or order of assembly of various parts. When necessary, a view of a large machine or device in its entirety may be broken and extended over several sheets if there is no loss in facility of understanding the view (the different parts should be identified by the same figure number but followed by the letters *a, b, c,* etc. for each part). The plane upon which a sectional view is taken should be indicated on the general view by a broken line, the ends of which should be designated by numerals corresponding to the figure number of the sectional view and which has arrows applied to indicate the direction in which the view is taken. A moved position may be shown by a broken line superimposed on a suitable figure if this can be done without crowding; otherwise, a separate figure must be used for this purpose. Modified forms of construction can only be shown in separate figures. Views should not be connected by projection lines, nor should centerlines be used.

Arrangement of Views

All views on the same sheet must stand in the same direction and should, if possible, stand so that they can be read with the sheet held in an upright position. If views longer than the width of the sheet are necessary for the clearest illustration of the invention, the sheet may be turned on its side so that the 2-in. (5.1-cm) margin is on the right-hand side. One figure must not be placed upon another or within the outline of another.

Figure for *Official Gazette*

The *Official Gazette*, the authorized publication of the Patent Office, is used to announce to the world the allowance and issuance of a patent. It is published biweekly and can be obtained for an annual subscription price of $89. Each patent issued is represented by a drawing and an abstract of the disclosure, a short verbal description.

When preparing the drawings for an application, the drafter should, as far as possible, prepare one of the views so that it will be suitable for publication in the *Official Gazette* as the illustration of the invention. See Fig. 10-14, which is a page from the *Official Gazette*.

Extraneous Matter

An inventor's, agent's, or attorney's name, signature, stamp, or address, or other extraneous matter, will not be permitted upon the face of a drawing, within or without the margin, except that identifying indicia (attorney's docket number, inventor's name, number of sheets, etc.) should be placed within ¾ in. (19.1 mm) of the top edge and between the hole locations defined earlier in this section under Size of Sheet and Margins. Authorized security markings may be placed on the drawings provided they be outside the illustrations and are removed when the material is declassified.

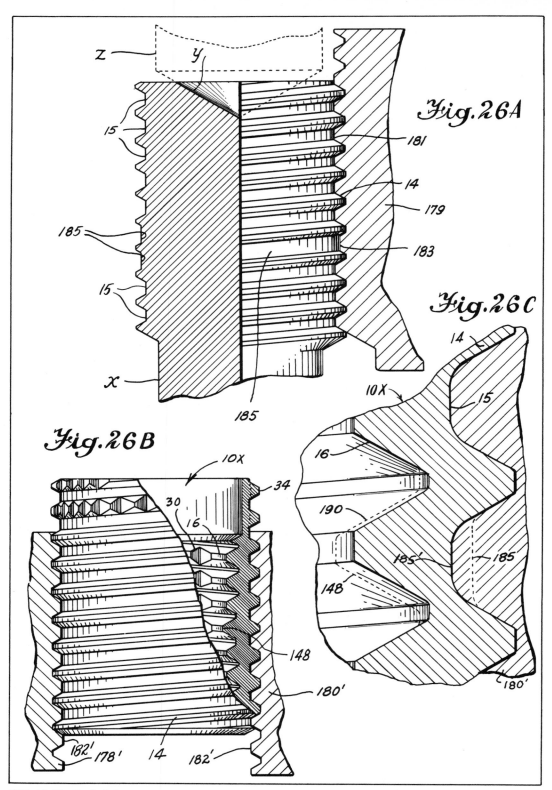

FIG. 10-13 Exploded view.

for supporting said upper frame on said lower frame generally parallel to said lower frame while permitting said upper frame to move between a high lift position above said lower frame and a lowered position below said high lift position; and, auxiliary scissors lift arms extending parallel to a vertical plane perpendicular to said horizontal direction; lift means for moving said upper frame between said positions; a pair of extendable feet mounted on said lower frame on each side of a median line through said lower frame parallel to said horizontal direction; and a single operating means for extending the feet of each pair to tilt said elevator horizontally when said upper frame is in said high lift position.

3,341,043
FOAMED PLASTIC ARTICLES
Thomas R. Santelli, Sylvania, Ohio, assignor to Owens-Illinois, Inc., a corporation of Ohio
Continuation of application Ser. No. 440,778, Mar. 18, 1965, which is a division of application Ser. No. 389,511, July 22, 1964. This application Dec. 19, 1966, Ser. No. 603,049
7 Claims. (Cl. 215—1)

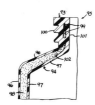

6. A plastic article comprising a body portion and a finish portion integral with said body portion, said body portion being essentially cellular between interior and exterior surface layers which are non-cellular, and said finish portion being substantially more dense than said body portion, said finish portion having dense, glazed interior and exterior surfaces confining therebetween any cellular material occurring in said finish portion.

3,341,044
SAFETY BOTTLE CAPS
Donald B. Valk, Madison, N.J., assignor to Wel-Kids, Inc., Madison, N.J., a corporation of New Jersey
Filed Aug. 18, 1966, Ser. No. 573,311
8 Claims. (Cl. 215—9)

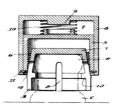

1. A safety bottle cap comprising a connector member having means for interlocking connection with a container, a locking member sleeved over the connector member and having means for causing interlocking of the connector member with the container, a cover member sleeved over the locking member and movable axially and rotatably relative thereto, and clutch means for interconnecting the cover member with the locking member upon said axial movement for thereafter causing turning movement of the locking member in clutch engagement.

3,341,045
HEAT INSULATED BOTTLE
Jack Sandler, Florham Park, N.J., assignor, by mesne assignments, to Air Reduction Company, Incorporated, New York, N.Y., a corporation of New York
Filed Aug. 13, 1963, Ser. No. 301,750
17 Claims. (Cl. 215—13)

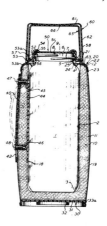

1. In combination, a rigid expanded plastic receptacle member, a flexible plastic skin by which the receptacle member is tightly surrounded, a rigid glass bottle fitting within the receptacle member and having at its top a restricted neck protruding from the top extremity of that member and below said neck a shoulder, and a flexible plastic inverted saucer centrally apertured for and tightly engaging said neck and peripherally secured to said skin, wherein said flexible plastic inverted saucer bears against said shoulder and forms a means retaining the bottle within the receptacle member.

3,341,046
FLUID-TIGHT BOTTLE CAP
André Bereziat, Lyon, and Guy Janssen, Chatou, France, assignors to Societe Astra de Bouchage, Surbouchage & Conditionnement, Societe Anonyme, Lyon, France, and Georges Lesieur & ses Fils, Societe Anonyme, Paris, France
Filed Mar. 29, 1966, Ser. No. 538,373
Claims priority, application France, Apr. 23, 1965, 14,383, Patent 1,453,863
2 Claims. (Cl. 215—41)

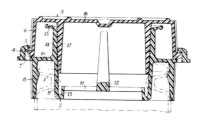

1. In a fluid tight cap for closing the neck portion of a container, said cap comprising a base element of flexible plastic material which is adapted to fit on the neck of the container, and a cap element which is adapted to close the top of said base element, wherein said base element comprises on the one hand two coaxial internal and external skirts fitting respectively to the inside and to the outside of the neck of said container, and on the other hand a substantially cylindrical sleeve projecting to the outside, the internal skirt defining an orifice, said cap

---

FIG. 10-14   Page from *Official Gazette.*

Transmission of Drawings

Drawings transmitted to the Patent Office should be sent flat, protected by a sheet of heavy binder's board, or they may be rolled for transmission in a suitable mailing tube; but drawings must never be folded. If they are received creased or mutilated, new drawings will be required.

Informal Drawings: Rule 185 R.P.P.C.

The requirements of patent rules relating to drawings will be strictly enforced. Everything described in the written specification must be shown in the drawings. A drawing not executed in conformity thereto, if suitable for reproduction, may be admitted, but in such cases, the drawing must be corrected or a new one furnished as required. The necessary corrections or mounting will be made by the Patent Office upon the applicant's request or permission and at his or her expense.

Plant Patents

A completely different category of patentable items is those described in 35 U.S.C. 161 as asexually reproduced new and distinct varieties of plants, including cultivated spores, mutants, hybrids, and newly found seedlings, other than a tuber-propagated plant or a plant found in an uncultivated state.

The drawings for such patentable plants are governed by a different set of rules than those prescribed for utility or design patents. These rules are found in 37 C.F.R. 1.165 and are summarized below.

Plant Patent Drawings

Plant patent drawings are not mechanical drawings and should be artistically and competently executed. Figure numbers and reference characters need not be employed unless required by the examiner. The drawing must disclose all the distinctive characteristics of the plant capable of visual representation.

The drawing may be in color, and when color is a distinguishing characteristic of the new variety, the drawing *must* be in color. Two copies of color drawings must be submitted. Color drawings may be made either in permanent watercolor or in oil, or in lieu thereof may be photographs made by color photography or properly colored on sensitized paper. Permanently mounted color photographs are acceptable. The paper in any case must correspond in size, weight, and quality to the paper required for other drawings. Nonpermanently mounted copies will be correctly mounted at the applicant's expense.

FOREIGN PATENT DRAWINGS

Patent Cooperation Treaty

Under the new Patent Cooperation Treaty (PCT) the rules for drawings are expected to be quite similar to those now in effect in the United States.

Since this treaty has only recently come into force, the final regulations have not been developed, although it is known that the sight size of the drawing will be 26.2 by 17 cm.

Foreign National Application

The patent laws of each country vary widely in regard to drawing rules and regulations. To provide the drafter with the information for each country would require printed material greater in length than this entire section. Over 10 different-sized drawing sheets are used in foreign practice, and materials vary from bristol board to kent unglazed paper, tracing cloth, and linen according to individual country laws. Most major cities of the United States have at least one firm specializing in foreign patent drawings, and it is recommended that the drafter or designer get in touch with such a firm to obtain any foreign patent drawings required.

Recent treaties creating a Common Market patent and a European patent are steps toward standardization of the patent process. It will be at least 10 years, though, until there is real progress in this direction.

REFERENCES

The publications listed here provide helpful additional information on subjects related to preparing patent drawings. They are available from the Superintendent of Documents, U.S. Government Printing Office, Washington, DC 20402.

Code of Federal Regulations, vol. 37: *Patents, Trademarks, and Copyrights*, July 1, 1977 (37 C.F.R.).
Manual of Patent Examining Procedure, sec. 608.02, July 1, 1977 (stock no. 0254-089).
U.S. Department of Commerce, *Guide for Patent Draftsmen*, 1975 (stock no. 003-004-00521).
U.S. Department of Commerce, *Patent Laws*, August 1976 (stock no. 003-004-00535-1).

Mathematical Graphics

Irving Dlugatch

Mathematical applications of graphics are of two types. The most common use is in the visualization or dramatization of statistics. The second application is in the solving of problems that are difficult to resolve otherwise or in which it is desirable to present all the solutions that exist simultaneously.

STATISTICAL CHARTS

Statistical charts are prepared to summarize data, to reveal relationships, and to interpret quantitative information. *Line charts* emphasize relationships. It is customary for the vertical axis to be measured in quantities such as dollars or pounds produced. The horizontal axis usually represents units of time. When the horizontal axis units are time units, the chart is sometimes called a *time series chart*. Figure 11-1 is an example of a *single-line chart*. Figure 11-2 is a *multiple-line chart* in which the three curves are plotted against the same base line. Figure 11-3 is a *component-part chart* in which the top curve represents the sum of the other three.

Statistical charts are sometimes distorted in order to accentuate some element of the statistics. Figure 11-4 represents the sales for a company. This chart indicates little increase in sales volume over a 5-year period. Rescaling the chart, in Fig. 11-5, shows a substantial increase.

Pie Charts

Pie charts consist of one or more circles divided into sectors for the purpose of showing the component parts of a whole. Figure 11-6 is a single-circle pie chart. Figure 11-7

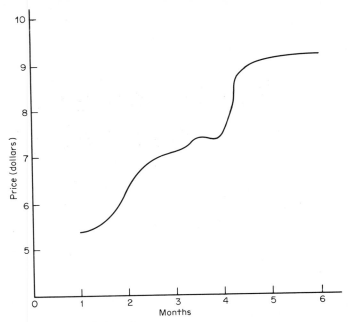

FIG. 11-1 Single-line chart.

demonstrates how multiple pie charts can illustrate changes in component parts over a period of time.

Bar Charts

In *bar charts* quantities are represented by the length of the bars. As with line charts, the vertical axis is measured in quantities. However, the horizontal axis need not be in time units. In fact, the horizontal axis need not be in any kind of units. Figures 11-8 and 11-9 are examples of bar charts. Figure 11-10 is a form of bar chart that has no horizontal axis.

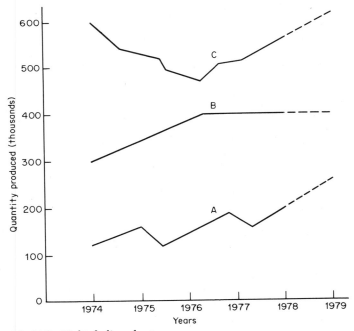

FIG. 11-2 Multiple-line chart.

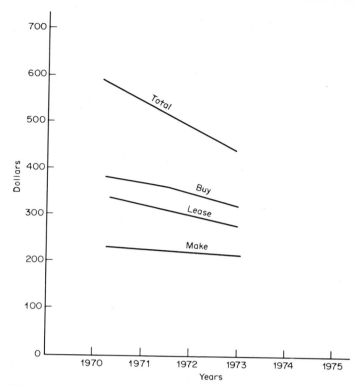

FIG. 11-3 Component-part chart.

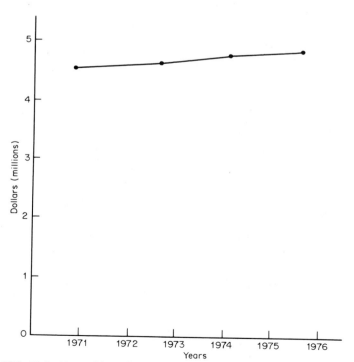

FIG. 11-4 Normal-line chart.

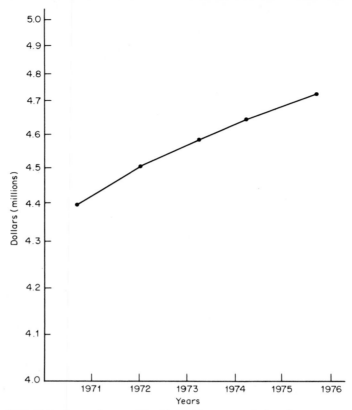

FIG. 11-5 Same chart as Fig. 11-4 with new vertical scale.

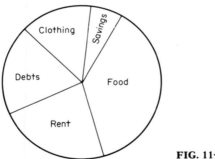

FIG. 11-6 Pie chart.

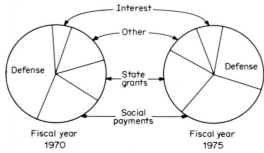

Fiscal year 1970

Fiscal year 1975

FIG. 11-7 Multiple pie chart.

11-4

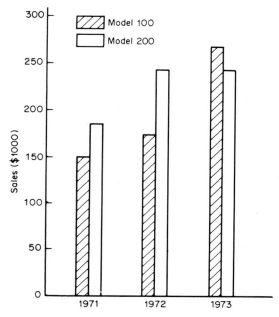

FIG. 11-8   Bar chart.

Histograms

A special form of the bar chart is the *histogram*. A histogram is a bar chart of a frequency distribution and is used to reveal trends or relationships that might be overlooked in a table. The purpose of a frequency distribution is to organize data items by classifying them according to some observable characteristic. *Frequency distribution* refers to the frequency of occurrence of values in the various classes. To construct a frequency distribution, it is necessary to determine the number of classes, the width of the classes, and the number of observations or frequencies in each class. Figure 11-11 is a typical histogram.

The bars in Fig. 11-11 are of equal width, representing equal class intervals. The height of each bar corresponds to the frequency of the class. The area of a bar is proportional to the frequencies in that class. If unequal classes are used, the height of the

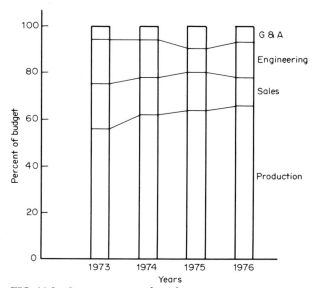

FIG. 11-9   Component-part bar chart.

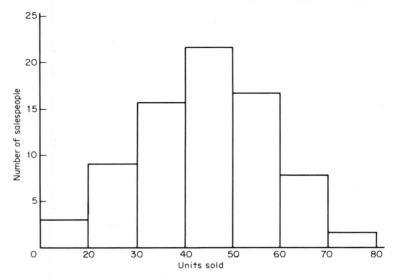

FIG. 11-10   Special bar chart.

bar must be adjusted to make the area correctly proportional to the frequencies in the bar. The width of the bars used in the histogram is calculated from

$$W = (L - S)/n$$

where W = width of class
   L = largest value
   S = smallest value
   n = number of classes

Typical values of n vary from a minimum of 5 to a maximum of 15. The value for a specific application depends on the purpose of the histogram. Classes must be selected so that both the smallest and the largest data items are included.

Frequency Polygons

A *frequency polygon* is another method of graphically presenting a frequency distribution. A frequency polygon is shown in Fig. 11-12. It is essentially a curve drawn

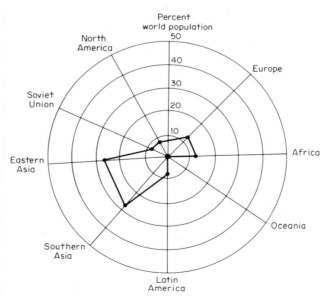

FIG. 11-11   Histogram.

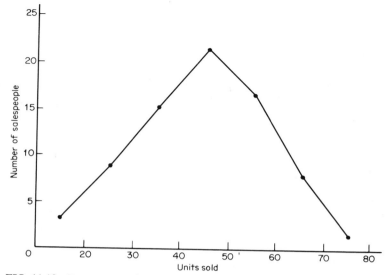

FIG. 11-12   Frequency polygon.

through the midpoints of the histogram bars in Fig. 11-11. The polygon is useful for making comparisons of two or more distributions. The curve can be smoothed by reducing the class size and increasing the number of items in the distribution. If this is carried to the extreme, the curve of Fig. 11-13 results. This bell-shaped *normal* curve is the generalized normal *population frequency distribution*. Whereas Fig. 11-12 represents only a small sample of all the available data, Fig. 11-13 describes the total data.

The frequency distribution can be converted to a *cumulative* frequency distribution as shown in Fig. 11-14. This graphical representation of a cumulative frequency distribution is called an *ogive*. This curve is useful in showing the number of values falling above or below a certain value, usually the *median* value. The median is the arithmetic mean or average value. In Fig. 11-14, the minimum value is 20 and the maximum is 70. The median value is therefore 45. The curve indicates that 38, or 54.2 percent, of the salespeople sold less than 45 units.

The purpose of averages is to give, in a single value, the typical size or location of a set of values. The *arithmetic mean* is the sum of the values of a group of items divided by the number of such items. The *median* is a measure of central tendency that occupies the middle position in an array of values. The array must be in either ascending or descending order. The *mode* is the most commonly occurring value in a series. In Fig. 11-11, the class of 40 to 50 would be the mode, with 45 the median of the class. The median is also the arithmetic mean in this case.

The mode is the most commonly occurring value. Frequency distributions do not show the actual values, however, and the mode must be approximated. The most commonly occurring value in the distribution is found in the largest class at the peak of the frequency polygon. The largest class is the *modal* class. Figure 11-15 illustrates how the mode can be graphically approximated from a histogram. This is based on the formula

$$M = A + W[d_1/(d_1 + d_2)]$$

where M = mode
    A = lower limit of modal class
    W = width of class
    d_1 = difference between frequency of modal class and frequency of class preceding it
    d_2 = difference between frequency of modal class and frequency of class following it

The relationship between the mean, median, and mode for a frequency distribution depends on the skewness of the distribution. *Skewed* curves differ from symmetrical in that some values are much higher or lower than the typical values found in the

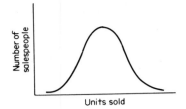

FIG. 11-13 Normal curve.

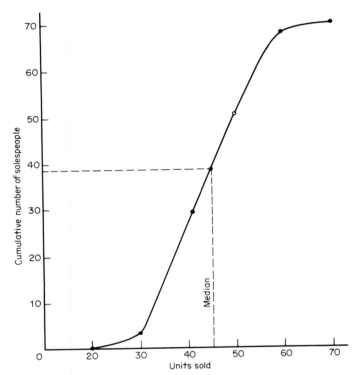

FIG. 11-14 Ogive.

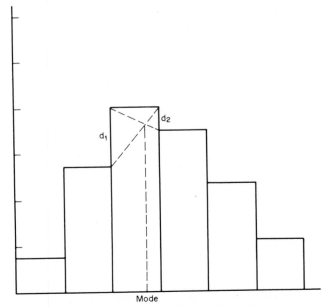

FIG. 11-15 Graphical approximation of the mode.

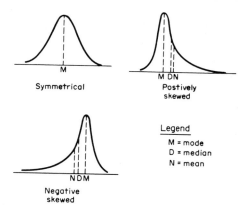

FIG. 11-16   Relationship between mean, median, and mode.

distribution. Figure 11-16 shows the relationships for the three types of distribution. It is seen that the mean, median, and mode are identical for the symmetrical distribution. For the *positively skewed* distribution, the mode has the lowest value, and the mean the highest. In the *negatively skewed* distribution, the mode has the highest value, and the mean the lowest. The median always lies between the mean and the mode.

TIME-SERIES ANALYSIS

Time-series analysis involves classifying and studying the patterns of movement of values over regular intervals of time. The four components of a time-series analysis are:

1. *Secular trend* is the smooth or regular long-term (i.e., secular) growth or decline of a series. The wavy line in Fig. 11-17 is the actual series. The smooth curve is the trend.

2. *Seasonal variations* are those that are periodic in nature and which recur regularly within a period of 1 year.

3. *Cyclical fluctuations* are periodic in nature and involve recurring up-and-down movement.

4. *Irregular movements* occur over varying but usually brief periods of time. They follow no regular pattern and therefore are not predictable.

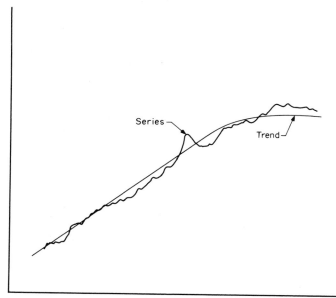

FIG. 11-17   Time-series: secular trend.

The time series in Fig. 11-17 includes all the components listed above. If a time series cannot be attributed to a trend, seasonal, or cyclical element, it is lumped into the *irregular* or *residual* category.

The trend component needs to be known in order to describe historical patterns or to predict persistent patterns in forecasting. Fundamental to the linear trend is the straight-line equation

$$Y = a + bX$$

where Y = trend value for a given time period
a = value of Y when $X = 0$
b = slope of the line

This is illustrated in Fig. 11-18.

If all the available data are plotted as points and no regular pattern can be detected, a linear trend can be measured by using a ruler to draw a straight line through the data by visually estimating the trend. A more reliable technique is the *method of least squares*, which mathematically fits the line to the data. Figure 11-19 shows a linear trend that has been fitted to time-series data.

There are two properties of a trend line computed by the method of least squares:

1. Sum of the deviations of the data values about the trend line will be zero. This is stated as

$$\Sigma(Y - Y_t) = 0$$

where Y_t = value on trend line (see Fig. 11-19)

2. The sum of the squares of the deviations is a minimum, or

$$\Sigma(Y - Y_t)^2 = \text{minimum}$$

The trend line equation is computed from

$$a = \Sigma Y/n$$

where n = number of data values

$$b = \Sigma(XY)/\Sigma(X^2)$$

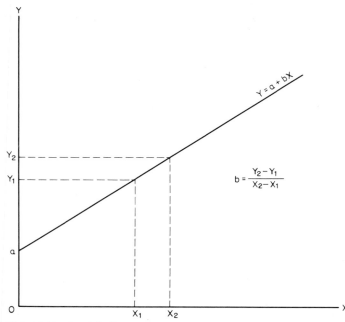

FIG. 11-18 Straight-line equation.

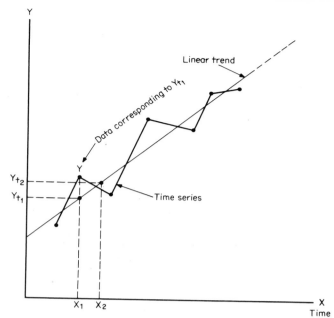

FIG. 11-19 Method of least squares.

and substituting in the equation

$$Y_t = a + bX$$

The dotted extension of the trend line is a projection of the trend to serve as a forecast.

REGRESSION ANALYSIS

Regression analysis is a technique for measuring and evaluating the relationship between two variables. In regression analysis, an estimating equation is needed to describe the pattern of the relationship. The estimating equation is called a *regression equation*. The variable to be estimated is called the *dependent variable* and is plotted on the Y, or vertical, axis. The *independent variable* exerts an influence on the dependent variable and is plotted on the horizontal, or X, axis.

A *scatter diagram* shows plotted points, each of which represents an item for which both a dependent and an independent variable exist. The diagram permits determining whether a useful relationship exists between the two variables and what type of equation is needed to describe the relationship. Figures 11-20, 11-21, and 11-22 illustrate scatter-diagram forms. The method used for linear regression is that of least squares. The computations for regression are facilitated if the formulas are stated in terms of deviations from the means of the X and Y variables. Thus,

$$x = (X - \overline{X})$$
$$y = (Y - \overline{Y})$$
$$xy = (X - \overline{X})(Y - \overline{Y})$$

and

$$a = \overline{Y} - b\overline{X}$$
$$b = \Sigma\,(xy)/\Sigma(x^2)$$

and

$$Y = a + bX$$

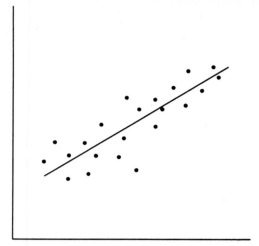

FIG. 11-20 Linear relationship.

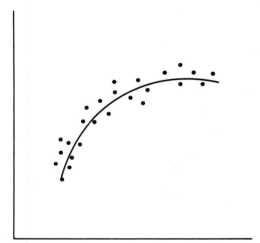

FIG. 11-21 Curvilinear relationship.

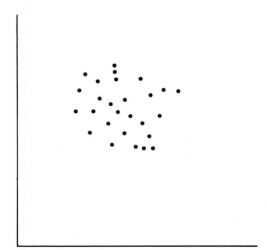

FIG. 11-22 No relationship.

GRAPHIC SOLUTION OF EQUATIONS

The real roots of an equation $f(X) = 0$ with real coefficients can be estimated approximately from the graph of the function $y = f(X)$, because they are the X coordinates of the points where the graph meets the X axis. This is shown in Fig. 11-23, which is the solution for the equation $2X^3 + 3X^2 - 5X - 6 = 0$. From the graph, the roots are -2, -1, and $+1.5$.

This technique is particularly useful in the design of servomechanisms. A *servomechanism* is a power-amplifying device in which the element driving the output is actuated by a function of the difference between the input and the output. The primary function of a servomechanism is to automatically control a given quantity or process in accordance with a given command. In the continuous-control type, a continuous action takes place to reduce the output error. The predominant characteristics of servo systems can be described by a differential equation.

Because the servo system uses feedback for control, it has a tendency to oscillate. To determine the stability of the system, the position of the roots of the characteristic equation must be determined in the complex plane. Figure 11-24 is a complex-plane plot of roots of a second-order system.

Servo system response is complex, so it is necessary to plot it for various degrees of

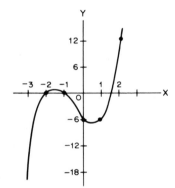

FIG. 11-23 Graphic solution of equation.

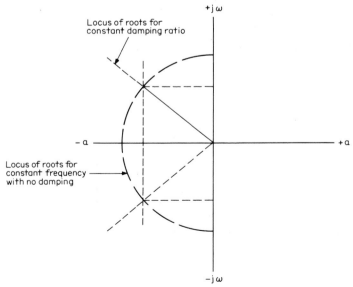

FIG. 11-24 Complex-plane plot of roots of a second-order equation.

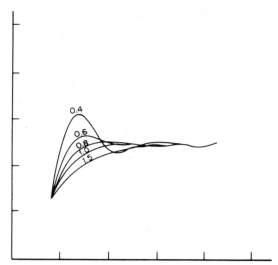

FIG. 11-25   Servo system: transient curves.

damping. Figure 11-25 is an example of such curves. These are dimensionless transient curves of the error response of a second-order system when subjected to a step velocity input.

GRAPHIC WAVEFORM ANALYSIS

A harmonic analysis can be made graphically from the graphic plot of a waveform. Figure 11-26 is such a waveform. The procedure is as follows:

1. Select one complete interval, T, of the function.
2. Divide the interval into k number of equal subintervals of width W.
3. Determine the coefficients of the sine terms in the equivalent Fourier series.

$$A_n = 2 \sum_{m=1}^{m=k} [(a_m \sin n\Theta_m)/k]$$

where A_n = peak amplitude of the nth harmonic
m = number of subinterval = 1, 2, 3, ... ,k

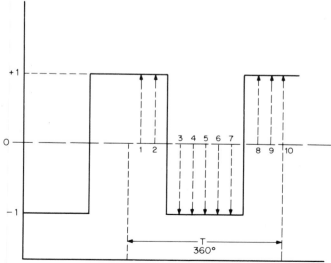

FIG. 11-26   Graphic waveform analysis.

a_m = amplitude of mth subinterval in graphic plot of waveform

Θ_m = phase angle of mth subinterval

= $2\pi m/k$ radians

4. Determine the coefficients of the cosine terms from

$$B_n = 2 \sum_{m=1}^{m=k} [(a_m \cos n \Theta_m)/k]$$

In Fig. 11-26, $k = 10$, $a_1 = +1$, and $a_3 = -1$.

The absence of a harmonic will be indicated by a zero amplitude for the coefficient. The accuracy of the analysis will increase as the number of intervals is increased. If the number of intervals is k, the highest harmonic n determined by this method should be less than $k/2$.

MAXIMA AND MINIMA

The function $f(X)$ has a maximum at $X = a$ if a constant can be found such that $f(X) < f(a)$ for $|X - a| <$ the constant, $X \neq a$. The constant must be different from zero so that there are points on the curve on both sides of a below the value of a.

Similarly, the function $f(X)$ has a minimum at $X = b$ if a positive constant can be found such that $f(X) > f(b)$ for $|X - b| <$ the constant, $X \neq b$. In this case there must be points on both sides of b above the level of b.

Figure 11-27 is a graph of the function $f(X)$. In this graph, a is the maximum of the function, and b and D are minima. For the curve between C and E, E is the *greatest* value, and b is the least value as well as being a minimum.

CURVE SKETCHING

Some problems require a clear idea of the shape of a curve as a whole. This is best obtained from a sketch that exhibits the major features of the curve. It is usually impractical to find this shape by plotting a large number of points on the curve. The clues to the sketch are:

1. Symmetry about the axes. If one variable can be replaced by its negative without changing the function, then the other variable is symmetrical about its axis.

2. Intersection with the axes. Determine the points where the curve crosses an axis and find the direction in which it does so.

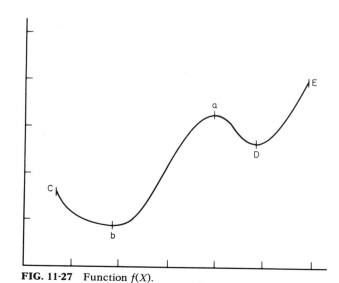

FIG. 11-27   Function $f(X)$.

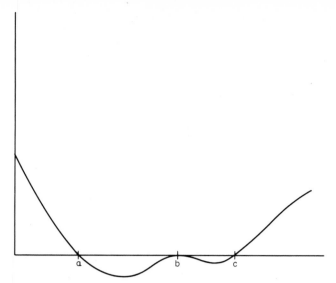

FIG. 11-28 $y = (x - a)(x - b)^2 (x - c)$.

3. Excluded regions. See if there is a range of values of X for which Y cannot have real values, and vice versa. There are no points of the curve with coordinates inside such ranges.

4. Behavior at infinity.

5. Behavior near the origin.

6. Singular points. These occur where dy/dx is indeterminate.

7. Asymptotes and their intersections.

8. Gradient of the curve.

9. The use of polar coordinates.

Figures 11-28 through 11-30 are examples of curve sketches.

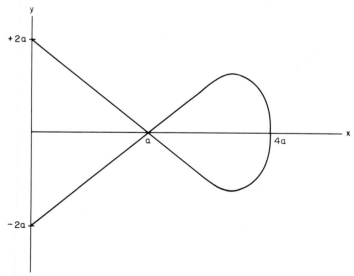

FIG. 11-29 $ay^2 = (x - a)^2 (4a - x)$.

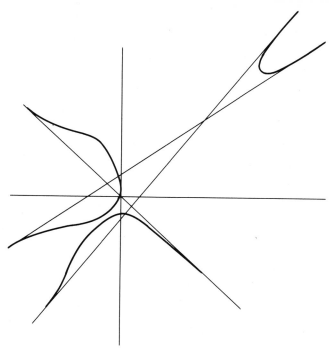

FIG. 11-30 $y^3 - 2xy^2 - x^2y + 2x^3 + 3x^2 + 6y^2 + 7x = 0$.

GRAPHIC INTEGRATION

Given the curve AC in Fig. 11-31. The area under the curve can be found by dividing the area into narrow vertical slices; the narrower these slices, the better the accuracy. The area of the slice is $y_1 \, \Delta_1 x$. The area under the curve is equal to the sum of all the slices. This is approximately equivalent to the integral of $y \, dx$ between the values zero and c for x.

GRAPHIC SOLUTION OF GAME MATRIX

Game-theory problems are solved by means of a payoff matrix like that in Fig. 11-32. The first-column numbers are the Blue player's strategies for playing the game. The

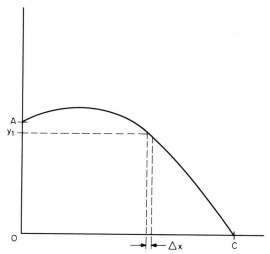

FIG. 11-31   Area under a curve.

Red Blue	1	2	Row minima
1	4	(2)	2*
2	2	1	1
Column maxima	4	2*	

FIG. 11-32 A 2 × 2 game theory matrix.

Red Blue	1	2	Row minima
1	0	2	0
2	4	1	1*
Column maxima	4	2*	

FIG. 11-33 A matrix without a saddle point.

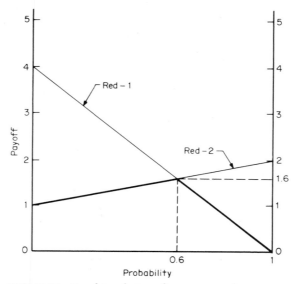

FIG. 11-34 Graphic solution of a game matrix.

numbers in the top row are the Red player's strategies. The number 4 represents the amount Red pays Blue for a number 1 strategy match. The number 2 is the payoff for a failure to match, and 1 is the payoff for a number 2 strategy match.

The bottom row gives the maximum value in each column. The last column is the minimum value in each row. Red will choose the smallest of the column maxima as the optimum strategy. This is called the *minmax*. Blue's optimum strategy is number 1, which gives Blue the *maxmin*. The minmax and the maxmin are both equal to 2 and are marked with an asterisk. When the maxmin and the minmax are equal, the game is said to have a *saddle point*. The saddle point, 2, is circled in Fig. 11-32.

If the matrix does not have a saddle point, it may be difficult to solve. Figure 11-33 is a sample of such a matrix. Notice that the row minima is equal to 1 and the column maxima is 2. Therefore there is no saddle point. Figure 11-34 is the graphic solution of the Fig. 11-33 matrix. The Red -1 line gives the payoff to Blue if Blue plays Blue -1 and Red plays the number 1 strategy. Blue's payoff when Red plays the number 2 strategy is shown by the other line. Blue's minimum gain for any value of the probability is given by the heavy lines. If Blue chooses a 0.6 probability (at the intersection of the two lines), Blue will maximize the minimum gain against either of Red's strategies. Hence this point is the optimum strategy, and the value of the game to Blue is 1.6.

PERT NETWORKS

The *Program Evaluation Review Technique (PERT)* is a method used in the planning and control of complex development and production programs in industry. It is used chiefly to resolve the problem of relating time needed to job costs. Figure 11-35 is an example of a PERT network.

The arrows, labeled with letters *a*, *b*, *c*, etc., are activities that are accompanied by a time-for-completion estimate. The squares at the beginning and end of the activity arrows are events or milestones. An event, unlike an activity, does not require the expenditure of resources. An event is used to identify the point in time at which an item is delivered.

The time required to accomplish each activity is estimated and noted on the network as shown. Thus, activity *a* is estimated to take 4 weeks. The earliest date that an event can be completed is the sum of the time for activities preceding it. For example, that date is 4 weeks from now for Event 1. For Event 2, both tasks *b* and *e* must be completed. Event *e* is the critical one, because it takes the longest. Therefore Event 2 will take 7 weeks (*a + e*). In this manner, the estimated time for completion of Event 4 is 11 weeks, because the longest path is *a + e + f* as designated by the heavy line in the figure. The longest path is called the *critical path*. It is also the path that has zero slack. *Slack* is the difference between any path and the critical path. It defines the allowable delay in the completion of a specific event.

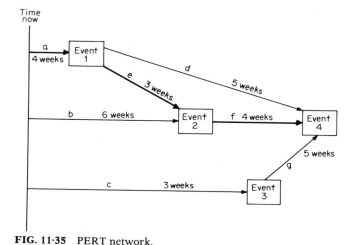

FIG. 11-35 PERT network.

GRAPHIC SOLUTION OF LINEAR PROGRAMMING

Mathematical programming is a quantitative technique for optimization that employs mathematical models. A typical problem for which mathematical programming is suitable is in the allocation of production capacity. The general nature of these problems is: a group of limited resources must be shared among a number of competing demands, and all decisions are interlocking because they have to be made under a common set of fixed limits.

Linear programming makes use of linear relationships among factors affecting the operation under study. It treats quantities that can be defined by adding multiples of certain other quantities. Linear programming makes use of inequalities because planning programs usually seek results that fall within certain limits, rather than a precise quantitative objective.

A set of inequalities stated as equations does not have a unique solution. There are many *feasible* solutions. Therefore, a final condition is imposed for choosing the *optimum* solution from those available. This is the criterion of minimizing a parameter or maximizing another. A solution obtained by mathematical programming constitutes a program or schedule of allocations. The optimum solution may be the least expensive.

The graphic solution of a linear programming problem is illustrated by Fig. 11-36. In this case, two products A and B are being processed by four departments: stamping, milling, drilling, and painting. Based on data concerning the time it takes to process each of the products in each department and the capacity of each department, the following equations are formed.

$$30A + 40B \leq 12{,}000 \quad \text{stamping}$$
$$60A + 20B \leq 12{,}000 \quad \text{milling}$$
$$40A + 25B \leq 10{,}000 \quad \text{drilling}$$
$$12A + 10B \leq 6000 \quad \text{painting}$$

All these constraints must be satisfied for any mixture of A and B chosen as a possible solution.

Each of the equations is plotted as a straight line in Fig. 11-36. The intersections of the axes are calculated as follows:

Stamping:

If $B = 0$, $A = 400$ (capacity of department)/(time to process 1 A unit)
If $A = 0$, $B = 300$

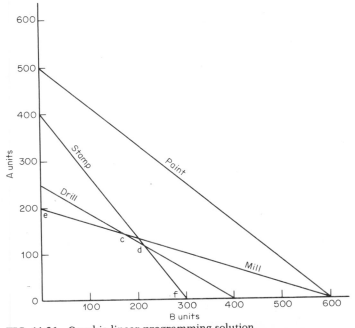

FIG. 11-36 Graphic linear-programming solution.

Any point on a line or below it is a feasible solution because it satisfies a constraint. Points below all the lines are feasible solutions that satisfy all the constraints. This is encompassed by the region outlined by the lines joining points 0, e, c, d, and f. The optimum solution lies on the boundary of the feasible solution space. In this case there are four possible optimum solutions: e, c, d, and f. It is known that product A yields a profit of \$5 and B a profit of \$4.

The solutions at each of the possible solutions are:

$$e: 5(200) + 4(0) \quad = \$1000$$
$$c: 5(140) + 4(175) = \$1400$$
$$d: 5(120) + 4(210) = \$1440$$
$$f: 5(0) \quad + 4(300) = \$1200$$

It is obvious that d is the optimum solution, with a maximum profit of \$1440. This calls for a program of 120 A units and 210 B units.

PERFORMANCE VARIABILITY

The performance of a device depends on the parameters of the component parts and on the particular set of values assigned to those parameters. Because of imperfect parts and subsequent time-dependent and environmental effects, system performance variability is inevitable. This concept is illustrated in Fig. 11-37, where a performance characteristic V of a device is plotted as a function of parameter P. Data for a plot of this type can be obtained by holding all parameters and environmental conditions except P constant at nominal values; P is varied over a range above and below its nominal value.

The nominal value of P falls at the point on the curve $V = f(P)$ at the design center of V, V_{nom}. The values of P are found to lie not exactly at P, but in the range indicated in the plot by the lower frequency distribution. The effect on V of this variability in P can be determined by projecting the P distribution up to the curve and over to the V axis as shown. If the curve is essentially linear, the distribution of V will have the same shape as that of P. If the curve is nonlinear in the range of interest, however, the distribution of V will be a distorted version of that of P.

Figure 11-38 shows the application of this technique to the case of the performance variability affected by several parameters simultaneously. The figure shows a positive

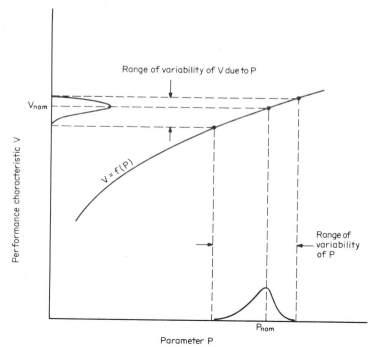

FIG. 11-37 Performance variability.

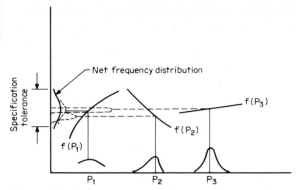

FIG. 11-38 Performance variability as a function of the variability of three parameters.

correspondence between $V_1 P_1$ and P_3, and a negative correspondence between V and P_2. It can be seen that V is highly dependent on P_1 and P_2, but only slightly dependent on P_3. The net variability of the performance characteristic is influenced by all three parameters.

DECISION TREES

Decision trees are graphic representations of probabilistic decision theory. The decision tree enables a planner to consider alternative courses of action and compare the results of the actions in terms of the expectations.

Figure 11-39 is the diagram of a simple decision-tree solution. The box at the extreme left is called the *decision point*. Here the manager is presented with two or more possible actions from which to select the one that promises the most desirable outcome. In this case, the manager must decide whether to use manufacturer's rep-

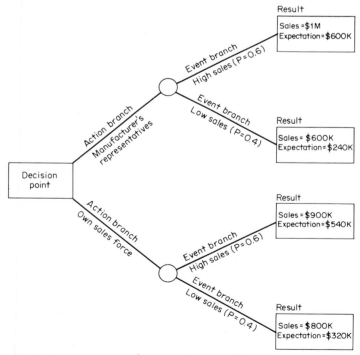

FIG. 11-39 Decision tree for a marketing problem.

resentatives or the company's own sales force to market a line of products. The two alternatives each make a branch of the tree originating at the decision point.

Each branch terminates at a node at which there are two outcomes (in this problem) of the action, called *event branches*. Each branch ends in a result, or *outcome*, such as profit or sales volume. The product of the outcome and the event probability gives the expectation.

In the illustrated problem, there is a 60 percent probability of achieving some high level of sales and a 40 percent probability of a lower level of sales. The manufacturer's representatives are more expensive because of their large commissions, but they are not always effective. Therefore, their high level of sales is estimated as $1 million and the low level as $600,000. For the company's sales force, the high-level estimate is $900,000 and the low level is $800,000.

The expectations are computed for each possible event. These are entered on the diagram. Adding the two expectations for each action gives the "value" for the representatives as

$$\$600,000 + \$240,000 = \$840,000$$

For the sales force,

$$\$540,000 + \$320,000 = \$860,000$$

Therefore, the best choice is to use the company sales force, because the "value" of this decision was $20,000 higher than that for the representatives.

BREAK-EVEN ANALYSIS

The *break-even point* is that point of a company's operations at which it passes from losses to profits or vice versa. Figure 11-40 depicts a linear break-even analysis chart. This is the simplest form of this type of analysis. The solid line is a plot of all the break-even conditions. The dotted lines are the plots of cost of producing a particular volume. Where any particular cost line crosses the solid line is the break-even point for that level of costs. A break-even point is where the costs exactly equal the sales volume. The dotted line represents the variable costs (they vary with volume), and the point at which the dotted line crosses the vertical scale marks the fixed costs. The fixed costs

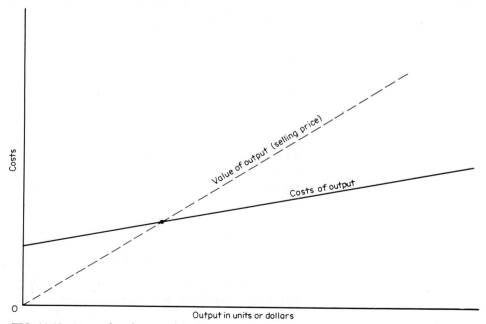

FIG. 11-40 Linear break-even analysis.

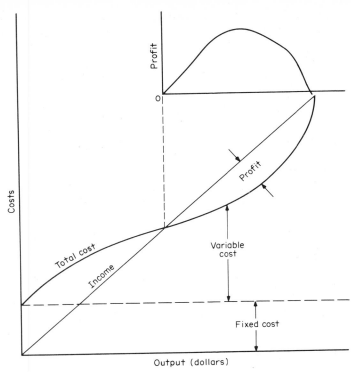

FIG. 11-41 Nonlinear break-even analysis.

are the same at any level of output. Thus the dotted line is giving the sum of the fixed and variable costs. The separation between the dotted line and the solid line is a measure of the profitability called the *profit path*.

Figure 11-41 is an example of a nonlinear analysis. In this case, the profit can go from zero to a peak and back to zero. The profit is plotted on an expanded scale in order to determine the peak more accurately.

Figure 11-42 is a special application of break-even analysis. It shows the relationship between carrying cost and ordering cost for the purpose of computing something called the *economic order quantity (EOQ)*. The EOQ is the best order size, i.e., that size

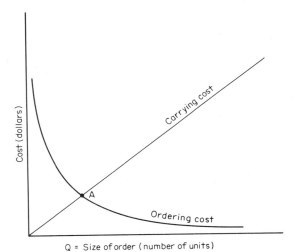

FIG. 11-42 Relationship between carrying cost and ordering cost.

at which the cost of carrying the inventory equals the cost of ordering. The crossing of the two curves at point A means the carrying cost is exactly equal to the ordering cost. This value of Q can then be used in the following:

$$EOQ = Q = [(2DJ)/(VE)]^{1/2}$$

where D = quantity of goods needed for a given period
   J = cost of placing an order
   V = unit value of items ordered
   E = percent of V, an estimate of the carrying costs

LEARNING CURVES

Learning curves relate the direct-labor hours required to perform a task to the number of times the task has been performed. For a large number of activities, the learning curve takes the form of Fig. 11-43. In essence, the curve stems from the fact that the time to perform an activity decreases by a constant percentage whenever the number of trials is doubled. That is, both the individual worker as well as the assembly line as a unit become more efficient with experience. Labor costs are directly proportional to the efficiency of the labor. Therefore, the curve in Fig. 11-43 plots dollars against volume instead of time against output. Typical curves are 80 percent curves, i.e., a 20 percent reduction in hours for each doubling of output. For the curve of Fig. 11-43, the cost goes down from $200 to $140 when the accumulated volume doubles from 40 to 80. This is a 30 percent reduction in costs, and the curve is a 70 percent curve.

The costs have a very strong relationship to the company's share of the competitve market. If cost declines with the number of units produced, the competitor that has produced the most units will probably have the lowest cost. Because the products of all competitors have essentially the same market price, the competitor with the most unit experience should enjoy the greatest profit. Conversely, large differences in cost

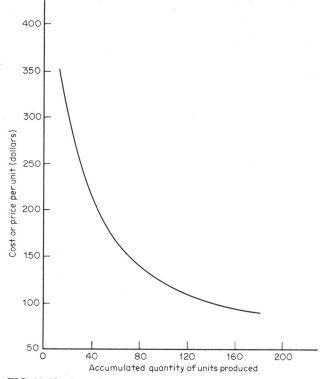

FIG. 11-43 Learning curve.

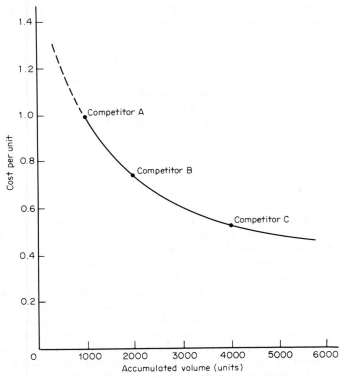

FIG. 11-44   Market share and costs for three competitors.

and profit can exist between competitors with widely different unit experiences. The relationship in market shares is shown in Fig. 11-44.

Figure 11-44 represents a stable market, which can occur only after a reasonable period of time. Competitor C dominates the market, with about 50 percent of the market. If the market is growing 15 percent or more annually, in units, then the dominant producer may have much more or less than 50 percent of the market.

Reprographics and Micrographics[1]

Regis L. Augusty

THE COMMUNICATION CRISIS

Many engineering managers have never really given serious thought to a most significant problem affecting their work and their company's profits. The problem is one of communications—the timely movement of design ideas to the end user. Today the process is cumbersome, slow, and, for the most part, a disaster.

[1]*Reprographics* is the internationally accepted term replacing *copying* or *reproduction*. It encompasses microcopying, photocopying, duplicating, in-plant printing, and reproduction by typewriters. *Micrographics* is a reprographics system that uses microimaging to achieve space savings, high-quality reproducibility, durability of records, and automated retrieval.

Outside Pressures

The pressure is on today's engineer as never before. The problems of energy, mass transportation, and major environmental issues rest squarely on the shoulders of the engineering community. The engineer is being asked to solve these problems with tools that have not changed in 40 years. Plus, the engineer now faces:

- Greater governmental controls
- Shorter lead time
- Higher costs
- Shortages of skilled personnel
- Keener competition

The Impact and Importance of Reprographics

The engineer deals daily with masses of information. How all this information is assimilated—and how effectively it is communicated—is tied directly to a company's ability to store, retrieve, and reproduce information. However, most of us have never really thought about the impact of reprographics services on our ability to get the job done—and done well.

Today, when a project is delayed, even a day, the effect can be quite serious. Consider the fact that when drawings are ready to be released, they usually are already late. Everyone needed them yesterday. Anything that further delays the distribution, be it people, machines, or communication systems, only aggravates an already intolerable situation.

A Twofold Problem

Time is our enemy. Engineering time is expensive. Since 1964 the cost of engineering has increased two or three times while productivity of the average engineer has leveled off. The engineer is told to become more productive; the question is, how? Hire more people? Develop better tools? Let's look at each of these separately.

Today, you would hire new people if you could get them—qualified drafters are difficult to come by. Moreover, hiring more people does not necessarily increase production. The truth is, production suffers because you must use your best people to train the new employees, and productivity drops.

The answer to increasing productivity in the engineering department is to supplement the labor force with labor-saving tools.

Which Tools to Use

There are many products on the market that are presumed to save time and help engineers become more productive; among them are better drafting tables, calculators to replace slide rules, modern writing instruments, better pencils, and, of course, computer-aided design (CAD) systems. All these tools are valid. A CAD system can save time. Although the system is expensive and the learning curve is long, improvements in productivity are realized. However, the major concern with a CAD system lies in its inability to produce prints in the quantity required to satisfy the needs of the end user. With CAD systems, we have a great deal of information tied up in digital form, and a limited means to secure the data. The system in other words becomes plotter bound.

REPROGRAPHICS: THE REAL COMMUNICATION CRISIS

We have said that the engineer deals daily with masses of information. The information comes from a variety of sources—customers, vendors, the sales department, manufacturing, and other locations. It comes in all forms—opaque copy, blue-line prints, reduced-size xerographic copies, microfilm aperture cards, microfiche, sepia copy, vellums, and polyester-based drawings. The engineer can be buried under a variety of input media.

The engineer is forced to sort and handle this diverse input and in turn develop data in the form of drawings, specifications, technical manuals, assembly drawings, tool drawings, and instruction manuals for installation, construction, and service. In reality, the engineer is tied irreversibly to reproduction processes that, for the most part, have not improved much during the last 40 to 50 years.

We have all heard engineers lament that it takes more time to get prints from drawings than it took to create the originals. The basic reproduction process in use today is the time-proven diazo system, which involves hand-feeding a variety of drawing sizes into a machine in which a strong light is passed through an original in order to create one print. This is a time-consuming process that has been known to damage originals if one is not careful in handling them. This is the primary reason why most drafters have chosen to draft on a polyester base rather than on vellum.

Reprographics: The Key to Engineering Productivity

If reference data are slow in leaving or arriving, the productivity of the engineer is drastically affected. We call this "engineering waiting time." No one calculates it, but it should be a factor in every job bid. If an engineer does not have timely, accurate, and legible data, his or her effectiveness is compromised, and productivity suffers. Consider this simple problem: You need six prints from 100 drawings. A third of the drawings are new polyester drawings, a third are vellum drawings that are one to five years old, and a third are paper or polyester sepias. You need 600 prints. You want them folded and collated into six sets. You take them to your reproduction center and ask that they be ready in one hour. You're told that will be impossible, and that's the problem.

Hurry Up and Wait

CAD systems do not help. Although CAD systems can help create engineering designs a great deal faster than conventional methods can, information is locked into a digital format, either on a disk or a magnetic tape. For prints to be made, a magnetic tape must be created and then taken to an off-line plotter like the Gerber Super Plotter or a Versatec electrostatic, where translucent originals are prepared. These translucent originals must then be run through the slow diazo process for initial distribution. This is a strange dichotomy. Here we are, using a highly sophisticated device to create drawings, and then using a decades-old diazo process to distribute the information. The result is frustration, waiting time, and lost opportunities. Engineering productivity suffers, and people and projects must wait.

The Scope of the Problem

To increase production in the engineering departments, one must review the entire process of idea development from initiation through completion. Ideas come from many sources—customers, vendors, management, manufacturing shops, and other areas where need exists. These ideas are evaluated against a base of existing information, personal experience, the drawing vault, change notices, customer input, and vendor input. The process is called *information retrieval.*

Once information is secured, it is acted upon. It is manipulated, changed, reworked, and recreated to the satisfaction of many parties. This is the *creative process.*

Completed data are passed on to engineering reproduction and control. The data must be documented, reproduced, and distributed to the shops, quality control departments, customers, vendors, management, etc. During this process automation and reliability are paramount. This is *reproduction and distribution.*

Once information is released, it must be refiled for future use. The file must be accessible, accurate, secure, up to date, and convenient and easy to use.

Simply stated, the total process consists of:

- Creation of information
- Reproduction of information

- Distribution of information
- Storage of information
- Retrieval of information

All the elements in the total process are interdependent. For example, it does little to have the greatest ideas in the world in your possession if you do not have an organized filing system that enables you to find those ideas. Also, it does little to be able to create drawings or ideas more rapidly only to lose time through the distribution process.

But that is in effect what is happening. Although industry at large is concentrating on how to make the engineer more productive, little attention is being paid to why he or she may not be productive. One of the major factors adversely affecting the engineer's productivity is the process of reproduction, distribution, storage, and retrieval of information.

THREE BASIC PROBLEMS

Reproduction is an essential ingredient in good engineering communications. If engineers are going to increase their productivity, they must begin by addressing three problems concerning reproduction. These three basic problems, which hamper the engineer's ability to communicate effectively and in a timely fashion with all parties involved in the complex engineering design activity, are:

- Lack of standardization of drawing sizes
- Alphanumerics on drawings
- Quality and age of originals

Drawing Sizes

There is no "standard-size" drawing in the United States today—there are a great many standard sizes. Consider the fact that the Military Specification MIL-D-1000 calls for engineering drawing sizes to be:

- 8½ × 11 in.
- 11 × 17 in.
- 17 × 22 in.
- 22 × 34 in.
- 34 × 44 in.

Yet, within the military, there is widespread use of drawings in sizes of 30 × 42 in., 28 × 40 in., and 24 × 36 in.

Industry sizes, as in the automotive industry, for example, range from 8½ × 11 in. up to 54 in. wide by any length. Other popular sizes are 42 in. wide, 36 in. wide, 30 in. wide by 12 ft. long, and there are many other unusual lengths and sizes. This random choice of sizes has severely hampered communications among engineers and drafters.

The Impact of Size on Engineering Communication

Because of the wide variety of sizes of original drawings, no manufacturer of reprographics equipment has been able or willing to spend the research dollars to develop a unit that can handle this wide array of sizes. Therefore, the engineer is left with the diazo system, which has a throat 42 in. wide and can accept this wide variety of input. This system is manually fed originals by a semiskilled operator at speeds that may have satisfied yesterday's engineers but are totally out of step with today's production demands and labor costs.

Microfilm is one excellent approach in dealing with this problem. The microfilm aperture card reduces all drawings to one convenient size on an electric accounting machine (EAM) card, which first allowed automation to be introduced to engineering reprographics systems. Microfilm placed an additional burden on the engineering department, however. We will discuss the advantages and disadvantages of microfilm later in this chapter. To truly automate engineering reproduction, though, the size standards of engineering drawings must be faced.

Consider the fact that the multibillion-dollar office copier industry would not be in existence were it not for the fact that most corporations communicate on a standard 8½ × 11-in. sheet of paper. Even the federal government is changing to the standard 8½ × 11-in. letter size in place of the 8½ × 10 in.-size. This has been done in order to gain the advantages of automation and improved communications through standardization.

Recommended Sizes

Because the 8½-in. letter width has been accepted as a standard size in automated equipment, drawings should be increments of this size. It is recommended that all drawings be created in the following sizes:

1. 8½ × 11 in.
2. 11 × 17 in.
3. 17 × 22 in.
4. 22 × 34 in.
5. 34 × 44 in.

The recommended sizes are spelled out in the DOD-STD-100A (formerly MIL-STD-100A). Every effort should be made to eliminate the use of drawings larger than 22 × 34 in.

Today it is possible to produce prints from engineering drawings at the rate of 1200 prints per hour on the Xerox 840 Engineering Print System, (called the EPS). This system, sold by Xerox Corporation of Rochester, New York, accepts drawings up to 22 × 34 in. The drawing is fed once into the machine. The number of prints required is selected, and prints are made at the push of a button. They are reduced to half their size, reproduced, folded, and collated in one single, automatic operation. The Xerox 840 brings duplicating speed to engineering documentation systems. Prints can be made on a wide variety of materials, including bond paper, vellum, and even polyester. However, the drawings, in order to be printed at this speed, must be reduced to an 11- × 17-in. format. This need for reduction brings us to the second constraint.

Alphanumerics on Drawings

Automation of engineering reproduction utilizing the Xerox 840 or microfilm necessitates reducing the prints to three-quarter size or half size. If the alphanumerics (dimensions and notes) on the drawing are not of sufficient height and spacing, the information is rendered useless because it cannot be read.

Alphanumeric Standards

The size of alphanumerics on a drawing is one of the most critical factors affecting the automation of the engineering documentation process. For drawing sizes 8½ × 11 in. and 11 × 17 in. the character heights on the drawing should be standardized at ⅛ in. minimum to ⁵⁄₃₂ in. maximum. These sizes are specified in DOD-STD-100A.

Drawing sizes from 17 × 22 in., 22 × 34 in., and larger must use alphanumeric sizes no less than ⁵⁄₃₂ in., or 0.156 in. This is imperative for a good half-size opaque or microfilm program.

In addition to having the proper height, it is advisable to have all characters or alphanumerics on a drawing typed. Some companies have found it inconvenient and time-consuming to have the dimensions typed. However, the notes can be typed and then spliced in or applied to a drawing. If hand lettering is used, the style should be a straight gothic. Spacing between lines of characters should equal half the character height. Uniform spacing between characters should be used. No guidelines should be drawn. Pencil or pen can be used. In some companies, notes are typed on a sheet of vellum using an IBM Executive or Selectric typewriter and the 10-pitch Orator type element. The notes are then spliced into the drawing with invisible drafting tape. Companies may also use an adhesive-backed drafting transparency to type their notes on, after which the transparency is applied to the drawing.

Fractions should be avoided on drawings. If fractions must be used, the numerator and the denominator should be the same height (for example, 5/32 in.) as the whole number.

Quality and Age of Originals

Drawings that have become faded and badly worn present the greatest challenge to an automated high-speed reprographics system. Sepias in particular fall into this category. Second-, third-, and fourth-generation sepias are the most difficult to deal with. If your company's drawing vault contains a wide variety of sepias and old, tired vellums, linens, etc., consider implementing a restoration program. When poor drawings lead to poor reproduction, the best thing to do is to pull the original, restore it, and issue a new microfilm. The Xerox 1860 and 2080 are both excellent labor-saving tools for the restoration of drawings (these units, and their capabilities, are discussed later).

The Constraint Summary

The problem of moving engineering data quickly—and the solution to that problem—rests with engineering and drafting departments. Good drafting practices are paramount. Just a few poor-quality drawings in a batch of a hundred can severely impair good communications. The American National Standards Institute (ANSI) has issued two standards (ANSI Y14.1-1975 and Y14.2-1973) which prescribe proper formats and styles. Again, drawings should not exceed 22 × 34 in. Lettering and notes on drawings must be of the proper height so that they can take the reductions required to implement an automatic reprographics system. All use of sepias should be eliminated.

TIME-SAVING TECHNIQUES

Drafting and engineering can effect considerable savings and time through the proper implementation of scissors-and-tape drafting. The key element is the proper selection of tools and mediums. The sepia, for example, has been used for years as a means of creating "similar-to" drawings. The problem with the sepia is that it tends to deteriorate in the file, eventually to the point at which it is no longer useful. Today there are alternatives.

Drawing Re-creation Systems

The photo wash-off system offers a high-quality alternative to the sepia. In the wash-off process, old drawings can be restored to like-new condition. Information can be reformatted and changes accomplished. The output is generally high-quality solid black lines on a white polyester substrate. This type of original provides excellent input for microfilm and other reproduction methods.

The drawbacks of the wash-off process are that it is a high-cost, slow process, and that usually the work must be sent out of the department or plant.

Creative Drawing Systems Using Xerographic Processes

As alternatives to the photo wash-off process and sepias, the Xerox Corporation offers two units to aid the drafter in creating, revising, and restoring drawings. The first is called the Xerox 1860 printer and is shown in Fig. 12-1. The 1860 handles originals up

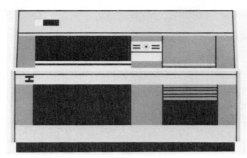

Fig. 12-1 Xerox 1860 printer.

to 36 in. wide and produces sharp black-on-white images onto vellum or plain bond paper. The output width is limited to 18 in. However, there is no finer tool for creating new inklike drawings from old sepias. The 1860 has the capability of dropping dirty background and smudges from originals while intensifying line definition. The major advantage of the 1860 printer is that it can be installed in the drafting room. There are no plumbing, venting, darkroom, or other construction requirements. The machine is simple to operate, and anyone can be trained to operate the unit in just a few minutes. With the 1860, scissors-and-tape drafting can be accomplished with ease and excellent results.

The second xerographic product is called the Xerox 2080, and is shown in Fig. 12-2. The 2080 promises to be one of the most revolutionary drawing-creation tools to be introduced in the past several decades. It has the potential to increase productivity in the creative drafting department by 30 percent or more. The 2080 will accept originals up to ⅛ in. thick, opaque or translucent, and will render the output on vellum, polyester-base paper, drafting stock, offset masters, or ordinary paper. The prints are of inklike quality, which makes them ideal for microfilm and other reproduction methods.

No Erasing

A unique feature of the machine is its nonfusing capability, which eliminates conventional erasing. Details, both large and small, can be removed quickly without damage to the base material. This permits a new drawing, with dimensional or scale changes, to be made, while the original remains intact. This capability is also important in restoring drawings. Scratches, blemishes, smudges, and old tape marks can be readily removed using the nonfusing capability and an elastic eraser. Once the image has been changed, it is then inserted in an off-line fuser and made permanent. The drafter can then add the new details to the drawing. The results are quicker turnaround time, better image quality, and total in-house control—and increased productivity.

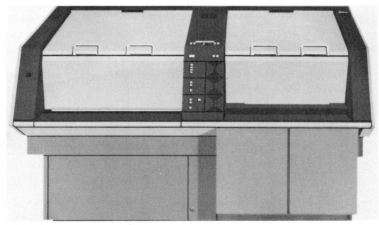

FIG. 12-2 Xerox 2080 printer.

Quality Enhancement Features

The 2080 allows a drafter to eliminate paste-up lines and shadow marks through the use of a simple lens-aperture control. For example, when a drafter needs to convert a decimal system on a drawing to its metric equivalent, the procedure would be:

- The drafter gives a print of the drawing with the metric equivalents marked on it to a typist.
- The typist types the metric dimensions on opaque gummed labels.
- The drafter removes the gummed labels and places them over the existing decimal dimensions.
- The drawing is fed into the 2080. The lens setting is set to burn out all the shadow lines of the opaque labels. A vellum or polyester print is made, and a new metric drawing is thus created.

Full "D" (22 × 34) Size Drawing Capabilities

The most significant feature of the 2080 from the drafter's viewpoint is its capability to create new drawings to scale up to 24 in. wide by any length. We recommend the 22 in. width that accomodates a 50 percent reduction which results in an 11-in. wide book-size output.

Scale Accuracy

The 2080 can also maintain scale accuracy. It will both reduce and enlarge. The parameters range from 45 to 141 percent. The zoom lens feature permits accurate scaling at all settings. Moreover, if a drawing is out of scale in the X coordinate but in scale in the Y coordinate, controls on the 2080 allow the operator to change the X dimension while maintaining the Y dimension. This is particularly important when distortion may have occurred in a microfilming process and the drafter must return to a full-size drawing from a microfilm print. The variable-enlargement feature, plus the XY control, allows the drafter to return the drawing to the original scale.

Microfilm Enhancement

The value of microfilm is enhanced by the 2080 and its enlargement capabilities. One constant complaint heard from engineers and others who use microfilm prints is that the prints obtained through a microfilm blowback process are not to scale. Now it is possible, with the variable-enlargement feature of the 2080, to bring the microfilm prints back to full scale with one or two passes through the 2080. Multiple regeneration of prints from prints does not have a negative effect on quality.

No Special Preparation

The 2080 requires no vending, plumbing, or special services. It operates on a 240-V line and is designed for use directly by engineers and drafters. It belongs in any drafting department with five or more drafters. Productivity gains of 2:1 and more are reported. The learning curve to effectively reach this level of productivity is four days.

Reproduction and Distribution Processes

Conventional distribution processes have for years been using the diazo system. However, there are several alternatives to diazo that have made great progress over the last few years.

In 1971 Xerox Corporation introduced the Xerox 840 Engineering Print System, called the EPS (see Fig. 12-3). This device accepts translucent or opaque originals up to 24 × 36 in. and produces from 20 to 40 bond-paper prints per minute. The prints are reduced 50 percent in size. The output size varies between 12 × 18 in. and 11 × 17 in. depending on input size. Up to 50 bins of sorters can be attached to the 840 EPS,

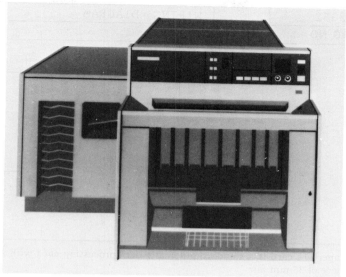

FIG. 12-3 Xerox 840 printer.

making it the most advanced form of total distribution for engineering drawings on the market today. The 840 will not reproduce from reverse-paper sepias, and the input size is limited to a maximum width of 24 in. For engineering drawings that are not up to microfilm standards, as mentioned earlier in this chapter, the prints could be difficult to read. Most users, however, find that the advantages of the low-cost, high-speed printing, folding, and collating capabilities far outweigh these disadvantages.

Unique Reproduction System

A West Coast electronics firm devised a unique use for the 840 EPS. Instead of using a microfilm half-size print program, the firm chose the 840. The company claims that it no longer has to worry about cameras, film, processing, checking, mounting, etc.; instead, reduced-size masters (11 × 17 in.) of all new drawings are made on buff paper. The buff master copy is then folded and placed in a standard 8½ × 11-in. file. When prints are required for release or reference, the 11- × 17-in. buff copy is pulled and duplicated on the 840 (same size). Once the 11- × 17-in. master is created, the original is never used to produce prints. The benefits to the user of this system are:

- The original drawing is undamaged.
- All drawings, regardless of size, are filed numerically in an 8½- × 11-in. file (folded).
- The expense of film conversion devices is avoided.
- No viewer is required—the reduced buff master is readable.

MICROFILM AND ENGINEERING DOCUMENTATION

Microfilm has been used by engineers for more than two decades. It is a proven system that provides the engineer with total print automation and eliminates wear and tear on expensive engineering drawings.

Types of Microfilm

There are basically three categories of microfilm: roll microfilm, the 16- and 35-mm film widths being the most commonly used; microfilm mounted in data processing cards, called aperture cards, containing a "window" that accommodates a 35-mm

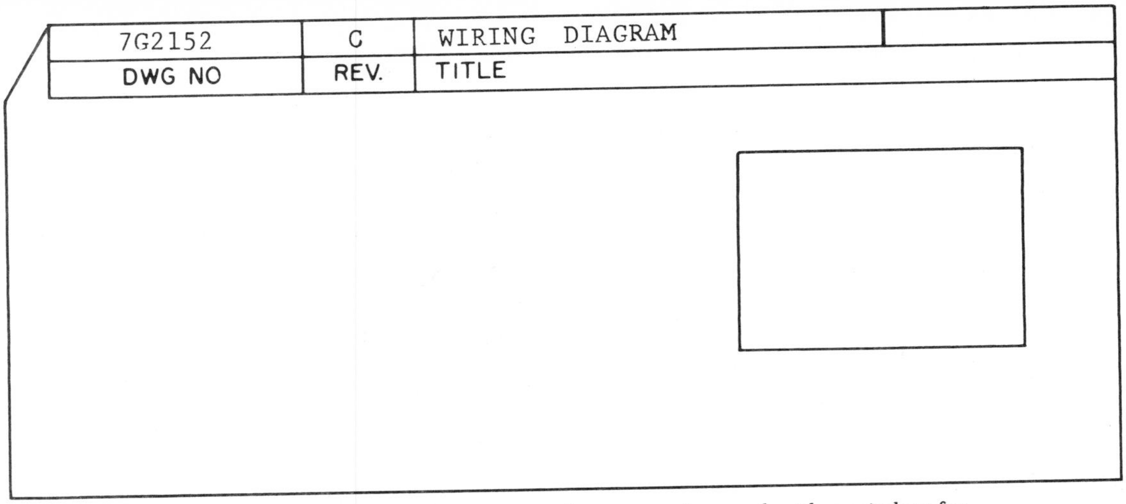

| 7G2152 | C | WIRING DIAGRAM | |
| DWG NO | REV. | TITLE | |

FIG. 12-4   An aperture card is a standard automatic data processing card with a window for insertion of a piece of 35-mm film.

frame (see Fig. 12-4); and microfiche, or fiche (see Fig. 12-5), 4- $\times$ 6-in. film format, with the images recorded on 105-mm film, that produces multiple exposures of 8½- $\times$ 11-in. documents in a disciplined arrangement that permits easy retrieval of engineering data of all kinds—specifications, bills of material, parts lists, computer analyses. All three types of microfilm are discussed below, with major emphasis on the aperture card as it applies to engineering documentation systems.

The Aperture Card

A sample aperture card is shown in Fig. 12-4. The aperture card is a single-unit format for all engineering drawings regardless of their size. The card can be updated manually or by keypunch and interpretation that print all the significant data about the drawing on the card. The 35-mm film chip is installed in the aperture by semiautomatic mounting equipment. The 3M Corporation of St. Paul, Minnesota, supplies the most popular brand of card, called Filmsort. This card has an adhesive polyester frame that holds the 35-mm chip firmly in place.

Multiple Prints

A key advantage of the aperture card system is the ability to duplicate many copies of the original card. These duplicated cards are created using a diazo-based film (3M Duplicard) on either a manual or an automatic card-to-card duplicator. Machines made by 3M, International Business Machines (IBM), and Keuffel & Esser (K&E) (the new 200 model) produce duplicate cards both manually and automatically. The average cost to create a silver duplicate (so called because of the silver halide film used in the duplication process) is approximately 30 cents. Duplicate cards (in which the silver duplicates are mounted) can be made at the cost of 3 to 5 cents per card. This is an economical means of distributing engineering information. The size and cost of the card permit the disbursal of drawings to satellite locations. When prints are required at these satellite locations, the card is manually pulled from the file, inserted into an aperture card printer like the Xerox 1824 (see Fig. 12-6), and a paper print is made in a matter of seconds. The prints are permanent, high-quality reproductions in black printing on a plain white bond paper. Vellum prints can also be obtained on the Xerox 1824. (See Table 12-1.)

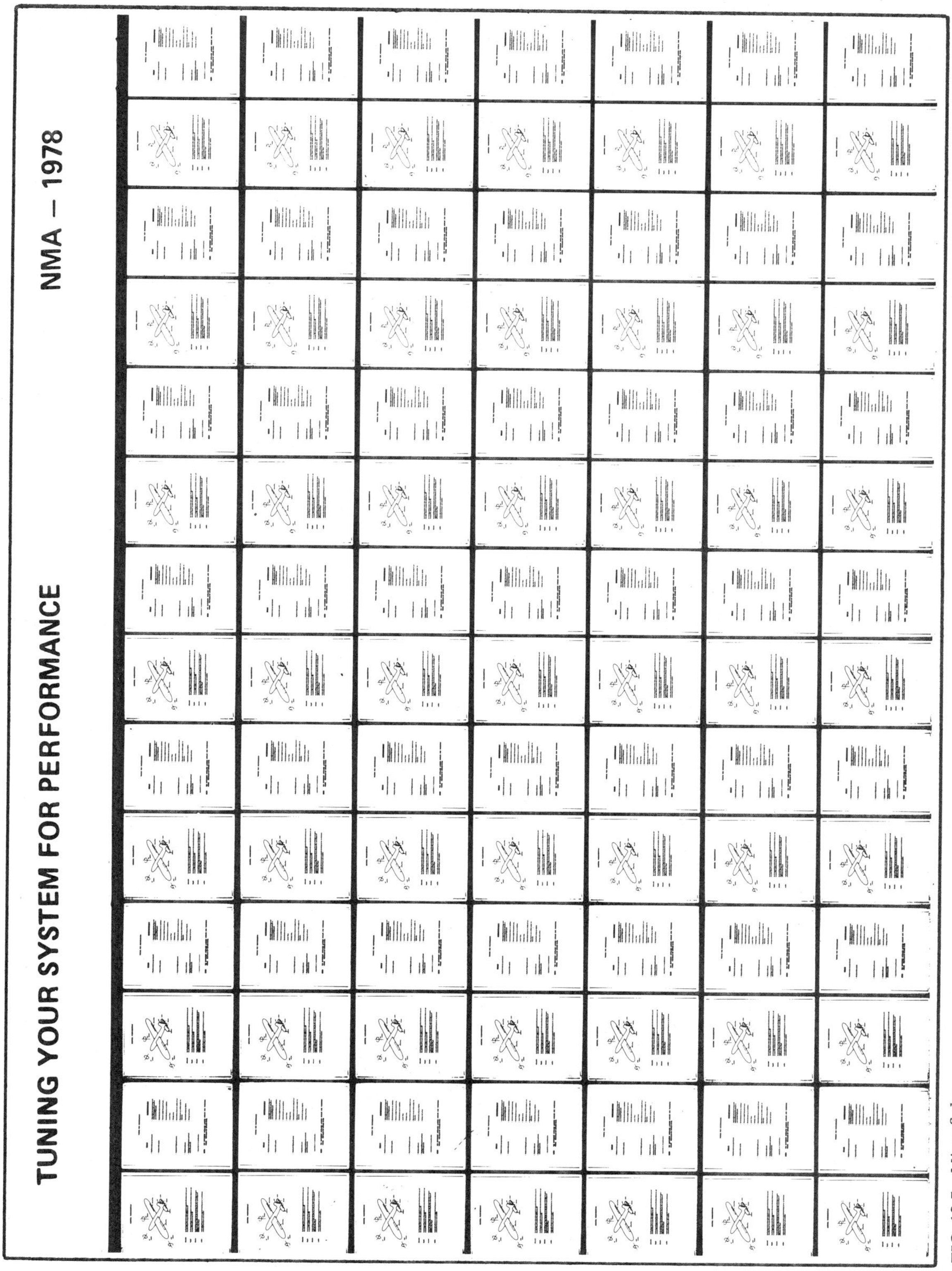

FIG. 12-5 Microfiche.

TABLE 12-1 Reduced-Print Image Sizes Using Government Reduction Ratios (MIL-M-9868D) and Magnification Ratio of Xerox 1824 Printer and 600 MEP

Original tracing size, inches	Reduced-print image size, inches			Paper print size
	Reduction = 16×*	Reduction = 24×*	Reduction = 30×	
8½ × 11†	7¾ × 10			8½ × 11
9 × 12	8¼ × 11			9 × 12
11 × 17†	10 × 15½			11 × 17
12 × 18	11 × 16½			12 × 18
17 × 22†	15½ × 20			17 × 22
18 × 24	16½ × 21¾			18 × 24
22 × 34†		13½ × 20½		17 × 22
24 × 36		14½ × 21¾		18 × 24
28 × 40			13½ × 19½	17 × 22
28 × 46			12½ × 22¼	18 × 24
30 × 44			14½ × 21¼	18 × 24
30 × 46			14½ × 22¼	18 × 24
34 × 44†			15½ × 21¼	18 × 24
36 × 30‡			17⅛ × 14½	18 × 24
36 × 48			17⅜ × 23¼	18 × 24

*Magnification = 14.5×; multiplier for reduction 16× = 0.90625; multiplier for reduction 24× = 0.60417.
† Recommended standard-size drawings for best quality, legibility, and utilization.
‡ Roll tracing 36 in. wide rotated 90° on camera board with 30-in. sections.

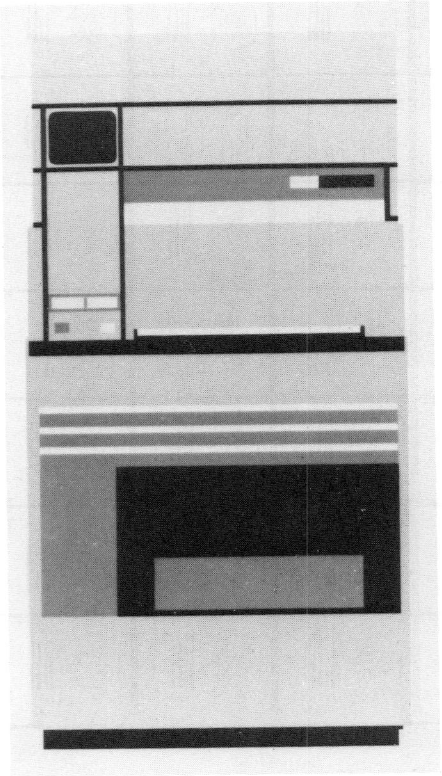

FIG. 12-6 Xerox 1824 printer.

FIG. 12-7 Xerox 600 printer.

The Advantages of Aperture Cards

There are many benefits to be gained in converting to an aperture card system. Here are a few:

- Easier and faster reference information
- Sequential and numerical filing of drawings regardless of size
- Preservation and longer life of the original drawing
- Elimination of paper-print files
- Reduced mailing costs
- Lower total cost
- More effective security
- Historical record of all drawing changes
- Automated drawing reproduction systems
- Automated storage and retrieval systems

The single biggest advantage of an aperture card system is its ability to handle large volumes of engineering drawings automatically. For example, the Xerox 600 Microfilm Enlarger Printer (MEP), pictured in Fig. 12-7, is designed to handle 200 engineering drawings—in the form of aperture cards—at a time. The 600 will automatically select the card, pull the paper out of the paper tray, print the image on the paper, and produce prints at the rate of 600 prints per hour. This system eliminates the use of the original as the print master, saves the drawing, and increases print production. (See Table 12-1.) 3M Company has a machine called the Quantimatic that has basically the same features as the Xerox 600 MEP.

Manual printers are also available for low-volume satellite locations. One such unit is the Xerox 1824 (see Fig. 12-6). The 1824 is designed to eliminate shop and print files. Aperture cards are sent to satellite locations, where production people or engineers can pull aperture cards from file and make prints at their convenience. Some manufacturers, such as OCE, supply reader-printers. The OCE 3600 is one such device that has a reading screen as well as printing capabilities.

Disadvantages of Microfilm

The disadvantages of microfilm can be summarized as follows.

- Aperture cards are not eye-readable—a special viewer must be used.
- Paper reproductions are normally reduced in size (see Fig. 12-6).
- Turnaround time on microfilm can be four hours to two weeks.
- Resistance of end users to the use of reduced-size prints can be strong.
- Prints are not to scale.

In the opinion of many people, the advantages far outweigh the disadvantages, and use of the aperture card system is recommended.

Steps in a Microfilm System

The steps in a typical microfilm system are illustrated in Fig. 12-8 and outlined below.

1. Original drawings are received and logged in from drafting.
2. The drafting log is given to a keypunch operator who punches in and prints the drawing detail, i.e., the number of sheets, the drawing number, the sheet number, the last sheet, the frame number, the drawing size, and the date of issue on the card.
3. The drawings are filmed on a 35-mm planetary camera.
4. The film is processed in an automatic film processor.
5. The processed film is checked on a densitometer to ensure that the density ranges from 0.9 to 1.2. (*Note:* Military specifications are 1.0 to 1.2.)
6. The drawings are returned to a secure vault.
7. The film is mounted by a semiautomatic mounter onto a keypunched card.
8. Duplicate diazo cards are created on a manual or automatic card-to-card duplicator.
9. Cards are distributed to satellite locations.
10. Prints are produced for distribution on an automatic microfilm enlarger printer.

To ensure that quality prints are always obtained, certain standards have been established. When these standards are properly applied, results are excellent. However, it is necessary to understand some of the techniques of photographic reduction to realize why these standards are important.

The camera can reproduce only what it sees. If exposure is set to pick up the lightest lines, it will "bone in" (make extremely heavy or fuzzy) the darker lines. Adjusted to the heaviest or darkest object, the camera will drop out those extremely light lines. It is therefore critical that all line work and lettering be of equal density. Letters and numbers that are not made clearly and with the proper density and spacing will "burn out" during the reduction process. In the printing process the reduced microfilm image is enlarged only 14.5 diameters. Small printed letters and numbers will not be legible. It is imperative to maintain proper size, spacing, and density of all alphanumerics (see Fig. 12-8).

Centering Marks

During filming, drawings must be centered so that they are established within the center of the aperture card. Centering arrows, to be used as register marks, must be put on each of the four sides of the drawing. The arrows should be centered on each side of the border that outlines the information area of the drawing. If they are not preprinted in the format, the drafter should add them.

Other Techniques Used to Ensure Quality Microfilming

As described in other sections of this handbook, good drafting practices are a must. However, to further improve the quality of engineering drawings for microfilming purposes, the following procedures should be applied to all drawings:

1. Plan your drawings and select an appropriate scale to reduce the clutter and to assure quality blowback.
2. When areas of the drawing are too crowded to be easily read at half scale, an enlarged view at an increased scale should be used.

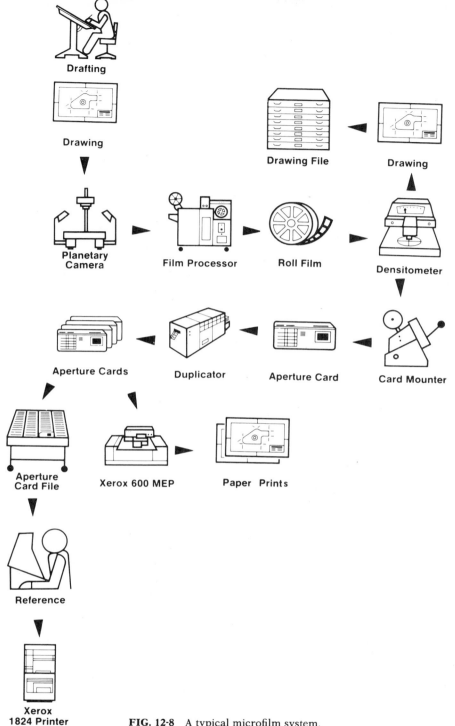

FIG. 12-8   A typical microfilm system.

3. Utilize good drafting materials and techniques to enhance the reprographics process.

4. Simplify drafting practices to eliminate or reduce elaborate artwork.

5. Shading should not be used on drawings that are to be microfilmed.

6. The use of material indicators or crosshatching should be minimal; if used, they must be open patterns.

7. Standard E-size drawings are the maximum recommended size for large drawings. The E-size drawings must be reduced at 30X reduction. Alphanumerics cannot be smaller than 5/32 in. on this size drawing.

8. A notation reading "Do not scale or half-scale print" should be placed near the title block.

9. All markings using a plastic or rubber stamp should always be placed on the front of the drawing.

10. If there is a reason to return a print to full scale, the words "Do not scale or half-scale print" should be removed from the enlarged print.

Other Benefits of an Aperture Card System

Once the drawings are reduced to a standard format, the aperture card, total automation can begin. For example, two manufacturers, Access Corporation of Cincinnati, Ohio, and Infodetics of Anaheim, California, have developed automated filing and retrieval systems for aperture cards.

Access Corporation makes a complete line of small-to-large automatic retrieval systems. Cards can be called from file automatically via a computer punch-tape or manually via a keyboard-entry system. Their largest model, referred to as the M system, can handle up to 200,000 aperture cards automatically. Infodetics' model 410-45 can store 200,000 aperture cards, request information via push-button phone, display information on a television monitor, produce duplicate film, and produce prints automatically on command.

In summary, the aperture card system is a powerful engineering and drafting tool. Good drafting practices are the key to its success.

Roll Film and Microfiche

Roll film has been in use for many years. Its use as an engineering tool is not recommended for a variety of reasons. The main reason is that engineering drawings change,

FIG. 12-9 Xerox 740 microfiche reader-printer.

and as drawings are filmed and stored via roll film, it becomes difficult to update the roll of film. After a time the process becomes unwieldy.

Microfiche is a fairly new development in the engineering environment. With the advent of computer output microfilm (COM), fiche has become another tool for the engineer. In some corporations microfiche from COM is being used for bills of material and change notices. Step-and-repeat cameras are being used to store engineering records and diaries. If prints are required at a later date, a Xerox 740 microfiche reader-printer (see Fig. 12-9) is used to produce dry plain-bond paper prints on demand. A new development of a special lens for the Xerox 740 now extends its usefulness to the engineer. This new feature now provides the engineer the capability of producing very legible plain paper prints from microfilmed engineering drawings on aperture cards.

The world of engineering graphics is being changed by advances in micrographics and reprographics. This chapter has tried to highlight some of those significant changes and alternatives.

SUMMARY

The demands of a changing market and engineering environment are precipitating a major revolution in the world of engineering graphics.

Computer aided design and computer aided manufacturing technologies are forcing engineering managers and reprographic managers to seek new ways to solve their communication problems. It is imperative to solve the problem of merging existing databases in the form of conventional drawing files with the new electronic drawing files. This is a considerable challenge. In most organizations the movement of electronic data is coupled with the movement of conventional hardcopy systems. The digitization of existing databases is an impossible task with today's technology. Restoring and retrieving digital data is becoming a massive problem. The new CAD technology and ability to create engineering data and drawings is outstripping the current ability of reprographics people to reproduce and disseminate the information. Therefore, the engineering manager in the future will have to focus on the total problem of managing engineering information. It will be necessary for him or her to balance and understand the significance of creating vast amounts of engineering data and the ability to move that data in a timely fashion to points of need.

Hardcopy and microfilmed reproduction systems needs will grow in the next decade. The use of paper prints, film, plotters, and high-speed duplicator-printers will expand. New tools will be introduced to the reprographics field that will enhance the engineering managers ability to focus on the total job of creation and distribution.

The engineering manager in the future who recognizes the many labor-saving facets available to him or her through modern reprographics applications will be the winner. The blending of digital creation with conventional drafting systems and understanding the trade-offs between different combinations of these two methods will be both a practical and necessary undertaking.

The world of engineering graphics is changing. This chapter has tried to highlight some of these changes, identify problems and shortfalls, and offer some possible alternatives and solutions.

REFERENCES

Corps of Engineers System for Microfilming and Printing Half-Size Construction Drawings, U.S. Army Corps of Engineers, Savannah, Ga., 1968.

Drawing Sheet Size and Format, American National Standards Institute, ANSI Y14.1-1975.

Engineering Drawing Practices, Military Specifications MIL-STD-100A (Oct. 1, 1967).

Line Conventions and Lettering, American National Standards Institute, ANSI Y14.2-1975.

"Modern Drafting Techniques for Quality Reproduction," National Microfilm Association, September 1971.

Requirements for Microfilming of Engineering Documents, Military Specification MIL-M-9868D (October 1, 1970).

Section  13

Retrieval Systems

Louise G. Schatzman

OBJECTIVES

Knowledge is a valued resource of people and companies, perhaps even a critical resource. It is a commodity in many sectors of our modern technocratic society, in which technologies themselves are often a mixed blessing. The intelligent application of technologies and the accumulation of reliable information are the crucial factors in the dissemination of useful information. The true value, then, rests in information itself and the knowledge to apply that information wisely, not merely in the publication of information. Yet it is these many millions of records and documents that are spotlighted in retrieval systems and record-management policies, rather than the more abstract but infinitely more valuable information they collectively convey.

One development of the information age will surely be more effective and more efficient systems for exploiting information resources without the waste of paper and the time-consuming paper-handling tasks. The essential purpose of this chapter is to point out some of the techniques and skills that are currently readily available to everyone in the design of retrieval systems for engineering data and information.

ADVANTAGES OF AN EFFICIENT RETRIEVAL SYSTEM

Realistically, an efficient information-retrieval system can present many advantages to any company. Among other things, such a system can:

1. Overcome resistance to change
2. Avoid interdepartmental conflicts
3. Cut delays to a minimum
4. Establish maximum control of information
5. Be responsive to the planning cycles
6. Keep flexible with the work schedule
7. Promote more effective functioning
8. Make critical resources more visible
9. Establish accountability for recording data
10. Create a dynamic environment for management decision making
11. Identify trends and inconsistencies in objectives
12. Integrate diverse groupings
13. Maintain reporting techniques that are flexible
14. Provide integrated systems so that different stages in the life of information can be coordinated
15. Maintain a "least-cost" concept
16. Enhance the most frequently used information
17. Create audit trails of information concept automatically
18. Advance with technology
19. Gather insight into the value of information
20. Provide insight for management strategic planning
21. Generate means of providing broad visibility to the information base
22. Prevent harmful knowledge monopolies
23. Prevent hardware traps
24. Improve constantly the efficiency of the transfer of information

THE TOTAL-SYSTEM APPROACH

Why design a retrieval system? Why, indeed, design a building or an industrial plant or some software system for your computer? Why not just let them develop at random? The analogy applies clearly.

The investment in information resources is immeasurable. Each data element as contained on some record or document is processed and reprocessed, generating new elements and new definitions. Without a cyclical system analysis—system synthesis in action—these expensive resources are easily lost, partly if not totally. Defining objectives, determining criteria for long-range development, and establishing self-auditing requirements for database evaluation can together provide the solid foundation and framework for the development of the retrieval system to fuller utilization of information resources.

Problems in Managing Information

Why design an information system? The problems of information management evolve from three main sources. The first and most important is the diversity of the problems of information storage and information retrieval. This is not one methodology but instead consists of two separate and very distinct ones.

Information can be stored on many mediums—paper, cloth, microfilm, magnetic tape, video tape, or magnetic disk. The choice may be determined by chance, by the technology available at that time and place, by the resources available, or by the characteristics of the information.

The physical components of most engineering information systems consist of sets of documents with intrarelationships as well as interrelationships between the various data elements that each contain. The physical nature of each document is only a part of the "address," or location, of the information.

Retrieval Problems

The retrieval of information is a problem separate from the storage and structure of information but necessarily connected to the physical location of the information. Retrieval of information comprises two processes. The first is the logical or intelligent retrieval of the informational data item—the object of the search. The second is the physical retrieval of the object containing the data item from its storage location.

The storage location is a piece of the information about an item to be retrieved, but it is only one small piece of that identifying information. The relative value placed in any total system design on the location of an object of information as compared to the object itself might well be a measurement of retrieval design efficiency.

Structure of Information

Therefore, since the information itself is more valuable than its location, the structure of the information itself might well be more important than the structure and interrelationships of their collective locations. If the form, type, and nomenclature of records are essentially only of use in locating the object of information, then the concentration of systems design effort on the information database structure can only lead to more effective and more efficient information retrieval.

THE TOTAL SYSTEM

The stacks of documentation and the objects of information that they contain do not constitute the total system. Users and user needs are vital parts of a total retrieval system. If the maximum usefulness is to be derived from an information base, good visibility of the contents is a necessity for the user. Many people will probably want to use the system, with a minimal amount of instruction, and use it with good understanding of what is or is not there. Therefore, the human side of planning should not be forgotten in a retrieval-system design. Information transfer is affected by all facets of human behavior—including individual as well as group behavior. This, of course, can vary from day to day and from occasion to occasion.

Organizing the total information system for retrieval will eventually include the following activities:

1. Analysis
2. Synthesis and development
3. Implementation
4. Maintenance
5. Standardization
6. Evaluation and auditing

None of these activities can be completely isolated from the others, nor can a step-by-step procedure be developed to complete one activity before initiating the next. They are interwoven and intermixed, but they are not just a vague jungle of activity. Each can be extracted mentally from the network of interwoven systems, after identification

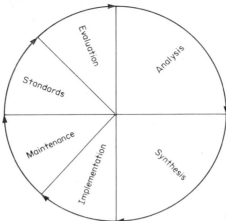

FIG. 13-1   Activities in information system cycle.

and isolation. The interconnections can then be examined, reviewed, improved, or eliminated as determined from the projected impact (see Fig. 13-1).

Analysis

The first activity requires at least two distinct investigations: (1) determining what information is available, and (2) projecting user needs. Determining what information is available is not the same as determining what documents are available, but the latter is probably an appropriate first step. From samples of each document, further information can be identified, such as:

1. What size and form?
2. What quantity and frequency?
3. Who issues it?
4. Who receives copies?
5. What items of information are contained?
6. How is the information monitored or audited?
7. What is it—a serial, update, one of a set?
8. Who should have access?
9. How is completeness controlled?
10. How should it be stored?
11. How long should it be stored?

Preparing forms for each document type with these and other questions answered will soon point out other questions to ask. Also, the interconnecting information-flow patterns will begin to be visible.

Data Items

Such tools as flowcharts, activity charts, and data matrix charts may assist in defining the data items contained on each document. Examination of these may indicate what is really needed, or what items are duplicates, and may even point out other items as they are needed. Flowcharting each information system as well as each document-handling system will emphasize the interactions and reactions and will also point out problem areas. As was mentioned before, this study of what is available is not a process that starts on day 1 and continues through until completion. A few items of information are noted each day and added to the total understanding of the information system as other studies are also in process. Because of exterior influences, there is usually continual change in information systems. Figure 13-2 illustrates a typical flow of documents.

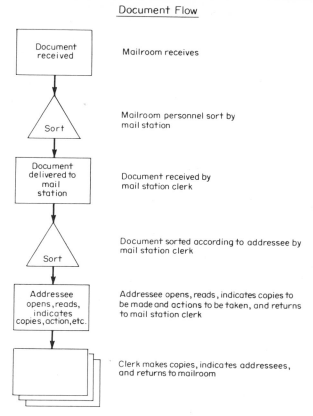

FIG. 13-2   Document flow.

User Needs

Investigation of user needs is a more subjective study. Some of the questions to ask are:

1. Who needs what kinds of information?
2. Where are these people located?
3. How fast do they need the object?
4. In what form do they need it?
5. How is the information used?
6. What items are most often retrieved?
7. What are the rates and volumes of retrieval?
8. How many people need access to the data base at one time?
9. Do any security problems exist?
10. How soon after generation is it needed?
11. Must the originals be retained, or can microfilm be used?
12. How many physical locations need duplicate files?
13. What hardware systems are available?
14. What is the relationship of numbers of retrievals to total database size?
15. What depth of auditing is required?
16. What are user trends expected to be after a given time span?

Conscientious cooperation with the user can only lead to a successful retrieval system, because retrieval is essentially part of the process of improving communication.

SYNTHESIS

The design of the retrieval system takes place whenever some of the decisions have been made or need to be made on collection, file structure, controls, automation, indexing, or hardware. Each decision will probably result in advantages and disadvantages in providing for every user's needs over the entire life of the information. Inasmuch as the system is attempting to meet the diverse needs of the researcher and auditor of tomorrow as well as the decision makers of today, and neither is completely predictable, the system must embody some compromises (see Table 13-1).

TABLE 13-1   Retrieval Patterns

String keyword
1. Select from system thesaurus the keywords nearest the desired information.
2. Attach data base on CRT using specific system procedures as defined by vendor.
3. Input desired keywords singly or in a series if system procedures allow.
4. Scan references to locate item desired.
5. Note file location of desired item.
6. Retrieve requested documents from file.

Tree method
1. Select general subject or category under which document would most likely be found.
2. Scan list of subdivisions under the general subject, selecting possible classifications, following system procedures.
3. Attach each subdivision, scanning whatever additional information is available.
4. Make note of desired references including file location.
5. Retrieve from stored location.

Parameters
1. Select index for whichever parameter you are most sure of.
2. Look for the series of entries that are within your requirements.
3. Scan the abstracts of the series to pinpoint the one document you desire.
4. Identify the date and microfiche location (if filmed) for the desired document.
5. Locate the microfiche or file for date and then locate the document.

AUTOMATION

A major decision must be made concerning automation. Is the total system to remain manual all the way through, is there to be some computer assistance at some of the stages, or is the system to be completely automated? Or is there a probability of changing at some point from one to another?

If there is an eventual possibility of automation, then the specifications for the system should be designed that way from the beginning. For example, a fully automated system that will handle coding, indexing, filing, numbering systems, and procedures, but with computer assistance on indexing only, can be planned for as a possible compromise that will give satisfactory results if the final step, computer assistance, never comes to fruition. However, it would not usually be possible to go from a fully computer-assisted system to a fully automated one without major backtracking. The same problem would result from changing from a fully manual system to a computer-assisted one. Unless the possibility had been considered in the original system specification, there would undoubtedly have to be major reworking of parts of the system. Obviously then, the overall philosophy in design might tend toward one step higher than the proposed one—toward overcollection, broader indexing, and more automation, until the delimiters are definable.

FILING SYSTEMS

The decision on the filing or storage system must be made fairly early in the synthesis process.

Storage or filing systems can be simple and random, or they can be extremely complex, requiring pages of explanations; or they can be somewhere between the two extremes. Random storage is, of course, the easiest for input, but it requires some appropriate tool if retrieval may be required. A unique document number may be used if the file is small, along with a log describing the information. A card index may be required if the number of retrievals is higher or goes back farther than the clerical memory.

One very common approach is file numbering. This structure consists of listing the names of the hundreds of types of documents that are anticipated, grouping them appropriately, and assigning a decimal code to each item. Subject organization might, of course, be used instead of document types. Combining the two could hardly be less than chaotic.

The four principal requirements of a document storage system in which efficient retrieval is also desirable are as follows:

1. *Simplicity of storage* so that filing is readily done hourly or daily
2. *Intelligible to all people* with a very few words of explanation
3. *Consistency of storage* so that the same document can be filed the same way by all people in all locations if they so choose
4. *Accessible* in a cost-effective time frame for most users, for reading or for generating copies

There are many commercially available hardware systems and color-coding systems that are space-saving and assist with retrieval efficiency for the small file. However, they may have some disadvantages. The purchaser becomes a captive of the system, and usually at a high price. For many types of records, the flexibility of folding or changing direction at a later date is almost, if not entirely, lost.

Simplicity

Keeping the filing system as simple as possible, using, for example, drawing number, date of document, or person's name, and then creating some manual or automated cross-indexes using standard hardware and simple color-coding to assist in file control, may be better compromises at least for an experimental model system. There are always special problems and unique situations that arise, however, to which these generalizations may not apply.

NUMBERING SYSTEMS

Another major problem area is numbering systems. Most companies, departments, and individuals use whatever appeals to them at the time without taking into account long-range design or control. Take the case of drawing numbers, for example.

Usually drawing numbers are used only to index drawings for filing and reference. Blocks of numbers may be assigned to contractors, offices, or disciplines. As time goes on and systems evolve, more use is usually made of drawing numbers as an index and as a control tool. For example, the computer-readable aperture card system uses the drawing number as the indexing medium. As systems are computerized, a greater need for a consistent, systematic assignment of numbers with minimum exceptions will arise.

Simplest Numbering System

The drawing number itself can be arbitrarily assigned, or it can contain coded information. The simplest form of drawing number would be to start at 0001 and assign numbers sequentially. The major advantage of this system would be:

1. Simplicity of assigning numbers, because the system is easy to describe and define.
2. The system is efficient, because few digits have to be copied, written, keypunched, referred to, etc.

3. Duplication errors are avoided, since a log is required.

4. Aperture cards are easily and economically sorted because there are fewer columns to sort on.

5. The system can apply to large or small sets of documents, but the input may be greater on smaller sets of documents because a detailed log would be required.

The greatest disadvantages of the simple system are that:

1. It requires a manual or computerized log of drawings related to the number. Many large companies have millions of drawings on aperture cards indexed by this type of numbering system. A central computerized log is used to maintain information about each drawing. Other companies, though, have millions of drawings on aperture cards, but no retrieval information exists about any drawing except for its number. Thus, retrieval by any other descriptor is not possible from these files.

2. The simple system may require changes in existing procedures.

3. To be most efficient, a central control clerk or an on-line computer system will be required to assign numbers as necessary. Preassigning blocks of numbers that must be given to contractors and/or disciplines would require more digits, thus leading to more inefficiency.

4. Standards and details might have to be renumbered to fit in with the scheme.

Complex Numbering System

At the other end of the spectrum, the drawing number could be completely coded by meaningful symbols to represent most of the needed information by means of a packed format of 10 to 12 columns, as long as each column has at least 26 possible symbols.

For example, inasmuch as 1 column alone can uniquely identify 10 different numerals and characters, then 4 columns can uniquely identify 10,000 different combinations of letters and numerals. Thus, a number format of only 5 or 6 alphanumeric characters is required for identification for most companies. Another 5 or 6 characters can be added to represent other frequently referenced information about the drawing, such as drawing type, commodity, or type of equipment. The advantages of the coded system are:

1. A simpler log can be maintained.

2. No control group is needed to assign numbers.

3. It is more efficient for people to reference the drawings, because the number gives more information.

4. Flexibility for different projects is possible by varying the code meanings.

5. Long-range retrievability is possible by discipline, equipment, type of drawing, etc.

6. Aperture cards are sortable into the most useful order, because similar drawings are automatically coded the same way.

7. All computer systems with the drawing number filed can be standardized.

The major disadvantages of a coded system are:

1. Procedural changes are usually required.

2. Computer program changes are usually required.

3. On aperture cards, if the number is lengthy, sorting and handling time may be excessive, and keypunching errors may be compounded. Also, the procedure may be difficult to explain to contractors, vendors, and other offices with input to the same database.

Numbering Documents

Some of the same problems exist for numbering of documents. The tendency seems to be for relatively random, individual-preference numbering systems. If some level of control, automation, or cross-indexing is desired, then designing the numbering system and assigning and controlling numbers can be useful and cost-effective. Keeping the number as short and simple as possible will usually be desirable.

INDEXING SYSTEMS

Another major decision is to decide which indexing system to use. As is the case with a filing system, once an index system is set up and going, changing it is expensive and time-consuming. Again, overdesign at the beginning is probably much better than underdesigning, because a compromising retreat from overdesign does not involve extra work.

Before deciding on whether to index, how to index, or how deeply to index, take another look at user needs as discussed earlier, under Analysis.

If different groups of people, different departments, or different disciplines require differing types of information or a differing structure of the same information, probably some cross-indexing is needed. Perhaps several different sorts of cross-indexing will be helpful, or several parameters should be selected, or specific ranges of key words will be required. For example, an engineering company working on various contracts will use different departments for executing various parts of the job. Electronics engineering, metals engineering, cost and scheduling, finance, structural engineering, civil engineering, permits, and documentation may all have their responsibilities on the contract. However, they may have different needs from the documentation and different requirements from the information retrieval system. The structural engineer may need to have indexes to the information based on the building numbers, or work breakdown structure, or physical areas. The electronics engineer needs to know the type of equipment, and the electrical engineer wants only the documents that contain electrical information. The cost and scheduling engineer needs to know the planned tasks for each area by each engineering group, as well as the projected costs. The finance department needs to know the projected costs for differing time periods and actual costs as they are incurred. Perhaps it will not be feasible to incorporate everyone's detailed needs into one indexing scheme, but by the use of codes, sorting parameters, a judicious keyword scheme, and/or abstracting, retrieval can be optimized for a maximum numbers of users.

Most storage systems, except for the very smallest, will—after a few years and a few changes in personnel—require some sort of index for efficient retrieval. The cost of logging, working out a few codes, and maintaining the information consistently as the documents are generated is not very high. Losing a few drawings, destroying some by poor handling, and redrawing some designs because no one remembered having done it before can all cost much more than overdesigning the retrieval system in the beginning.

Index Design

In designing an index, there are few facts to go on—only conjecture from past experience and insights into future trends. Value judgments as applied to information cannot be eliminated, but they are necessarily more subjective than objective. Therefore, keeping the retrieval indexing flexible is probably one of the best decisions that can be made.

Although managers and budget makers want certainty, "answers" on costs of indexing based on estimated numbers of retrievals, updates, accesses, or other manipulations will be mostly conjecture. In determining the objectives of the system, only one aspect can be optimized in reality. Therefore, by setting the goal as optimal retrieval per dollar of investment or optimal retrieval time per database item—or whatever suits the purposes best—the "best guess" on details and costs should have some relationship to reality.

It is easy in our technological age to be mesmerized by precision. Psychologists say most people's thinking is one-half analytical and one-half objective. Therefore, in designing retrieval systems, it is important to keep in view at all times the behavioral side of the problem—how people learn, what triggers memory, and all the other intangibles involved with dealing with people.

Perhaps the person designing the retrieval system should be a different person from the analyst who gathers data on probable input and users' needs. Certainly a different psychological profile is required for extended periods of time. The methodologies are different—one deals in facts of today, the other in conjecture for tomorrow. One deals in objective facts, the other in subjective and intuitive value judgments.

Operations Research

The operations researcher attempts to apply scientific methods to management problems, thus using only mathematical criteria to derive answers. This approach omits value judgments, emotional dimensions, and the intuitive, social, and political aspects of human behavioral patterns. Are scientific methods any more rational than value judgments and intuition?

Designing a system is a problem-solving procedure that involves much more than just getting a handle on data. The system is not the totality; the index to the information is not the information itself; and the model is not the reality. A methodology to principles, whether scientific, financial, personal, or social, is a process of examining the whole integrated system in all its dimensions.

There are many books and articles on systems analysis and systems design. All may have some value for the uninitiated, but probably none will provide all the answers. Ideally, several reference sources will be available and will be used by any systems designer just to provide a new approach and to help in keeping an open mind (see Suggested Readings at the end of this section).

Because the indexing system is essentially the heart of a retrieval system, the design must necessarily reflect the objectives of the overall system and optimize on those criteria. A few charts or graphs using the numbers projected from the analysis process should provide some answers to start with on a model system. The computer-assisted index opens all sorts of doors for options, reporting methods, and management information systems (see Fig. 13-3).

Index Specification

A minimum specification for an index to an information base of 20,000 items might include at least five descriptors about the document and the information contained therein. These might be the:

1. Document identification.
2. Document date.
3. Building area, or equipment type affected.
4. Engineering discipline or department affected.
5. A key word, probably derived from the subjects discussed. If several keywords per document are needed, duplicate entries can be prepared, or a string keyword program can be developed if the funds are available.

Assuming a fixed-field, fixed-length computer database, then for 20,000 entries a very economical set of programs will update the database and generate several programs as desired. The report, sorted by document identification, will list all the items in logical order. The user, knowing the drawing number for a drawing index, has only to look at this list to determine whether it has been sent to the file. Ideally, other information is also given in the report, such as revision number, title, issue date, or building or area affected. For all items about a certain area or breakdown structure, the area manager would go to that index. For all items about a certain discipline or department, the appropriate codes can have been entered, and thus an index can exist for full or partial reporting.

Action Items

Action items—requiring answers by scheduled dates by assigned persons or groups—can be monitored by means of a spinoff system from this kind of database. Peripheral

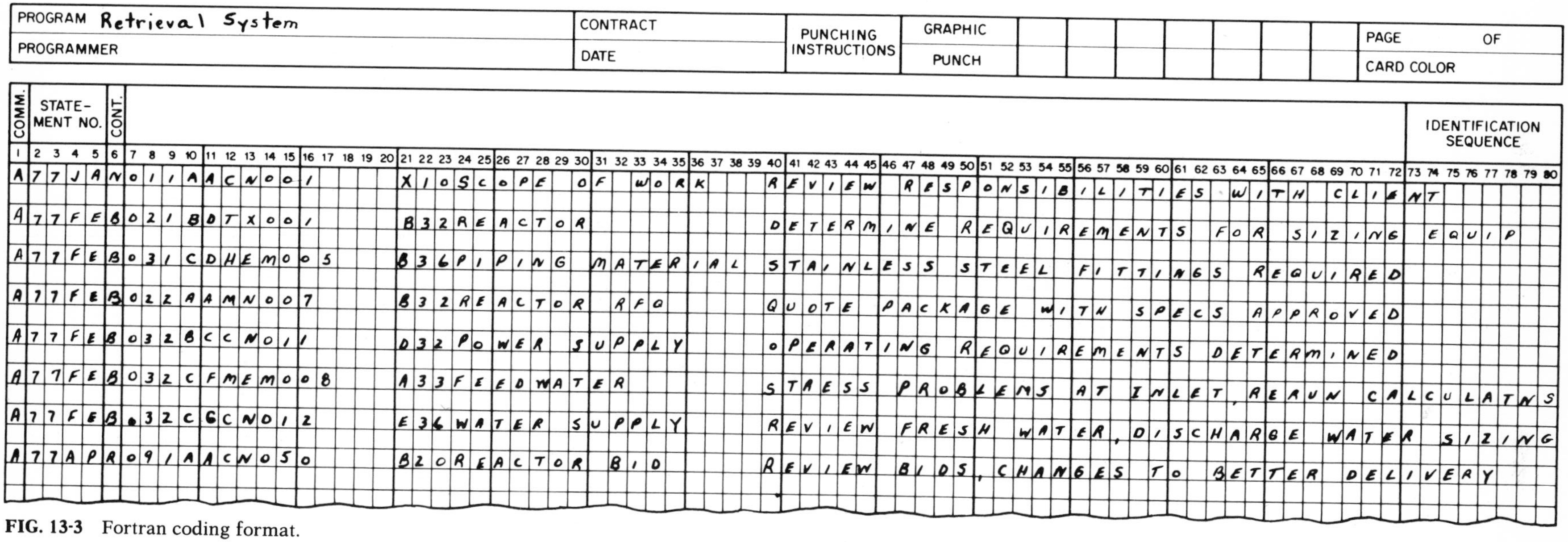

PROGRAM Retrieval System
PROGRAMMER

CONTRACT		GRAPHIC		PUNCHING INSTRUCTIONS		PAGE	OF
DATE		PUNCH				CARD COLOR	

STATE-MENT NO. (1-5)	CONT. (6)	(7-72)	IDENTIFICATION SEQUENCE (73-80)
		A77JAN01/AACN001 X10SCOPE OF WORK REVIEW RESPONSIBILITIES WITH CLIENT	
		A77FEB021 BDTX001 B32REACTOR DETERMINE REQUIREMENTS FOR SIZING EQUIP	
		A77FEB031 CDHEM005 B36PIPING MATERIAL STAINLESS STEEL FITTINGS REQUIRED	
		A77FEB0ZZAAMN007 B32REACTOR RFQ QUOTE PACKAGE WITH SPECS APPROVED	
		A77FEB0328CCNO1/ D32POWER SUPPLY OPERATING REQUIREMENTS DETERMINED	
		A77FEB032CFMEM008 A33FEEDWATER STRESS PROBLEMS AT INLET, RERUN CALCULATNS	
		A77FEB032CGCNO12 E36WATER SUPPLY REVIEW FRESH WATER, DISCHARGE WATER SIZING	
		A77APR09/AACN050 B2OREACTOR BID REVIEW BLDS, CHANGES TO BETTER DELIVERY	

FIG. 13-3 Fortran coding format.

report capability is only one of the advantages of having computer-assisted information retrieval.

Another report might consist of a list of overdue reports, which can be generated by searching for document identification numbers between date parameters. Reports received last month or last week might be useful. Once an index is set up on the computer, many uses can be found for it that had not been anticipated.

IMPLEMENTATION

How to get started is usually a problem. Starting with a small sample from one department or group that is already interested in the system might pave the way for faster progress later. There will be a few problems with procedures that are not quite complete or fully understood. There may be information items that were not communicated at the analysis stage. Also, once credibility and trust in the system have been established, additional items are frequently discovered and must be entered into the system. Generating a few reports, circulating these reports to users, making presentations, giving seminars, and quickly solving minor problems with users as they come up can develop user acceptance with one group at a time. Good communication with each user may build a whole group of people who will "sell" the system to the next department or project.

One of the problems that may be encountered is the resistance of some departments to trust someone else with their documents. Most managers carry an umbrella policy over whatever documents they generate or receive. Convincing them that a group whose principal responsibility is storage and retrieval of documents can function as reliably as the originating group may take some personal explanation. Most managers will, however, appreciate consistency in systems, deeper communication levels, and efficient service.

Working Files

Another problem may be that the working files are not always organized in the way that seems most feasible for a general file. For example, the engineer may file telexes by the company that sent them; the purchasing office may file purchase orders by number, the telex office by telex number, the traffic office by geographic area. A central information office might use none of those numbers for filing but may instead index each of them so that retrieval is efficient for each department, and so that all parties have full visibility of the desired information in some index.

MAINTENANCE

Consistent updating of the database on schedule is of prime importance to the credibility of the system. The process of reading, classifying, and coding the information on all data sheets, drawings, or documents is part of the routine updating. Also important are verifying accuracy, keeping statistics, and massaging keywords for consistency as well as for meeting user needs. Maintaining a high level of integrity of the database is an absolute essential. If one entry is wrong, that may be the one that will be requested!

Keeping up with personal contacts, asking for comments, and following up on problem areas and bottlenecks are all necessities. Good communication is essential every step of the way in a retrieval system and will result in improved efficiency.

STANDARDIZATION

The use of codes and the development of some standards are necessary for long-range efficiency. They are also helpful for generating specific reports, such as exception reports. It is especially important that numbering systems have been well designed, that they are monitored for errors, and that logs are maintained by issuing departments, so that the central handling group can audit their records and have assurance of completeness. Keyword standardization will be covered in the following section on keywords.

There are extensive standards that have been developed nationally by each engineering discipline and by most national associations. How each of these standards affects the database manipulating an individual company's information is unique to that company.

KEYWORDS

Keyword indexing is becoming more and more popular. There are many different approaches and much difference of opinion as to whether to use open keywords or to have a limited glossary of keywords. If the documentation is limited to one discipline, such as electrical engineering, then using only the keywords advocated by an organization such as the Institute of Electrical and Electronics Engineers (IEEE) may be logical. Whenever the database covers parts of many disciplines and departments, such a practice may create other problems in its effort to be relatively inflexible.

There are many advantages in using an open keywording system. First of all, the words that the originator used and that the receiver has read are going to be the ones they remember. Second, the judgmental differences in assigning information to one glossary word or another will never be a problem. Fewer procedures, instructions, and guidelines are required, so that if there are several locations inputting data to the same database, there will be fewer problems and conflicts. Third, the meaning of a particular word may vary from standard keyword set to another, depending on the developer of the set.

Another reason for open keywords is that, in using the words of the originators, the managers of the information system are more sensitive to the phases and modulations in the information flow. Instead of the users' subjects being forced into the information system structure, the subjects, as structured by user-selected keywords, are reflected. Experimenting both ways by the database manager will develop the best approach for each application.

ABSTRACTING

One way of generating more efficient retrieval from a document database—as distinct from a drawing database—is to compose very short abstracts of the information on each document. The cost and effort of generating a letter, for instance, are expended only if there is a specific decision, question, definition, or report to be communicated that is worth the investment. If the essence of the information in each document can be precisely abstracted in 8 to 10 words, this abstract can greatly enhance the retrievability of that information. Lengthy abstracts require a much more sophisticated computer function than the average company can consider. Retrieval would involve optical scanning and on-line retrieval that would require cost-effectiveness of thousands of dollars per retrieval to pay for the generation and maintenance at present-day costs.

EVALUATION AND AUDITING

Both evaluation of the total information system and auditing of its functions are important to management, information specialists, and users. Objective self-evaluation can be a routine procedure. A subjective self-evaluation is another problem, but it is probably of more benefit to the information specialist than to any other person. With limited budgets—when whatever is needed the most is all that gets done—there will be constant inadequacies and frustrations. If these problems are documented with costs and real value, then personnel needs are usually recognized and realistic budgets usually appear. Time is money in information retrieval, as it is in design efforts (see Table 13-2).

TABLE 13-2   A Retrieval System

Analyze the information you have . . .
And . . .
Make conjectures about the user needs of tomorrow.
Then . . .
Look at the equipment available, the costs of the equipment and supplies, the tools needed to learn how to use the equipment, and the cost of those tools.
Make a model . . .
Analyze again: Did the system accomplish the task economically?
Get started, even in the middle . . .
Maintain the best possible quality!

Audits by external groups can be beneficial, because the information specialist, more than anyone else, must be constantly aware of the total system. Self-audits are a "must," but seeing a system as others see it is another important criterion of evaluation.

FEASIBILITY STUDIES

If the evaluation is to be thorough in examining user needs and in gathering experience statistics, feasibility studies become almost a formula. A hundred hours spent in encoding is certainly more productive than a hundred hours spent in random searching. However, that hundred hours in encoding would not be justified to save the time spent on one or two 2-day retrievals. The study can include the value of all of the management information reports subjectively, even if no dollar value can be computed readily.

Defining the scope and objectives, describing each step, and finally preparing the final report can be a preliminary discussion and implementation tool.

MICROFILM

Should microfilm be used? How does one determine what to film and when? Each situation must probably be examined for its own values. There are many advantages in using microfilm, but there are also disadvantages. If the document is to be retained more than 15 years, then the cost of raw document storage alone justifies the cost of filming. If the document is to be retained less than 15 years, and retrievals are approximately one per every 100 pages, then it will probably be cost-feasible to film each document as it is received. It is harder to justify filming after a document has passed out of active use.

The advantages of using microfilm probably outweigh the disadvantages in the following situations:

1. For large drawings that require extensive and expensive floor space to store
2. For documents with high retrieval rates—more than one retrieval per hundred pages
3. For sets of documents with more than 15 pages from which more than five copies are made, providing the users will use the film for the information and not to generate hard copies
4. For sets of documents that are mailed, again depending on user needs
5. In cases where file integrity and rapid retrieval are more valuable than the cost of filming

The disadvantages of filming are the equipment requirements and the need to educate users as to the advantages of filming. People seem to gain so much psychological satisfaction from stacks of paper that reeducation often requires much time and special motivation. With the improvement of updatable microfilm and automated micro-fiche-handling equipment, the trend may be changing.

CENTRAL INFORMATION COORDINATION

There have long been controversy and differences of philosophy about centralized versus decentralized systems. Each has some advantages in some situations. There are probably some range of volume and complexity of information over a certain time span that make one system more cost-effective than the other in information centers. Other conditions, such as physical location, equipment availability, level of computer technology, and archival requirements, may effect the design of information retrieval system. However, as new disciplines—such as seismic engineering and environmental engineering—expand, new controls and regulations will affect the conduct of businesses and their records-management philosophies. It is more important than ever that information specialists with professional training in organization of information for retrieval play an integral part in the planning and management of information retrieval systems.

SUMMARY AND CONCLUSIONS

There can be devastating consequences to having inadequate information. Management of information has not always been recognized and accepted as a prime and valuable tool. The need for specialists in the information-science field is fostering the formation of such departments in more and more institutions.

The information specialist can use strategic planning to provide management with expanded pictures of what information is available, and can also provide users with efficient economical retrieval. Whether the system is manual, computer-assisted, or totally automated, a total system design, with an emphasis on retrieval of information and not merely storage of documents, is important. There are masses of information at our disposal. More than 32,000 different journals in science alone were published last year, and there will be even more information in years to come.

TABLE 13-3   Automation

Equipment and documentation	Number needed	Equipment and documentation	Number needed
Equipment		Vessel	5
Vessels	2	Insulation	1
Mechanical	15	Painting	1
Instruments	100	Instrument	5
Vendor drawings		Purchase orders	
Vessels	50	Mechanical	5
Mechanical	300	Vessel	1
Instruments	100	Piping material	1
Design drawings		Structural material	2
Plot plans	1	Electrical material	2
Piping plans	2	Instrument	5
Piping isometrics	30	Paint and coatings	1
Structural	5	Insulation	1
Electrical	15	Miscellaneous documents	
Vessel foundations	2	Letters, including interoffice memorandums	
Mechanical flow diagrams	5	Conference and meeting notes	
Process flow diagrams	2	Purchase orders, expediting notes, inspection information	
Mechanical foundations	15	Finance records	
Schedules	20	Procedures	
Specifications		Employment records	
Mechanical	10		
Piping	3		
Electrical	5		
Structural	10		

A CASE STUDY

Assume that a client, the ABC Oil Company, has contracted for a naphtha stabilizer. The project manager expects that design and construction will require 12 months to complete and provides the estimates shown in Table 13-3.

This is a rather small project and probably would not require sophisticated techniques for information retrieval, but it may demonstrate some of the principles presented in the chapter.

The information specialist, using the input of Table 13-3, would assist in the design of the numbering systems. This may require coordination with such departments as finance, cost and scheduling, procurement, inspection, and all the engineering disciplines, as well as with the client. Dividing the project into logical geographic areas

TABLE 13-4

A	Upstream side
B	Reactor
C	Downstream side
D	Power system
E	Water and utilities
X	Applies to everything

might give six area codes as shown in Table 13-4. Discipline codes can be worked out that incorporate existing cost codes, if possible, or the codes may be unique, as shown in Table 13-5.

TABLE 13-5

08		Personnel and employment
10		Project management
20		Procurement
	28	Inspection
30		Engineering
	31	Process
	32	Vessel
	33	Mechanical
	34	Instrument
	35	Electrical
	36	Piping
	37	Coating and insulation
	38	Civil and structural
40		Finance
50		Cost and scheduling

A drawing numbering system might be designed as follows:

AAA-BCCDEEEF

where AAA is the contract number
    B is the area code
    CC is the discipline
    D is the type of drawing
    EEE is the numerical or alphabetical sequence
    F is the alphabetical revision number

An encoding format would then be developed for correspondence and miscellaneous documentation. Computer programs would be written to generate a database of records using this format. Programs would also be required to update and produce various reports as desired by users. Document identification indexes and date indexes, which are usually keyword indexes, are more frequently required than are area and discipline indexes. Tables 13-6 through 13-11 illustrate the various forms that might be used.

TABLE 13-6

Column numbers	Entry
1	Computer program control column
2–8	Date of document
9–11	Fiche location
12–20	Document identification
21	Area code
22–23	Discipline code
24–39	Keyword
40–80	Abstract, title, or as desired

TABLE 13-7

Keyword	Date	Fiche DOC ID	Area	Disc	Abstract
Feed water	03 FEB 77	2CF MEM008	A	33	Stress problems at inlet, to rerun calculations
Piping material	03 FEB 77	ICD MEM005	B	36	Stainless steel fitting required
Power supply	03 FEB 77	2BC CN011	D	32	Operating requirements determined
Reactor	02 FEB 77	IBD TX001	B	32	Determine requirements for sizing equipment
Reactor bid	09 APR 77	1AA CN050	B	20	Review bids, changes to better delivery
Reactor RFQ	02 FEB 77	2AA MN007	B	32	Quote package with specs approved
Scope of work	01 JAN 77	1AA CN001	X	10	Review responsibilities with client
Water supply	03 FEB 77	2CG CN012	E	36	Review fresh-water and discharge-water sizing

TABLE 13-8

Number of pages	10,110	
Number of copies	4	
Total pages	40,440	
Cost @ 10¢ per page		$4,044
Proposed method on microfiche:		
41 fiche $3.00 per COM fiche	$123.00	
4 duplicates @50¢ per copy	20.50	
Total per microfiche system		143.50
SAVINGS		$3,900.50

TABLE 13-9*

	Estimated no. of documents	Paper copy and covers	Jacket fiche	Duplicate set fiche
Manuals, data and reference	2,000	7 manuals 550.00	35 fiche @$2.27	35 fiche @$0.17
Material			$79.45	$5.95
Drawings, originals	500	$500.00	85 fiche @$1.10 $93.50	85 fiche @$0.17 $14.45
Totals		$1050.00	$171.95	$20.40

*To generate a working copy of all backup data and documents required in operating and maintaining the plant, the cost would be $1050 for paper but only $171.95 for jacket microfiche. A duplicate set on fiche is then $20.40. With microfilm, readers and a printer are necessary. With paper copies only a printer is needed.

Updating requires new prints for paper files; updating jacket plus some new duplicate fiche would be required for the jacket fiche system.

TABLE 13-10

ASSUMPTIONS:

Estimated number of specifications	40
Average number of pages per specification	20
Average number of revisions	1.85
Average number of copies per revision	30

Costs using hard copy only:
40 specs, 20 pages each, 30 copies with
1.85 revisions @10¢ per page $4,440.00

Costs using microfiche:
1 original of 40 specs, 20 pages with
1.85 revisions @6¢ per frame 88.80
30 duplicated of 40 specs @30¢ 666.00
Total cost using microfiche 754.80

TOTAL SAVINGS USING MICROFICHE $3,685.20

TABLE 13-11

Item description	Frequency of issuance	Type of copy	Remarks	Control Number
Drawing progress report	Monthly	Hard copy	Start 9-1	T142
Engineering drawings	As approved	Aperture card	Controlled by drawing register	Varies with drawing
Specifications index	Periodic revisions	Microfiche	Same as item 2	T128
Equipment list	Monthly	COM	Same as item 1	T146
Vendor drawings	As released for construction	Aperture Card	Controlled by vendor register	As noted
Vendor drawing register	Periodic revisions	Hard copy	Same as item 1	T148

SUGGESTED READINGS

Peter Brophy, Michael K. Buckland, and Anthony Hindle (eds.), *Reader in Operations Research for Libraries*, Information Handling Services, Englewood, CO, 1976.

Alice Chamis, "Design of Information Systems: The Use of System Analysis," *Special Libraries*, vol. 60, January 1969, pp. 21–31.

Edward A. Chapman, "Planning for Systems Study and Systems Development," *Library Trends*, vol. 21, April 1973, pp. 465–478.

Edward A. Chapman, Paul L. St. Pierre, and John Lubans Jr., *Library Systems Analysis Guidelines*, Wiley-Interscience, New York, 1970.

C. West Churchman, *The Systems Approach*, Dell Publishing Company, New York, 1968.

James F. Corey and Fred L. Bellomy, "Determining Requirements for a New System," *Library Trends*, vol. 21, April 1973, pp. 533–552.

Paul J. Fasana, "Systems Analysis," *Library Trends*, vol. 21, April 1973, pp. 465–478.

John M. Fitgerald and Ardra F. Fitzgerald, *Fundamentals of System Analysis*, John Wiley & Sons, New York, 1973.

Audrey N. Grosch, "Application of Systems Analysis to the Special Library," in Eugene B. Jackson (ed.), *Special Librarianship*, Scarecrow Press, Metuchen, NJ, 1980.

Robert M. Hayes and Joseph Becker, *Handbook of Data Processing for Libraries*, 2d ed., Melville Publishing, Los Angeles, 1974.

Gerald F. Hice, William S. Turner, and Leslie F. Cashwell, *System Development Methodology*, rev. ed., North-Holland Publishing, Amsterdam, 1978.

Frank G Kirk, *Total System Development for Information Sytems*, John Wiley & Sons, New York, 1973.

F. Wilfred Lancaster (ed.), "Systems of Design and Analysis for Libraries,"*Library Trends*, vol. 21, April 1973.

Henry C. Lucas, Jr., *Toward Creative Systems Design*, Columbia University Press, New York, 1974.

M. Mahapatra, "Systems Analysis as a Tool for Research in Scientific Management of Libraries: A State of the Art Review," *Libri*, vol. 30, June 1980, pp. 141–149.

Thomas Minder, "Application of Systems Analysis in Designing a New System," *Library Trends*, vol. 21, April 1973, pp. 465–478.

Edythe Moore, "Systems Analysis: An Overview," *Special Libraries*, vol. 58, February 1967, pp. 87–90.

Perry E. Resove (ed.), *Developing Computer-Based Information Systems*, John Wiley & Sons, New York, 1967.

Ludwig Von Bertalanffy, *General System Theory: Foundations, Development, Applications*, George Braziller, New York, 1969.

Gerald M. Weinberg, *An Introduction to General Systems Thinking*, Wiley-Interscience, New York, 1975.

Ronald E. Wyllys, "System Design—Principles and Techniques," in *Annual Review of Information Science and Technology*, vol. 14, Wiley-Interscience, New York, 1979, pp. 3–35.

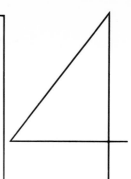

Section 14

Creating an Effective Design/Drafting Manual

George E. Rowbotham

When drafting skills go to waste or are allowed to deteriorate, the fault is often the lack of a design/drafting manual (D/DM)—or use of a poor one. Putting together an effective D/DM nowadays involves a lot more than just collecting and printing the necessary data. Instead, these more sophisticated manuals must be designed, tested, and maintained with the same care given to the technology and practices they describe. They must be carefully structured and organized to clearly describe gears, castings, forgings, dimensioning and tolerancing, datums, surface texture, computerized parts lists, printed-wiring drawings, symbology, and other technology and practices—along with an effective method of updating the manuals.

In general, there are nine primary goals of a D/DM, as follows:

1. Establish uniformity of practice and achieve correct drawing interpretation both within the company and on a national level. (This conserves time, reduces costs, and curtails mistakes.)

2. Simplify and promote the proper use and application of customer standards. (Conformance with contractual requirements results in considerable savings, because it often avoids the need for redrawing.)

3. Reduce supervision. (Personnel will not require verbal step-by-step guidance.)

4. Reduce manufacturing costs. (Practices are included in the D/DM that will promote ease of manufacture and assembly.)

5. Hasten the execution of work. (The decision-making process becomes complete, and therefore more accurate and much faster.)

6. Enable more thought to be expended on new and advanced drafting problems. (Solved problems are documented.)

7. Minimize operating problems that result from divergent drafting practices. (Differences of opinion are reconciled.)

8. Train drafters. (Drafters learn the latest technology and practices and profit from the experience of others.)

9. Avoid redundant drafting.

With engineering and technology becoming more and more complex, the ability to communicate clearly is of critical importance. The D/DM should be prepared on the premise that an engineer's expression of design requirements, through specifications and drawings, is only as valuable as the ability of those documents to communicate such requirements to manufacturing, procurement, testing, and quality assurance for clear and specific action. It will provide instruction, direction, and guidance. It will be an assurance that engineering drawing output is reasonably close to the latest state of the art, conforms to contractual obligations, and supports company operating methods. Also, a D/DM should be effectively indexed.

QUALITY DRAWINGS—THE ULTIMATE OUTPUT

When drawings are easily understandable, complete, and accurate, they contribute greatly to the avoidance of errors, lost time, and scrapped parts, as well as to the successful functioning of the hardware. Preplanning the drawing structure can also aid substantially in producing an economical and adequate design package. In summary, this is the prime purpose of the D/DM.

STANDARDS

American National Standards Institute (ANSI) Standards

Practices included in the D/DM should comply with ANSI standards, tailored to internal operational requirements. By doing this, drafting practices will comply with practices used on a national basis, making drafting output fully understandable to customers, vendors, and suppliers.

Military Standards

If necessary, the D/DM will support the requirements of MIL-D-1000 and MIL-Std-100. The approach will, of course, be tailored to internal operational requirements.

WRITING—A PRIME CONSIDERATION

The greatest fault of most D/DMs is that they are simply not readable. Design and drafting personnel are too often puzzled because of shoddy communication. They're confused by vague and inconclusive answers. Modern civilization's most important product is information, such as a D/DM. This vital product, information, is useless unless it is clearly understandable to the people who can make use of it. Many truly great ideas have laid dormant simply because they weren't effectively communicated.

The prime myth underlying the writing problem is that it's easy to teach people to write. It is extremely naive to propose that all that has to be done is to enroll a company's professional people in English or in technical report writing classes at the local college and the problem will be solved. That might be a good start, but it's only a start—in low gear, at that. Experience proves beyond a doubt that people cannot be taught to write easily. They can learn, provided that they have sensitivity for words and style and a good understanding of what is acceptable English usage—and, most important, provided that they are willing to dedicate many of their leisure hours to careful reading and evaulation, and a concerted effort to imitate good writing. Few people can be great writers, but most everyone can learn to write intelligently. A great deal depends on a well-organized mind.

Clear writing starts in an author's mind. After an author knows what he or she wants to say, and whom he or she is saying it to, clear writing should generally follow. This preliminary thinking leads to the main idea the writer wants to get across. Every effort should be made to get that idea across in the lead sentence. The lead sentence should be a direct, positive statement, and should not be weighed down with qualifications and detail. A good lead prepares the reader's mind for what follows—to expect the direction the message will move in and to grasp it quickly. If sentences shuffle along, the reader will give up in boredom; if they dart one way, change direction, and then start off again, the reader will give up in exhaustion.

Writing, particularly in a technical document such as a D/DM, should include clear, concise statements that are free of ambiguity. The reader should be able to extract the proper information quickly and with minimum effort. Clear writing is the knack for telling the facts clearly, simply, and briefly. Brevity conserves the reader's time. For example, this sentence prolongs reading time: "It is used to rupture missile frames while in flight in order to initiate disintegration." This statement conserves reading time: "It makes missiles explode in flight." The chief fault of many D/DM writers is generally "overwriting." In an effort to be exact, they clutter writing with "fat" words and phrases that could probably have been omitted or replaced by a short word. Long words are good only if no shorter word will do the job. Technical terms can be useful as shorthand, but only if the reader understands them. "Thermal equilibrium" is not, for example, good shorthand for "stable temperature."

Toward the end of the seventeenth century, John Dryden (British essayist, satirist, and poet laureate) said: "The chief aim of the writer is to be understood." Two centuries later, Robert Louis Stevenson expanded on Dryden, saying "Don't write merely to be understood. Write so that you cannot possibly be misunderstood." Another fault of most writers is the excessive use of long sentences. Short sentences (and short words) decrease the mental effort needed to elevate ideas from the page into the reader's mind. One way to boost readability is to keep the average number of words per sentence under 20. Two thousand years ago Pliny the Elder wrote: "I apologize for this long letter; I didn't have time to shorten it."

To sum it up: the D/DM should be a model of clarity and thoroughness.

ORGANIZATION OF CONTENTS

Subsections (subjects) should be carefully organized in a readable and logical manner. Subjects and individual standards should not be buried in an avalanche of unrelated detail. For example, the standard on drawing scale should not be included in the subsection covering arrangement of views. It should, in the interest of clarity, be presented as a separate subsection, because its scope is different. It defines a specific detailed practice that deals with the prevailing drawing scale and, where applicable, multiple scales. Likewise, such unique and complex subjects as line conventions, sectioning, projection, and dimensioning and tolerancing should also be clearly presented as separate sections or subsections. Both third-angle projection (the American practice) and first-angle projection (a confusing practice) should be thoroughly but briefly defined in the D/DM, along with appropriate illustrations. Both forms of projection should be clearly defined for three significant reasons, namely: (1) the drafter will understand third-angle projection when it is clearly defined; (2) the drafter will understand better how to avoid first-angle projection, which should never be used; and (3) personnel encountering first-angle projection on drawings prepared outside the United States will understand how to deal with this confusing practice.

Survey of Needs

An in-house, in-depth survey may be conducted to identify internal needs. As part of this effort, the needs of key personnel, including both those responsible for drawing preparation and the users of those drawings, should be determined. An objective questionnaire may be used for this. (See Fig. 14-1*a* through *f* for an example.) Such a questionnaire may be unnecessary, of course, if the D/DM writer has a thorough knowledge of the product line or if a D/DM committee is organized.

Table of Contents

A typical table of contents is illustrated in Fig. 14-2. Contents will vary according to product line, but they should always be reduced to essentials.

Index

The purpose of an index is to make the D/DM more useful. The D/DM should be indexed so that the reader can find the needed information quickly and without difficulty. An index must be selective, however. No effort should be made to index a subject under every form by which a reader might possibly look for it. For example, entries under *surface texture* should not be duplicated under *surface finish*, nor should the entries under *screw threads* be duplicated under *threads*, and so on. To help the reader, cross-references can be used; for example, *threads (see also screw threads)* can direct the index user to related subjects without unnecessary repetition.

Page Numbering

The D/DM should be divided into sections titled according to subject matter and identified by a letter of the alphabet. Every page in each section will carry this letter as a prefix to the page number. In order to facilitate classification of subject matter and to allow for insertion of additional pages, each section is divided into subsections numbered by one or more digits as required, and with decimals used to designate individual pages. Each subsection should begin on a right-hand page.

Each page number will consist of a letter, a subsection number, a decimal point, and a page number. A typical D/DM page number, therefore, would be A1.01, which may be followed by A1.02, A1.03, etc., with odd numbers always on right-hand pages and even numbers on left-hand pages. If it becomes necessary to insert pages between existing numbers, these are given a third digit, i.e., A1.021, A1.022. In this case the page sequence would appear as in Fig. 14-3.

Paragraph Numbering

Paragraphs should be numbered in the D/DM to facilitate reference and to indicate topical organization of the material. These numbers begin anew with each subsection and bear no relation to page numbers. Each prime paragraph is identified by a whole

DRAFTING ROOM MANUAL QUESTIONNAIRE

Date:_____

This questionnaire should be completed prior to the initiation of the drafting room manual preparation effort.

Name and title of individual interviewed:_____

_____

Department:_____

A check mark ☑ should be entered in the appropriate blocks.

☐ DRAWINGS TYPES

Which of the following drawing (dwgs) types are used:

☐ Design layouts
☐ Sketches
☐ Detail dwgs
☐ Assembly (assy) dwgs
☐ Tabulated detail dwgs
☐ Standard parts dwgs
☐ Inseparable parts dwgs
☐ Same-as dwgs
☐ Installation dwgs
☐ Envelope dwgs
☐ Kit dwgs
☐ Book form dwgs
☐ Cut and paste (scissor) dwgs
☐ Photo dwgs
☐ Undimensioned dwgs
☐ Vendor control dwgs
  ☐ Specification control
  ☐ Source control
☐ Wiring harness assy dwgs
☐ Wiring data dwgs
☐ Scribe coat dwgs
☐ Lofting dwgs

☐ Printed wiring (PC) dwgs
☐ Flexible printed wiring dwgs
☐ Electronic schematics
☐ Logic diagrams
☐ Wiring diagrams
☐ Block diagrams
☐ Interconnection diagrams
☐ Flow schematics
☐ Mechanical schematics
☐ Piping dwgs
☐ Tubing dwgs
☐ Structural dwgs
☐ Pictorial dwgs
  ☐ Isometric
  ☐ Trimetric
  ☐ Axometric
  ☐ Dimetric
  ☐ Perspective
  ☐ Exploded views
☐ Other _____
☐ Other _____
☐ Other _____

☐ INTERACTIVE GRAPHICS

☐ Are any of the foregoing drawings produced by interactive graphics? If yes, underline the applicable drawings.

☐ If interactive graphics is used, what savings resulted?_____

☐ What hardware is used?_____

_____
_____
_____
_____
_____

FIG. 14-1(a)   A sample questionnaire.

☐ CONTRACTURAL REQUIREMENTS

    ☐ Are MIL-STD-100 and MIL-D-1000 requirements involved?
    ☐ Are any contractural requirements involved? If so, what?_____
    _____

☐ RESTORATION PROBLEMS

    ☐ Are there drawing restoration problems: If yes, to what extent?
      ☐ Excessive -- Comments_____
      ☐ Minor -- Comments_____

☐ MICROFILMING

    ☐ Are drawings microfilmed?
      ☐ If yes, is there any loss of detail in microfilming?
      ☐ Is microfont lettering used?
      ☐ Are alignment arrowheads used?

☐ CONVENTIONAL REPRESENTATION

    ☐ Is conventional representation used?

☐ Threads	☐ Minimized section lining
☐ Gear teeth	☐ Minimized projected views
☐ Piping	☐ Descriptions instead of views
☐ Tubing	☐ Partial views for symmetry
☐ Repeated features	☐ Springs
☐ Circles instead of ellipses	☐ Plumbing hardware
☐ No underlining	☐ Standard parts
☐ Pictorial drafting avoided	☐ Symbols
☐ Nameplates	☐ Shadowing avoided
☐ Knurling	☐ Other_____
☐ Nuts	☐ Other_____
☐ Serrations	☐ Other_____
☐ Electrical Terminals	☐ Other_____
☐ Electrical receptales	☐ Other_____
☐ Bearings	☐ Other_____

☐ DRAWING TITLES

    ☐ Do drawing titles consist of a noun or a noun phrase and sufficient modifiers
      to distinguish the part from other similar parts?
      ☐ If no, what is the practice?_____

☐ PATENT PROCEDURE

    ☐ Is there an outline or procedure for the handling of items of a patentable nature?

☐ TRADEMARKS

    ☐ Are there any trademarks used to identify parts and assemblies?

☐ CHECKING PRACTICE

    ☐ Is there an established checking practice?
      ☐ Are dwgs checked 100%?
      ☐ If no, what are the limitations?_____

FIG. 14-1(b) A sample questionnaire, page 2.

□ <u>WELDING</u>

    □ Are welding symbols used?

□ <u>ELECTRICAL</u>

    □ Are electrical symbols used?

□ <u>DESIGN STANDARDS</u>

    □ Are design standards established?

□ <u>TABLES AND CHARTS</u>

    □ Are there any particular tables and charts that are frequently used?
      If yes, what?_____
      _____
      _____

□ <u>DRAWING SIZES</u>

    □ What drawing sizes are used?

□ A-8.5 x 11	□ E-34 x 44	□ J-34 x 48
□ B-11 x 17	□ F-28 x 40	□ K-40 x 48
□ C-17 x 22	□ G-11 x 42	□ Other_____
□ D-22 x 34	□ H-28 x 48	□ Other_____
□ A-9 x 12	□ F-36 x 48	□ J-36 x 50
□ B-12 x 18	□ L-18 x 50	□ K-42 x 50
□ C-18 x 24	□ P-24 x 50	□ Other_____
□ D-24 x 36	□ H-30 x 50	□ Other_____

    □ Any restrictions involved on the use of large size drawings?
      If yes, describe:_____
      _____
      _____

□ <u>REVISIONS</u>

    □ Are revision records completely recorded in detail in the drawing revision block?
      If no, is reference made to the Engineering Order or Change Notice?

□ <u>TECHNIQUES</u>

    □ Are drawings prepared in pencil?
    □ Are drawings prepared in ink?
    □ Are drawing notes typed?
    □ Are dimensions and tolerances typed?
    □ Are adhesive-backed preprinted drafting aids used?
      If yes, what?_____
    □ Are modified reproducibles used to produce new drawings?

□ <u>NUMBERING</u>

    □ Is there a part number identification procedure?
    □ Is a significant part number system used?
    □ Are item numbers (dash numbers) used?

FIG. 14-1(c)   A sample questionnaire, page 3.

☐ <u>DISTRIBUTION</u>

☐ What will the drafting room manual distribution amount to approximately?

☐ Draftsmen____	☐ Management____	☐ Sales____
☐ Designers____	☐ Quality Control____	☐ Accounting____
☐ Checkers____	☐ Purchasing____	☐ Other____
☐ Engineers____	☐ Vendors____	☐ Other____
☐ Manufacturing____	☐ Customers____	☐ Other____

☐ <u>NEXT ASSEMBLY</u>

☐ Is the next assembly drawing numbered recorded on the dwg?
   If no, where?_____

☐ <u>MULTI-SHEET DWGS</u>

☐ Are multi-sheet drawings used?

☐ <u>TOLERANCING</u>

☐ Is unilateral tolerancing used?
☐ Is bilateral tolerancing used?
☐ Is plus or minus tolerancing used?
☐ Are shop run tolerances established?
☐ Are general tolerances established in the drawing title format?
   If yes, what?_____
☐ Is positional tolerancing used?
   If yes, are the following concepts applied:
   ☐ Maximum material condition (MMC)?
   ☐ Regardless of feature size (RFS)?
   ☐ Radius concept?
   ☐ Diameter concept?
   ☐ Projected tolerance zone?
   ☐ If any departures, explain:_____
   _____
☐ Is tolerancing of form (geometric) and runout applied?
☐ Is symbology used?

☐ <u>DIMENSIONING</u>

☐ Are decimals used exclusively?
☐ Are fractions used exclusively?
☐ Are fractions and decimals used together?
☐ Is datum referencing used?
☐ Is limit dimensioning used?
☐ Is free state variation dimensioning used?
☐ Is the metric system used?
   ☐ If yes, is dual dimensioning used?
      (Are inch and millimeter dimensions combined on one drawing?)

☐ <u>DRAWING SCALES</u>

☐ What drawing scales are used?

☐ Full size	☐ Four times size	☐ Other____
☐ One half size	☐ Ten times size	☐ Other____
☐ Twice size	☐ Other_____	☐ Other____
☐ One fourth size	☐ Other_____	☐ Other____

FIG. 14-1(d)   A sample questionnaire, page 4.

☐GEARS, SPLINES AND SERRATIONS

    ☐Are gears used?
        If so, which type?

☐Spur   ☐Hypoid   ☐Crown teeth
☐Helical   ☐Worm   ☐Gear racks
☐Bevel   ☐Internal   ☐Other_____

    ☐Are splines used?
        If so, which type?

☐Involute
☐Parallel sides

    ☐Are serrations used?

☐SPRINGS

    ☐Are springs used?
        If yes, which type?

☐Compression   ☐Leaf
☐Extension   ☐Flat
☐Torision coil   ☐Flat spiral
☐Torision bar   ☐Bellevile
☐Volute   ☐Other_____

☐PRODUCT LINE

    Which types of parts, or parts elements, are used?

☐Machined parts   ☐Extrusions
☐Stampings   ☐Rubber parts
☐Threaded parts   ☐Weldments
☐Pipe threaded parts   ☐Electronic packaging
☐Plastic parts   ☐Metal powder parts
☐Forgings   ☐Optical parts (glass)
    If yes, are separate forging   ☐Torch cutted parts
    drawings prepared?   ☐Brazed parts
    ☐Yes   ☐No   ☐Broached parts
☐Castings   ☐Tubing
    If yes, are separate casting   ☐Purchased parts
    drawings prepared?   ☐Modified purchased parts
    ☐Yes   ☐No   ☐Gaskets
    Which type of casting is used? ☐Cams
    ☐Sand   ☐N/C produced parts
    ☐Die   ☐Ceramic parts
    ☐Permanent mold   ☐Metal powder parts
    ☐Plaster mold   ☐Sheet metal parts
    ☐Investment castings   ☐Riveting
      (Precision-loss wax proc.) ☐Other_____

☐SECURITY MARKINGS

    ☐Are any security markings required?
    ☐Confidential   ☐Top Secret
    ☐Secret   ☐Other_____

FIG. 14-1(e)A sample questionnaire, page 5.

☐ DRAWING NOTES

    ☐ Are general notes standardized?
    ☐ Are local notes standardized?
    ☐ Are they preprinted?
    ☐ Are they typed?

☐ SURFACE TEXTURE

    Are surface texture requirements defined?
    ☐ If yes, are waviness requirements defined?
    ☐ If yes, is lay restriction defined?
    ☐ Is symbology used?

☐ PRODUCEABILITY

    ☐ Are drawings reviewed and approved by a produceability engineer?

☐ DESIGN REVIEWS

    ☐ Are design reviews held?

☐ SAFEGUARDING DRAWINGS

    ☐ Are procedures established to assure the safety of drawings?
    ☐ Is a fire proof vault used?

☐ PARTS LISTS

    ☐ Are parts lists included directly on the drawing?
    ☐ Are separate parts lists prepared?
    ☐ Are parts lists computerized?

☐ UNIQUE PRACTICES

    ☐ Are unique drafting practices applied?
    If yes, describe:_____
    _____
    _____

☐ REMARKS

    _____
    _____
    _____
    _____
    _____
    _____
    _____
    _____

FIG. 14-1(f) A sample questionnaire, page 6.

TYPICAL TABLE OF CONTENTS

FIG. 14-2   Table of contents.

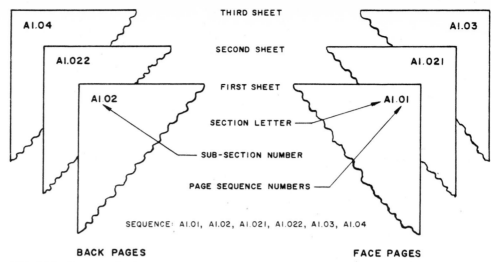

FIG. 14-3   Page numbering.

integer followed by a decimal point. Subordinate paragraphs are numbered by adding integers, separated by decimal points, such as 1.1.1.

Figure Numbering

All figures, whether they consist of illustrations, tables, or charts, should be consecutively numbered on each page according to the following system: the first part of the number will consist of the page number, i.e., the portion of the page designation that follows the decimal point, and the second part of the number will consist of a digit, separated from the first part by a decimal point, denoting the sequence on that page. Thus, three figures on page A7.01 will be numbered 01.1, 01.2, and 01.3. This also applies to pages that have been added. Therefore, figures on page A7.011 will be numbered 011.1, 011.2, and 011.3.

Illustrations

The inclusion of ample illustrations will eliminate lengthy explanations. The Chinese are credited with coining the phrase "A picture is worth a thousand words." To a drafter a picture is easily worth ten times that many words. Illustrations in a D/DM should be clear and complete, but only so far as is necessary to clearly illustrate the point under discussion. An illustration should be placed as near as possible to its first reference in the text but should not break into the middle of a paragraph. Care should be taken to spare the reader the necessity of turning a page to refer to an illustration. Linework, decimal points, and lettering should be sharp enough and thick enough to reproduce well. Scale and proportion should be uniform. Spacing of letters and numerals should appear approximately equal (see Fig. 14-4). Illustrations that will be reduced in size for inclusion in the D/DM should have ample space between lines, letters, and numerals in order to avoid overcrowded or smudged appearance (see Fig. 14-5). Border lines should enclose illustrations, and all figure numbers should be placed outside the border.

Location of Illustrations

Where correlation to text material is not affected, illustrations should be located in the top right-hand column of the first page of a section or subsection. Figure 14-6 shows (a) a page with a single-column illustration and (b) a page with a double-column table.

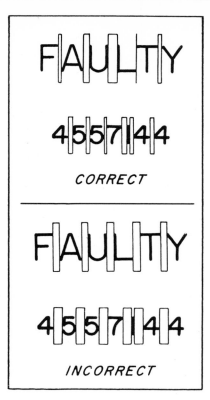

FIG. 14-4 Spacing of lettering and numerals.

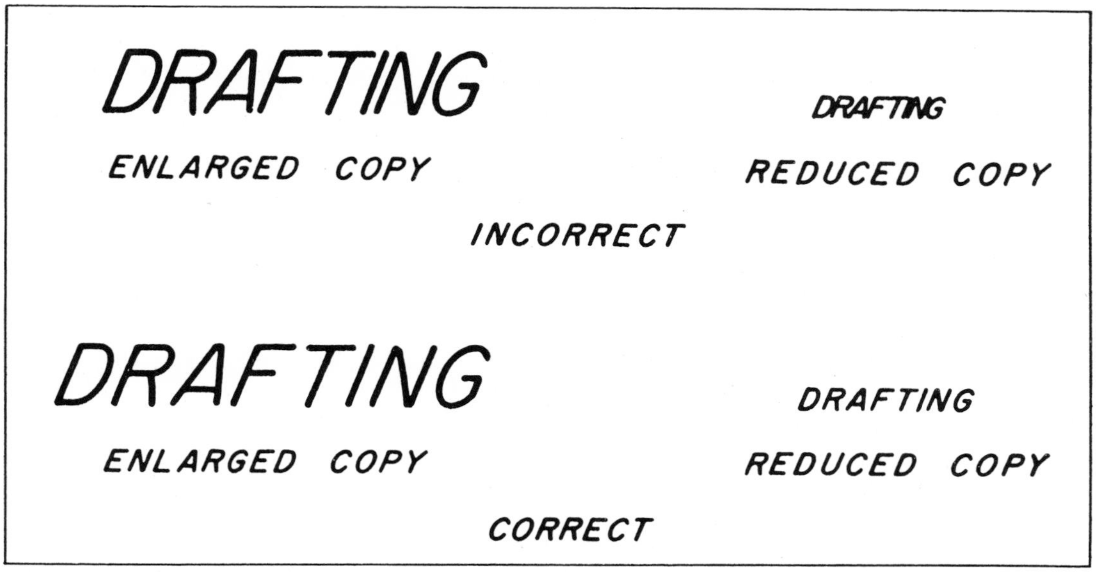

FIG. 14-5 Reduction (size reduction of detail).

Pinpointing "Know-How" Requirements

As illustrated in a book titled *Value Engineering: A Systematic Approach*, by Arthur E. Mudge (McGraw-Hill; 1971), a system of "know-how" requirements provides the reader of a D/DM with a reliable way of reducing the amount of work to be done in performing specific daily activities, thereby increasing efficiency and saving time and money. An especially valuable feature of this system is the incorporation of "reading

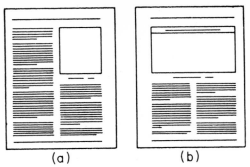

FIG. 14-6 Location of illustrations.

levels" into the table of contents, allowing the reader to select the information of greatest importance on the basis of job function or responsibility. Reading levels pinpoint know-how requirements. For example, four different levels of study can be coded and described as follows:

A = To be committed to memory

B = To be read and studied for complete understanding

C = To be read for familiarization and general understanding

D = To be scanned for basic understanding of contents for future reference

With such coded levels of study, drafters, checkers, designers, engineers, and so forth can quickly isolate and study those contents of the D/DM which relate most frequently to their daily work. It should not be intended that every page of the D/DM be committed to memory by everyone. It should not be necessary for a drafter to refer to the D/DM every time he or she draws, letters, or revises a drawing. If this happened, drafting would proceed at a snail's pace. In time, the contents of the D/DM will become part of everyone's know-how through constant association and use. For some people, however, certain pages will require concentrated study from the start. An example of how D/DM know-how requirements can be pinpointed is shown in Fig. 14-7. By helping each individual understand what know-how is expected of him or her, goals can be reached by the best way, in the shortest time, and at the lowest cost.

Two-Part D/DM

A D/DM could, for practical reasons, be divided into two parts. Part 1 could stick strictly to drafting practices, and Part 2 could explain Part 1. With this approach, the D/DM user can study the explanatory notes and thereby be able to refer to the concise drafting practice presentation with full understanding.

FORMAT AND ARRANGEMENT

In designing the D/DM, every effort should be made to save the user's time by ensuring that what is presented will be easily referenced. A reader normally enters the page visually at the upper left-hand corner, and the layout should proceed from that point on. Adequate cross-referencing should be employed to avoid duplication of information. Tables, preferable to graphs, should be reduced to the simplest form consistent with the degree of sensitivity required so as to avoid making the reader interpolate between steps. Text and illustrations should be printed on D/DM standards forms. Text matter should be arranged in two columns, with adequate space between the columns. Typing should be performed directly on a printed worksheet form that dimensions the spaced columns, format arrangement, and so forth in nonreproducible blue lines. Figure 14-8 shows an example of such a printed worksheet form.

KNOW-HOW CHART-DRAFTING ROOM MANUALS

SUBJECT	PAGE	EXECUTIVE	INSPECTOR	MANUFACTURING ENGINEER	MANAGER-ENGINEERING	ENGINEER-ELECTRICAL	ENGINEER-MECHANICAL	DESIGN AND DRAFTING SUPERVISOR	DESIGNER-ELECTRICAL	DESIGNER-MECHANICAL	CHECKER	DRAFTSMAN-SENIOR	DRAFTSMAN	FILE CLERK
FOREWORD (PURPOSE OF MANUAL)	1.0	A	B	B	A	A	A	A	A	A	A	A	A	A
ARRANGEMENT OF VIEWS	5.0	D	B	B	B	B	B	A	B	B	A	A	A	C
CARE AND HANDLING OF DRAWINGS	14.0	A	A	A	A	A	A	A	A	A	A	A	A	A
CASTING DRAWINGS	22.0	D	B	B	C	C	B	B	C	B	A	B	B	D
CHECKING PRACTICE	19.0	B	C	C	B	C	C	B	C	C	A	B	B	C
COMPUTER GRAPHICS	32.0	B	D	D	B	B	B	B	B	B	C	C	C	D
CONVENTIONAL REPRESENTATION	9.0	D	D	C	C	C	C	B	C	C	A	A	A	D
COST REDUCTION TECHNIQUES	10.0	C	C	A	A	B	B	A	B	B	A	A	A	C
DESIGN LAYOUT PRACTICE	20.0	B	C	B	B	A	A	A	A	A	B	B	B	D
DIMENSIONING AND TOLERANCING	16.0	C	A	A	B	B	B	A	B	B	A	B	B	D
DRAWING INTERPRETATION	15.0	B	A	A	B	B	B	A	B	B	A	A	B	D
DRAWING LEGIBILITY AND REPRODUCEABILITY	12.0	C	B	B	B	B	B	A	B	B	A	A	A	A
DRAWING MEDIUMS AND REPRODUCTIONS	11.0	C	C	C	B	C	C	A	C	C	B	B	C	A
DRAWING NOTES	13.0	D	D	D	D	D	D	B	C	C	B	B	C	C
DRAWING REVISIONS	17.0	C	B	B	B	B	B	B	B	B	B	B	B	B
DRAWING SHEET SIZE AND FORMAT	4.0	D	C	C	C	C	C	C	C	B	A	A	B	B
ELECTRONIC PACKAGING DESIGN	27.0	C	C	C	C	A	C	A	A	C	B	C	D	C
ELECTRONIC SCHEMATIC DIAGRAMS	24.0	C	C	C	C	A	C	B	A	C	B	C	D	C
FORGING DRAWINGS	23.0	C	C	C	C	C	B	B	C	B	B	B	C	C
LETTERING	8.0	D	C	C	C	C	C	C	C	C	A	A	A	C
LINE CONVENTIONS	6.0	D	C	C	C	B	B	B	B	B	A	A	A	D
NUMBERING OF DRAWINGS AND PARTS	18.0	C	B	B	B	B	B	B	B	B	A	A	B	C
MECHANICAL SCHEMATIC DIAGRAMS	26.0	D	D	C	C	D	A	B	C	A	B	B	C	D
ORGANIZATION	3.0	A	D	C	A	D	D	A	C	C	C	C	C	C
PARTS LISTS	21.0	C	C	C	B	B	B	B	B	B	A	B	C	C
PRINTED CIRCUIT ARTMASTERS	25.0	C	B	B	C	A	C	A	A	D	B	B	C	C
PRODUCEABILITY (MANUFACTURING CONSIDERATIONS)	30.0	B	A	A	B	B	B	B	B	B	A	B	B	D
SECTIONING	7.0	C	B	B	C	C	C	B	C	C	B	A	A	B
SHOP RUN TOLERANCES	31.0	C	A	A	B	B	B	B	C	B	A	B	C	D
SCREW THREADS	29.0	D	B	B	C	C	B	B	C	B	A	B	C	D
UNDIMENSIONED DRAWINGS	28.0	D	B	B	C	B	B	A	B	B	A	B	C	C
WORK AUTHORIZATION	2.0	A	C	C	C	B	B	A	B	B	B	B	C	D

FIG. 14-7   Pinpointing "know-how" requirements.

Ideal Page Layout

Figure 14-9 illustrates an example of an ideal page layout. The text and related illustrations are grouped close together and are enclosed with linework. This type of page layout should be used whenever possible. Paragraphs that carry over to the following page should be subheaded again, followed by the abbreviation "cont'd."

Tables

The lines of type in a table should be in groups of three to five lines, with space between each group. On comparatively large tables, lines may be drawn in the spaces between

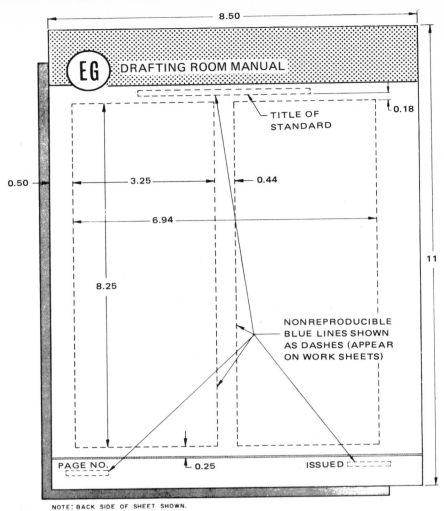

FIG. 14-8   Worksheet.

each group for ease of reading. Vertical column lines should always be used, and column headings should be boxed (see Fig. 14-10). Large solid blocks of unbroken type should be avoided. Lengthy paragraphs should be broken into smaller paragraphs. An illustration attracts the eye; a page of solid, unbroken type wearies it. Beyond the virtue of attractiveness is the greater contribution of clarification through well-chosen and well-prepared illustrations that conform to accepted drafting practice.

Position of Text and Illustrations

Text and illustrations should be presented in a normal book format, as shown in Fig. 14-11; if necessary, tables and illustrations may be presented so as to be read from the right-hand side of the manual when it is turned 90°, as shown in Fig. 14-12. Effort should be taken to minimize the need for turning the manual to refer to text, illustrations, or charts.

Drawing Notes

A list of acceptable drawing notes should be included in the D/DM, along with other pertinent details concerning dimensioning and tolerancing, lettering, projection, and so forth. All too often, shop personnel will read two drawings that originated in the

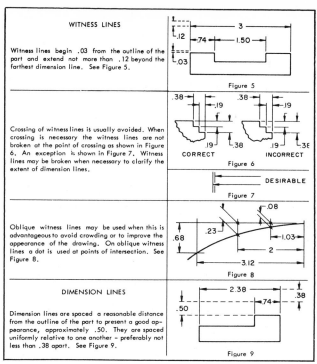

WITNESS LINES	
Witness lines begin .03 from the outline of the part and extend not more than .12 beyond the farthest dimension line. See Figure 5.	Figure 5
Crossing of witness lines is usually avoided. When crossing is necessary the witness lines are not broken at the point of crossing as shown in Figure 6. An exception is shown in Figure 7. Witness lines may be broken when necessary to clarify the extent of dimension lines.	CORRECT INCORRECT Figure 6 DESIRABLE Figure 7
Oblique witness lines may be used when this is advantageous to avoid crowding or to improve the appearance of the drawing. On oblique witness lines a dot is used at points of intersection. See Figure 8.	Figure 8
DIMENSION LINES Dimension lines are spaced a reasonable distance from the outline of the part to present a good appearance, approximately .50. They are spaced uniformly relative to one another – preferably not less than .38 apart. See Figure 9.	Figure 9

FIG. 14-9 Ideal page layout.

CAPTION				
XX	XXX	X	XXXXX	XXX
XXXX	XXXXXX	XXX	XXXXXXX	XXXXX
XXXX	XXXXXX	XXX	XXXXXXX	XXXXX
XXXX	XXXXXX	XXX	XXXXXXX	XXXXX
XXXX	XXXXXX	XXX	XXXXXXX	XXXXX
XXXX	XXXXXX	XXX	XXXXXXX	XXXXX
XXXX	XXXXXX	XXX	XXXXXXX	XXXXX
XXXX	XXXXXX	XXX	XXXXXXX	XXXXX
XXXX	XXXXXX	XXX	XXXXXXX	XXXXX
XXXX	XXXXXX	XXX	XXXXXXX	XXXXX

FIG. 14-10 Tables.

READING POSITION

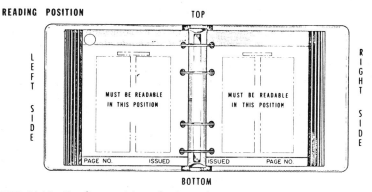

FIG. 14-11 Reading position—horizontal.

READING POSITION

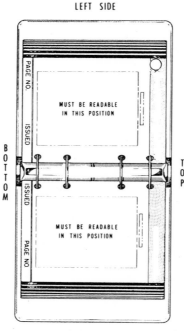

FIG. 14-12 Reading position—vertical.

same drafting room only to discover that what should have been identical notes on the drawings are so different that they appear to have been drawn by two different drafters in two different companies. Different notes can take on different meanings and may result in a scrapped part. The way to avoid this problem is to include a complete set of acceptable drawing notes in the D/DM. A desirable method of presenting drawing notes is to list them by category—for example, bonding, casting, machining, welding, etc.—and also by drawing type, such as formed sheetmetal drawing, printed-wiring-board assembly drawing, schematic drawing, or tubing assembly drawing.

Uniformity of Interpretation with the Least Effort

The chief goal should always be to make it possible for the reader to extract from a drawing what he or she needs quickly, with a minimum of effort, and without ambiguity. This can be accomplished by showing two abbreviated related illustrations. The first illustration will include the caption "drawing callout," and the other illustration will include "interpretation." This editorial technique is highly recommended and should be widely used. It can eliminate pages of cumbersome text. Examples of this technique are illustrated in Figs. 14-13 and 14-14.

Typical Examples of Sound Practices

A typical example of the type of drafting practice that will greatly enhance drawing clarity is illustrated in Fig. 14-15. In contrast, the poor practice shown in Fig. 14-16 will probably result in a great deal of costly scrap on the shop floor. Proper drafting practice should be presented in the D/DM.

Drafting Economy

The D/DM should stress that drawings should be reduced to their essentials. In the interest of speed and drafting economy, examples of conventional representations should be included to illustrate clear delineation in the drawing of screw threads, gear teeth, welding, piping, electrical components, and similiar detail (see Figs. 14-17, 14-

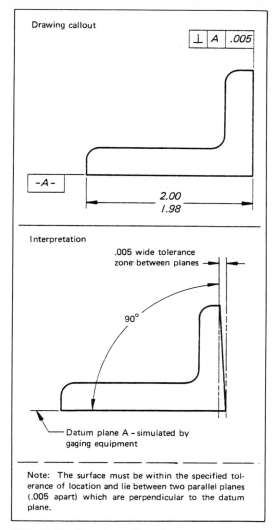

Drawing callout

⊥ | A | .005

-A-

2.00
1.98

Interpretation

.005 wide tolerance
zone between planes

90°

Datum plane A – simulated by
gaging equipment

Note: The surface must be within the specified tolerance of location and lie between two parallel planes (.005 apart) which are perpendicular to the datum plane.

FIG. 14-13   Drawing callout; interpretation: perpendicularity.

18, and 14-19. Instructions should be included in the D/DM that will prevent drafters from drawing any detail lacking intrinsic value, such as shading, artistic frills, redundancy, fancy lettering, and other ornamental detail. Proven techniques should also be described and recommended, such as interactive graphics, cut-and-paste drawings, automated parts lists, photodrawings, microfilming, typewritten lettering, and other cost-saving innovations. It should be emphasized that drawings should be prepared with the thought that shop personnel will not generally be able to visualize the intent of design and drafting personnel without a thoroughly clear presentation. It doesn't make sense to save a few dollars in drawing preparation and then to scrap hundreds of dollars worth of material in the shop because the drawings weren't clear enough. There should never be any compromise between drafting costs and sound drafting practices.

CONVENTIONAL REPRESENTATION

In many instances, portrayal of an object by exact projection of every surface boundary is both laborious and unnecessary. In fact, a drawing so executed may contain such a maze of unnecessary lines that vital details are obscured or deemphasized. For this reason, it is common practice to deviate from strict projection in the interest of economy and drawing clarity. For example, lines may be drawn which do not exist; lines which do exist may be omitted; a feature may be drawn as though displaced from its true position; shapes may be approximated; standardized codes or symbols may be used. Such practices, if they are widely recognized and accepted, are referred to as *conventional representations* and should be included in the D/DM.

Conventional representation, to be a valid means of communication, must include only practices that will be clearly understood by those using the drawing. The various sections of the D/DM should include examples of those drawing conventions as a guide to the accepted practices that can save drawing time without sacrificing clarity. Expe-

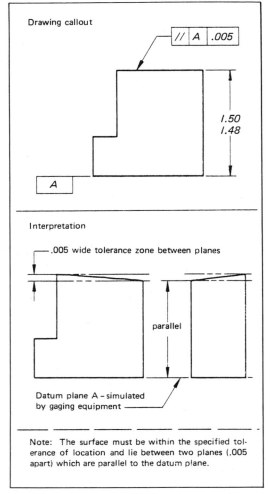

FIG. 14-14   Drawing callout; interpretation: parallelism.

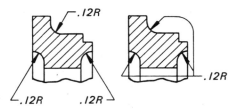

Recommended   Not recommended   **FIG. 14-15**   Optimum practices.

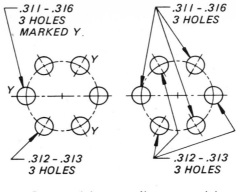

FIG. 14-16 Optimum practices—hole location.

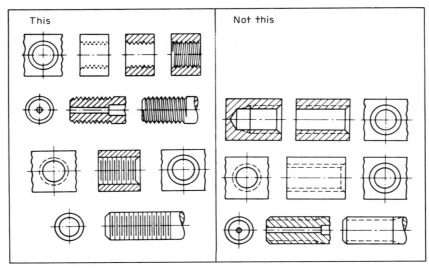

FIG. 14-17 Conventional representation.

rience proves over and over again that widely used and readily understood conventional practices result in substantial savings.

The fact that proven and accepted conventions are beneficial, however, does not justify extreme forms of simplification where uncertain intent may cause misinterpretation in manufacture, procurement, or inspection. Known as *simplified drafting*, this practice is often referred to as a "drafting menace." Simplified drafting advocates such practices as eliminating dimension lines, radii, arrow heads, circles to delineate holes, extension lines, and an endless amount of other essential engineering detail. This can conceivably reduce the cost of preparation of an engineering drawing by perhaps 25 percent, but it can easily increase overall operational costs by a staggering amount in the form of an endless parade of scrapped parts, debates, memorandums, adverse customer relations, and wrecked schedules. The D/DM should expressly prohibit this practice.

PATENT PROCEDURE

An outline or procedure for the handling of items of a patentable nature should be included in the D/DM. Or, as an alternative, reference could be made to Section 10 in this handbook.

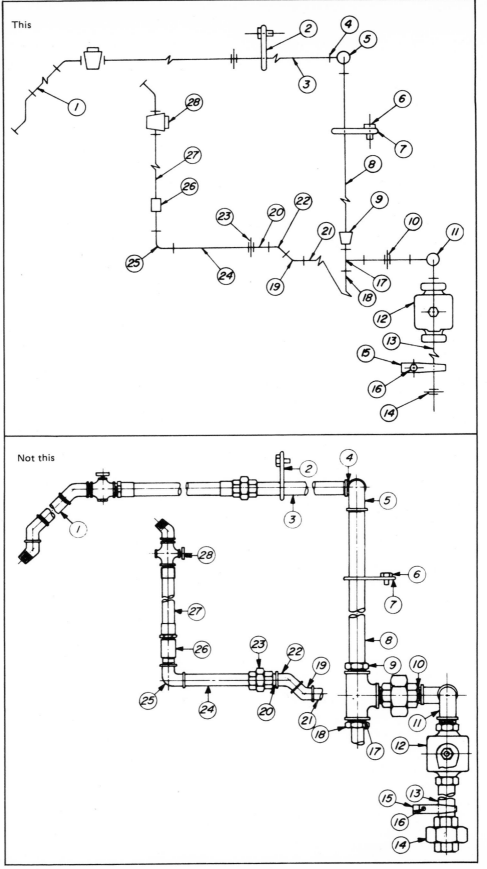

FIG. 14-18 Conventional representation.

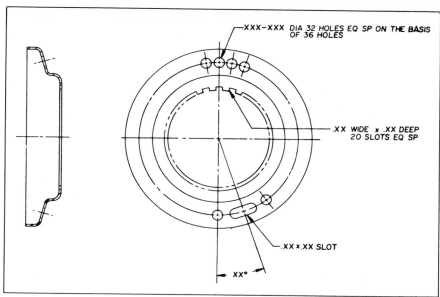

FIG. 14-19   Conventional representation.

TABLES AND CHARTS

A selection of useful tables and charts should be included, such as flat pattern development; conversion charts; twist-drill data; plate, sheet, and strip stock sizes; wrench clearances; sheet-metal bend data; and other pertinent information.

ABBREVIATIONS

Abbreviations should generally be avoided, as they may obscure the meaning for some readers. Certain commonly used and generally understood abbreviations, such as psi, rpm, dia, R, lb, and D/DM are acceptable. Mathematical symbols are an important part of the engineering language, but such symbols as # should be avoided.

BINDING THE D/DM

Many D/DMs are designed as paperback throwaways for which periodic replacements of sections will be distributed on a predetermined schedule. This guarantees that upon distribution of new material the manual holders will be able to directly insert the latest information into their D/DMs. Some D/DMs are divided into several specialized, separately bound books. Each book has its own introduction, list of contents, technical information, glossary, and index. This is a well-conceived plan providing for the revision of a specialized section without republishing the entire manual. Also, it allows specialists to keep on hand only those sections of the manual necessary for their specific job. For example, a drafter who prepares only electrical or electronic schematics probably has no need for instructions on separate parts listing, mechanical drawings, or dimensioning and positional tolerancing.

A paperback D/DM may be more suitable for large drafting departments consisting of fifty or more drafters. The hard cover ring-binder type may be more practical for small organizations because of cost considerations.

Updating

Generally, the D/DM is subjected to constant and frequent change. It is vital that manual users apply the updated information. Failure to receive and comply with the latest information can result in losses in both time and money; drawings may have to be redrawn or costly parts scrap may result. Various methods of manual maintenance have been tried by different companies. The most common, and rather undependable, method is to have the drafter, designer, checker, and engineer fully maintain their own manuals. The trouble with this method is that these busy people rarely have time to perform the total manual maintenance function. Besides, technical people are not usually adapted to clerical work. Consequently, new sheets may not be added; obsolete sheets may not be removed, and often data may be misfiled. On the other hand, manual maintenance may be delegated to a clerk. This method also has critical drawbacks. Locating the manuals and finding out if each one is complete can take up a great deal of a clerk's time. Another pitfall with this method is the possible delay in updating. When an extensive number of sheets are involved and a large quantity of manuals must be serviced, it may take weeks to complete the updating task. Some companies, probably out of sheer desperation, send the revised and new sheets to the secretary of the manager of the department affected. The busy secretary faces the same problems confronted by the clerk, and more. Annoying as these problems may be, though, they are not any justification for making clerks out of design and drafting personnel or for trying to make a "Wizard of Oz" out of some clerk. The paperback, though not always practical, is one solution. The alternative, despite its faults, is to delegate the updating task to design and drafting personnel under close supervision.

Updating Checklist

Another way to make sure that the D/DM is kept up to date is to publish a checklist. Often, updates and revisions may not be inserted, or new sheets will be inadvertently inserted in the wrong place. The checklist is issued periodically to indicate additions and revisions to the D/DM.

One way to help prevent misfiling of pages in a manual is to color-code the pages. For example, all pages intended for the D/DM can be identified by a green stripe on the left side of the sheet; a blue stripe can be used for the design manual, a brown stripe for the procedures manual, and so on.

A portion of a checklist released in March 1980 is shown below.

A1.01	February 1979	B2.02	March 1980
A1.02	January 1979	B2.02	March 1980
A1.03	January 1979	C1.01	March 1980
A1.04	June 1979	C1.02	April 1979
A2.01	December 1978	C2.01	April 1979
A2.02	June 1979	C2.01	January 1980
A2.03	June 1979	C3.01	January 1980
A3.01	February 1980	C3.01	February 1980

DESIGN INFORMATION

Generally, it is advisable to exclude design guidance from the D/DM as completely as possible, allowing the use of this subject only when it is inseparable from drafting practice. Although the D/DM is essentially a drafting practice handbook, however, certain design, manufacturing, and processing information cannot be completely divorced from drafting practice, and consequently, should be included where appropriate. For example, it would not be practical to describe how a casting drawing should be prepared without getting into design considerations such as draft, parting lines, cores, section blending, fillets, shrinkage allowances, porosity, locating points, ribs, and the various casting methods. Besides, many drafters are called upon to make minor design decisions, such as how close a pierced hole should be to the edge of a surface, what radii should be specified, what size of chamfer or countersink should be used, what tolerances should be used, or what washer or screw should be used.

Drafting practice standards and design standards are distinctly different concepts.

Drafting practice instructs drafters how to put information on drawings. A drafting practice provides an improved means of communication, its basic purpose being to aid in defining the design configuration and communicate this clearly to the fabricator and inspector. A design standard is a documentation of physical requirements intended to specify a combination of materials, shape, dimensions, and other characteristics so as to yield a product with functional performance and reliability. A design standard involves considerations of strength, stress, functional performance, safety, reliability, quality, weight, size, aesthetics, circuitry, etc.; a drafting standard is concerned with communicating the finished design to the persons who will make the product. Also, the D/DM should include guidance as to the producibility of parts from the drawings. Figure 14-20 illustrates a typical example of the kind of information that deserves attention.

THE METRIC SYSTEM

The use of the metric system is bound to be extended and will demand wide attention. This vital subject, described in Section 8 of this handbook, should be discussed in any D/DM.

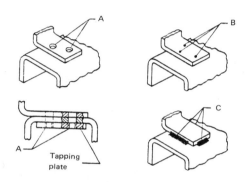

ASSEMBLY METHODS COST ANALYSIS CHART									
Operation or Material	Cost Factor	Qty & Pos	COST COMPARISONS						
			Spot Welds	Proj Welds	Rivets	Arc Welds	Bolts & Nuts	Bolts & Tapping Plate	Blind Rivets
Method			1	2	3	4	5	6	7
Spotwelding	100	3B	300					200	
Projection Welding	100	3B		300					
Forming Weld Proj	89	3B		267					
Punching Hole	89	4A			356		356	356	356
Rivet	70	2A			140				
Driving Rivets	96	2A			192				192
Arc Welding (1 inch)	250	3C				750			
Bolt	115	2A					230	230	
Nut	106	2A					212		
Lockwasher	18	2A					36	36	
Assembling Bolts	136	2A					272	272	
Tapping Plate (Matl)	321	1A						321	
Drilling Hole	89	2A						178	
Tapping Hole	89	2A						178	
Blind Rivet	742	2A							1484
TOTAL COST			300	567	688	750	1106	1771	2032

The above table is for illustrative purposes only and its application should be adjusted to costs prevailing at the time of its use. Cost comparisons are based on spotwelding as Unit 100. The table is not intended to indicate that the least costly method is the best; function and strength of assembly must also be considered.

FIG. 14-20 Assembly methods—cost comparisons.

THE IMPACT OF COMPUTER TECHNOLOGY ON D/DMs

It is widely believed that the need for D/DMs will expire in the future because of the new computer technology used in the preparation of drawings. Nothing could be further from the truth. The requirements for quality drawings will remain for machine output. Interactive graphics (IAG) will probably even increase the need for D/DMs. The cost of IAG systems ranges from $250,000 to $500,000. Consequently, it will be more important than ever before to do the job right the first time. If anything, it may be necessary to include a section in the D/DM on computer-aided manufacturing (CAM). IAG/CAM systems are automating more and more design, drafting, and manufacturing processes. And most experts concur that whatever influence the computer has had on design, drafting, and manufacturing will be minor compared to what's ahead. The need for a D/DM will continue.

AUDITING

An audit should be taken occasionally among drafters, designers, and checkers to obtain their evaluation and viewpoints of the D/DM. The questions asked should include: Does the D/DM include the needed information? Is it written in a style that is easily understood? How could it be improved? What's missing?

THE D/DM COMMITTEE

Every group substantially concerned with the D/DM should be offered the opportunity to participate in deciding what the provisions of the D/DM will be, so that the finished D/DM represents a consensus. Every effort should be made to thrash matters out so thoroughly that the decisions that are reached are unanimous, or nearly so. To accomplish this, a D/DM committee should be organized. This committee should consist of representatives from engineering, manufacturing, quality control, and purchasing. The D/DM committee reviews, comments on, and approves or rejects proposed drafting standards. The committee should be led by a director or chairperson, preferably a drafting supervisor. Also, a D/DM coordinator should be appointed whose responsibilities will be to handle administrative detail, including the arrangement of meetings, recording minutes of meetings, processing D/DM project questionnaires and voting summaries, and the incorporation of new and revised standards in the D/DM.

D/DM Projects

The committee work should be accomplished through assigned D/DM projects. The need for a D/DM project may result from established schedules, or it may result from suggestions or inquiries received from a committee member of other responsible personnel. Each proposed standard (project) should be assigned a D/DM project number. This will aid in scheduling and effective follow-up.

A D/DM project should be undertaken only after careful consideration to determine the need for and suitability of the proposed development. D/DM projects can be separated into two types, namely:

Type 1—Study Projects   A study project is an exploratory effort to determine the need for a D/DM project in a specific area, to report on related technical advances, or to establish the ramifications of suggested projects. This practice conserves time, for it may prevent the preparation of something that is not needed. Study projects should be assigned a D/DM project letter. A D/DM study project may result in the assignment of an active project.

Type 2—Active Projects   An active project is one for which an initial determination of need has been established and which is actively being developed for publication in the D/DM.

Circulation

Two copies of the proposed project or standard should be circulated to each member of the D/DM committee. These copies should be accompanied by a questionnaire and a ballot for committee members' votes concerning the proposed project or standard. A due date for the questionnaire and ballot, allowing ten days from the date of mailing to the return date, should be indicated on the forms. The ballot will permit the committee member to vote yes, no, or waive. Failure to reply should be taken to indicate approval. Ample space should be provided on the forms for comments. Figure 14-21 illustrates a questionnaire form.

Committee Responsibilities

It is essential that all committee members give serious consideration to the proposed D/DM standard. It is at this point that the proposed development should be questioned or disapproved if for any reason it appears unsuitable for its intended purpose or is contrary to management objectives. Every effort must be made to avoid unnecessary expenditure in developing and completing such a project. At the same time, in consideration of the fact that the circulated project represents an effort resulting from an

---

Date_____

Please complete one copy of this questionnaire and return it on or before

_____ (Allow 10 days.)

To: (Enter name and location
of coordinator here.)

QUESTIONNAIRE

STANDARDS COMMITTEE DRM PROJECT NO.

Title:_____

☐ APPROVE   ☐ DISAPPROVE   ☐ WAIVE

COMMENTS: (In case of a negative vote, include reasons for such vote and
suggest action, revised wording or figure revisions to overcome the specific objections. If an answer is not received
on the due date noted above, it will be considered that the
project is approved.)

_____
Name - Committee Member

Date_____   _____
Department

FIG. 14-21 D/DM project questionnaire.

expressed need, all negative votes must be accompanied by an explanation and suggestions as to how the objections may be overcome.

Summary of Replies

A summary of the committee vote on the circulated questionnaire, indicating those approving, disapproving, and waiving, and a summary of the comments, should be prepared by the coordinator and circulated to each member. Those not replying should be recorded under "approve" for purposes of determination of committee ballot majority. A sample of summary preparation is illustrated in Fig. 14-22.

The D/DM coordinator should prepare the summary of comments in accordance with the following:

1. The summary format should be based on the D/DM project proposal paragraphs receiving comment, with comments being recorded in the numerical order of the paragraphs to which they apply.

2. The committee member commenting should be recorded and his or her vote indicated.

```
        SUMMARY OF VOTE, COMMENTS AND PROPOSED DISPOSITIONS OF
                       QUESTIONNAIRE DATED_____

              STANDARDS PROJECT NO. XX  (PROJECT TITLE)

                     Date  _____

                              MEMBERS

   APPROVE - 6              DISAPPROVE -- 1                   WAIVE

   S. Smith *               W. Sterling *
   J. Jones **
   A. Cox *
   N. Brown *
   P. White *
   P. Freeman *

                       *   With Comments
                      **   No Ballot Returned

                           COMMENTS

   Paragraph 1

   S. Smith                 Revise first sentence to read  _____
   (Approve)                REASON:  To stress that  _____
   Coordinators             Disagree -- This point is covered in
                                       paragraph 2.5.1, page C1.05

   W. Sterling              Delete first paragraph
   (Disapprove)             REASON:  Not true because _____
   Coordinator              Agree -- will change accordingly.

   P. Freeman               Change first sentence to read _____
   (Approved)               REASON:  Improvement of clarity_____
   Coordinator              Agree -- Will revise to read  _____
```

FIG. 14-22 Summary of questionnaire votes.

3. The line of the paragraph to which the comment applies (if applicable) should be recorded.

4. Comments should be reported verbatim, not paraphrased. It is imperative that committee members be able to weigh the full import of each comment.

5. The coordinator should analyze each comment, consider all factors, and list proposed disposition beneath each comment, reporting also the reason for the disposition.

6. In determining disposition of comments received, it should be the responsibility of the coordinator to evaluate the comments and propose disposition based on committee consensus, which is 75 percent approval.

7. One copy of the summary of vote, comments, and proposed dispositions should be circulated to each member.

Meeting Review

At committee meetings, review and discussion of projects, comment summaries, and dispositions should be confined to those points of controversy that affect the technical content, intent, or interpretation of the project involved. Although it is mandatory to hear any point a member wishes to make, discussion must always be kept as brief and to-the-point as possible. The conservation of committee meeting time is imperative in order that all items on an often crowded agenda may receive committee action and the development of all D/DM projects can be facilitated.

Coordinator Preparation

It is the responsibility of the D/DM coordinator to analyze and plan projects in preparation for a committee meeting. The coordinator should endeavor to evaluate comments received in terms of their importance, and on the basis of each item's need for committee resolution. The review should be confined to those points that require committee resolution. By this means, considerations that have already been agreed upon, as well as purely editorial items, can be eliminated, and meeting time can be devoted to more important discussion.

Members' Preparation

It is the responsibility of D/DM committee members, when they review a circulated project or summary, to earmark those items to which they feel committee attention should be directed at the next meeting. This will ensure that no point of importance will be neglected in committee discussion.

Minutes of Meetings

It is the responsibility of the D/DM coordinator to record and circulate the minutes of each meeting. The minutes will be approved or disapproved at the next committee meeting.

Attendance and Participation

Regular attendance on the part of committee members is of the utmost importance. It is only through a cooperative effort that D/DM committee objectives can be fully achieved. Members should attend or be represented by an alternate at no less than 70 percent of the meetings. Prolonged inattendance and lack of representation places an inequitable burden on regularly attending members. Although company duties and conditions may make it impossible for a member to attend regularly, it is undesirable for a person to retain membership and consistently send an alternate to meetings. The chairperson may appoint subcommittees for accomplishment of special assignments or to study specialized subject in order to develop preliminary drafts for general committee review and approval. A committee member's or an individual's specialized knowledge in a technical field may be considered the prime requisite for appointment to a special subcommittee. Those selected for subcommittee work need not be members of the D/DM committee. Unless otherwise authorized by the chairperson, special subcommittees are automatically dissolved when they have completed their assigned task.

Committee Approval

In order that the recommendations presented in a proposed standard or project can be said to truly represent all segments of the organization, every effort shall be made to achieve unanimity. A project shall be considered finally approved by the D/DM committee when all major objectives have been resolved by a 75 percent approval vote of the committee convened.

In the event a committee member cannot attend a meeting and has a major objection to a project nearing completion, it will be his or her responsibility to outline the objection with justifying reasons in a memo to the committee chairperson. Such absentia objections shall be subject to resolution by the 75 percent approval vote of the committee convened.

Revision Authority

Revisions to drafting standards and projects should be authorized by committee consensus. When major comments or objections have been received from a circulation that affects the intent or interpretation, the coordinator must refer the matter to the committee for discussion and resolution. Revision authority should be the consensus of the convened committee. The coordinator may proceed with a revision under either of the following conditions:

1. When the project circulation results in a 75 percent approval vote.

2. When only minor comments are received that can be resolved through the coordinator's disposition.

Proposal Revision

If the D/DM project has been discussed and revision authorized, the D/DM coordinator will prepare a revised proposal according to instructions established by committee consensus. There should seldom be any need for a D/DM project to be held up until the next meeting for resolution of comments and proposed disposition. Prompt resolution will be a definite contribution toward early publication.

Revision Circulation

Two copies of the revised project should be circulated to each committee member. A questionnaire ballot should accompany each copy, with a ten-day return period indicated.

Classes of Revisions

A project may be revised as a result of questionnaire vote, summary of disposition, or by direction of committee convened. When the project is revised, the revision will fall into one of the following classes, which shall serve as a guide to the coordinator:

Class 1 revisions are technical revisions of text or illustrations resulting from questionnaires or meeting discussions. They shall be recirculated to the committee on a ten-day ballot circulation.

Class 2 revisions are semitechnical or minor corrections that do not materially affect the intent or interpretation of the text or illustrations. This class of revision might include rearrangement of paragraphs, rephrasing of paragraphs of text, and changes to illustrations with objective of added clarity. During the course of a project development, such revisions may result from questionnaire comment or by committee direction. In such cases, by direction of the committee convened, the project may be revised without requiring the ten-day ballot.

Class 3 revisions are those revisions of strictly editorial nature that affect spelling, punctuation, conformance to standards of preparation, or merely correct drafting technique to conform with accepted practice. Changes in this case have no effect on the intent or interpretation of the standard. These revisions may be made by the D/DM coordinator without prior circulation to the committee.

Disagreements

Controversial D/DM projects, as well as points on which agreement cannot be reached, should be set aside temporarily. After a number of these issues have accumulated, any committee member who is particularly interested will be offered the opportunity of reactivating the controversial D/DM issue of his or her choice.

Publication

The D/DM coordinator should, after committee approval has been given, prepare the final copy and arrange for printing and distribution.

Section 15

Checker's Guide

George E. Rowbotham

A checker's guide is indispensable. We have the design/drafting manual (D/DM), design manual, product assurance manual, military and industry standards, endless specifications, PERT charts, standard parts books, procedures manuals, organization manuals, and so on. Nevertheless, experience proves the need for an organized checker's guide in order to achieve better understanding, planning, monitoring, controlling, and auditing in the checking effort.

The purpose of the checker's guide is to outline specific checking functions that are essential for an engineering checker to thoroughly perform his or her job. Further, a checker's guide provides other company or customer personnel with a description of the checking effort. It is documented to provide requirements rather than to increase checking skills. It is not intended to replace the knowledge, experience, ability, and good judgment of checking personnel in fully exploring the adequacy and accuracy of engineering drawings. No document alone can develop skills—checkers learn their trade for the most part on the job, not by legislative procedures.

This does not mean, however, that the checker's guide is not practical or useful. Careful study of the contents makes the acquisition of familiarity with checking requirements much easier. The checker's guide provides the key steps the checker must follow on each drawing, engineering release, or revision to ensure complete conformance to customer and internal requirements. The checker's guide emphasizes the value of checking. Recommendations regarding the contents of the checker's guide, and related procedure, follow.

OBJECTIVE

Drawings shall be carefully reviewed and checked not just once but repeatedly, to ensure functional fitness, accuracy, conformance with design layout, schematic completeness, clarity of presentation, ease of manufacture and assembly, and conformance with applicable contractual specifications and directives.

RESPONSIBILITIES

To establish an understanding of job responsibilities, the following paragraphs briefly describe how these jobs interrelate.

General Responsibility

First, quality design and drafting are the responsibility of every designer, drafter, and checker. Every designer and drafter must carefully review his or her own work and not depend upon the checker or others to catch errors. The fact that the drawing is checked by a checker does not relieve the designer or drafter of responsibility.

The Designer's Responsibility

The designer is responsible for the preparation of quality design layouts. Design layout consists of the design of a component or a group of components complete to the point from which detail and assembly drawings can then be effectively generated. The design layout must be to scale, accurate, and complete, showing the detail design configuration and other essential engineering requirements, including envelope parameters, strength and hardness requirements, finishes, and materials.

Careful consideration must be given to conformance with customer requirements, pertinent specifications, operation, reliability, circuitry, manufacturability, adjustment, weight, assembly, installation, clearances, maintainability, adherence to standards, and such.

The Drafter's Responsibility

The drafter is responsible for quality drafting. Detail and assembly drawings must meet such requirements as conformance with the design layout, accuracy, clarity of presentation, conformance with standards, reproduction legibility, ease of manufacture, proper assembly, adequate clearances, and proper part names and numbers. Drawings must include correctness in dimensions and tolerances, materials, finishes, heat treatments, sufficient wall thicknesses, and other data required for castings, wiring diagrams, gears, sheet-metal parts, springs, and such, as defined in Section 3.

The Checker's Responsibility

The checker is responsible for quality checking. Quality checking ensures quality drafting.

Suggestions

Constructive criticism or suggestions for improvement of the checker's guide must always be welcome. They should be directed to the attention of the checking supervisor.

CHECKER'S NEEDS

The following shall be provided to the checker:

1. Check prints and the original drawings (to freeze drawings during checking). It is mandatory to use a print, rather than the original, for checking purposes. This avoids defacing the original drawing, provides a record of the alterations that will be required, and tests reproduction legibility. After checking, the marked check

prints and original tracings are turned over to the drafter for correction. All changes are evaluated by the drafter, and any objection to the change must be discussed with the checker. No changes shall be made after the drawing has been signed by the checker without his or her knowledge or approval. The check prints should be filed for six months and then destroyed.

2. It is the responsibility of the originator to assemble the complete documentation package before submitting drawings to checking. The complete package includes: check prints and drawings, reference prints, marked prints, calculations, pertinent memos, related drawings, approved schematics, block diagrams, layouts, special catalogs, release authorization, and any other pertinent data that may assist the checker. Incomplete packages may be returned to the originator for completion. When the calculations are included, the checker shall refer to them but derive his or her own calculations. Incomplete checking information results in delays, duplication of effort, and wasted motion, and it also increases the possibility of error.

3. Assembly drawings should be included. Generally, any job to be checked should be shown completed at least to the next assembly.

4. The cognizant design supervisor, or a designee, shall examine the package for completeness and the drawings for obvious errors or omissions. He or she shall ensure that any rough layouts or schematics are approved, signed, and dated by the responsible engineer so that checking can be assured of the preliminary requirements on which the design was based. The originator shall also make certain that marked print(s) are the latest issue. Following the design supervisor review, the originator shall deliver the package to checking.

5. If overtime will be required to complete a checking job within established schedules, a memo authorizing the amount of overtime needed should be signed by the responsible engineer and submitted to the checking supervisor for action as far in advance of the period the overtime is to be worked as possible. This requirement is essential, because checking must be able to justify its overtime expenditures.

DESIGN REQUIREMENTS

The checker's knowledge of the design requirements and the manufacturing processes involved, and his or her evaluation of drawing practices, can have a significant influence on the accuracy and cost of parts or assemblies. Layouts of new designs should be carefully studied and discussed with the designer or engineer to ensure that the checker will have full knowledge of the function and application of the design. If the checker believes that improvement in design should be made, he or she must review the matter with the responsible designer. Completed drawings must reflect the objective thinking of such coordination. In reviewing drawings, it is good practice to refer to all layouts pertinent to a part, as well as prints of associated parts. If the part is similar to parts in service, prints of such parts should also be obtained for reference. Layouts must be scrutinized for complexity to determine if the preparation of an additional check layout is warranted.

EXCESSIVE ERRORS

If a general review of the drawing reveals excessive errors that can be attributed to carelessness, the checker will return the drawing to the drafter without further check, and the drafter will then take corrective measures.

CHECKER'S INFLUENCE

A checker who ably performs his or her tasks gains the confidence of designers and engineers and may have a definite educational influence on the drafter. A checker's conclusions, when tactfully presented, can be beneficial in promoting coordination throughout the various phases of job development and in reducing the occurrence of errors.

DECISIONS

In some instances, the checker may be required to make a decision between the practice employed by the drafter and what he or she may consider to be preferable when an established practice is lacking. If the practice is clear and acceptable, the checker should approve the drawing, disregarding personal preference. On the other hand, the drafter will be required to change the drawing if the practice is faulty. In either case, measures must be promptly taken to deal with any controversial practice in the D/DM. In any disagreement between a checker and a drafter, the checker must take the initiative to bring about a workable understanding. The checker should understand and appreciate the drafter's feelings and work attitude based upon his or her own experience as a drafter. The drafter's feelings are easily understood; it is human to dislike being shown one's errors. The checker must exercise extreme understanding and patience with the drafter and be willing to explain the reason for each change. The checker's experience, if properly applied, will become part of the drafter's "know-how." If the checker and drafter fail to reach an agreement on a point of controversy, the matter must be directed to the attention of the checking supervisor for resolution. If he or she cannot resolve the problem, then it will be brought to the attention of the design and drafting section supervisor or a higher level for final disposition.

COMPROMISE WITH PERFECTION

Checking, like anything else, can be overdone. A drive for absolute perfection would usher in excessively high costs and would curtail work output. A constant drive for drafting economy is essential. It is proper to compromise with perfection when something nearly as good will do the job almost as well at less cost in time and effort. In other words, the checker must exercise good judgment and not be overconcerned with such trivia as slightly oversized arrowheads, slightly overextended lines, and the like. This does not mean that poor or mediocre performance can be condoned, let alone encouraged. The drafter who sets low goals or who consistently fails in performance must not be allowed to remain on the job. This does not mean that drafters should always be penalized for mistakes. Remember the old saw "He who does nothing makes no mistakes." The ultimate goal of checking is to avoid costly errors, maintain conformance with documentation requirements, and provide reproduction legibility. This is drawing quality control assurance.

Note: Credit for portions of this section is extended to *Graphic Science* Magazine.

Section 16

Numbering Systems

George E. Rowbotham

GENERAL

Despite their crucial importance, high visibility, and continual use, numbering systems are often the least known and least thoroughly understood drafting practices, and often their impact is grossly underrated compared to other management and engineering functions. In broad terms, numbering systems provide the basis of engineering documentation. They are vital engineering tools. They are not an end in themselves, but a means. The right question to ask in respect to them is not just what are they, but what are their functions, what is the optimum numbering system for a specific application, how are the numbers controlled, or what are the proven numbering principles.

In this era of rapidly advancing technology, organizations are compelled to come up with endless innovations if they are to progress or even survive. Generally, improvement begins with optimum practices (many of which are included in this handbook), with a new engineering image, and with advanced manufacturing technology.

The disciplines of numbering systems command the close attention of management, because they cover almost all aspects of an engineering and manufacturing organization, and they also concern customers and vendors.

BASIC NUMBERING SYSTEMS

There are three basic types of numbering systems: *nonsignificant*, *significant*, and *semisignificant*.

1. The nonsignificant system (purely serial numbering) employs numbers or a combination of numbers and letters, none of which have any significance in describing the part.

2. The significant system (mnemonic and devised codes) employs numbers or a combination of numbers and letters, all of which have significance in describing the part (see Figs. 16-1 and 16-2).

3. The semisignificant system employs numbers or a combination of numbers and letters, some of which are significant and some of which are nonsignificant. This system combines some features of both the significant and nonsignificant systems. Such numbers are used where both significant information, such as model, size, and finish, and nonsignificant information, such as a serial number, are both included in the part or drawing number (see Fig. 16-3).

The common requirement of these practices is that they provide accurate, orderly, and retrievable records of each part and assembly. Also, their implementation and maintenance must be effectively managed, not just once but always.

One of the main goals of significant and semisignificant numbering is to provide a simple and foolproof means of controlling variety at the source—i.e., to prevent a new part from being designed when a suitable one already exists under a different name and number. A second goal is to facilitate the work of all departments in quickly locating closely similar parts so as to achieve better use of existing tooling, rationalize manufacturing methods, enable quick cost comparisons to be made, and facilitate planning to obtain the maximum benefits from electronic data processing. These numbering systems have often become unworkable, however, because they have not followed sound fundamental principles, have failed to allow for expansion, or have grown too complicated or unwieldy. Before such systems are implemented, a study should be conducted to determine what other companies are doing. There are scores and scores of versions of these systems. Many of them are controversial.

A considerable number of the proven fundamental numbering principles are presented in this section of the handbook.

Which System?

In any theoretical study or rational commentary, some preliminary concepts must be established and accepted as self-evident on the basis of their intrinsic merit. Examples of this are the U.S. Constitution, the Euclidean geometric system, and most popular classical philosophical systems. Therefore, theoretical numbering activity could be developed on the basis of three axioms:

AXIOM 1.   It is essential to establish certain theorems on the basis of their self-evidence.

AXIOM 2.   Basic terms must be carefully distinguished and defined before misunderstandings develop.

AXIOM 3.   All numbering systems, with their various versions, are not ideal for any given organization. (Francis Beaumont and John Fletcher, English poets and dramatists of the early seventeenth century, wisely concluded: "What's one man's poison is another's meat or drink.")

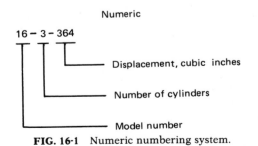

FIG. 16-1   Numeric numbering system.

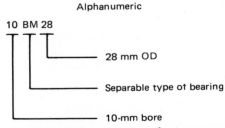

FIG. 16-2   Alphanumeric numbering system.

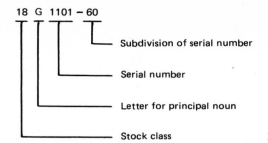

FIG. 16-3 Semisignificant numbering system.

DRAWINGS AND PART NUMBERS

It is recommended that the part number be the same as, or include, the drawing number.

DRAWING SIZE

The sheet-size letter of a drawing may be placed before or after the drawing number, if desired, to provide additional reference information. However, this drawing sheet size shall not be considered an integral part of the drawing or part number and should occupy a separate block adjacent to the drawing-number block.

DRAWING-NUMBER STRUCTURE

The drawing number should not exceed 15 characters. These characters include numbers, letters, and dashes, with the following limitations:

1. Letters I, O, Q, S, X, and Z shall not be used. This will avoid confusion, because these letters can be mistaken as certain numbers or symbols.
2. Numbers shall be arabic numerals. Fractional, decimal, and Roman numerals shall not be used.
3. Blank spaces shall be avoided.
4. Symbols and signs such as parentheses, asterisks, solidus (slash), degree, plus, and minus shall not be used, except when referencing a government or industry document whose identification contains such a symbol.
5. Drawing and part numbers shall not include as an integral part revision letters or numbers.
6. Letters shall be uppercase (capitals).

PARTS LIST NUMBERS

Each parts list (PL) shall be assigned the same number as has been assigned to the drawing to which it applies. Parts list numbers may be prefixed with "PL."

REPORT AND SPECIFICATION NUMBERING

Reports, specifications, and similar documents prepared in the engineering department require identifying numbers. Various schemes have been devised for report numbering, including elaborate coding to indicate the nature of the document. Unless careful planning that traces each possible condition to its ultimate conclusion is used in preparing these schemes, complications and contradictions will arise during employment of the numbering system.

The prime purpose of a report number is as a means of identifying and filing the document, and it is recommended that simple consecutive numbering be employed. Report numbers therefore begin at 1 for the first report issued and are then assigned in consecutive order, regardless of the nature of the document. When several divisions of a company are involved, a letter prefix should be used to identify reports originating in each division, such as "E-10," "D-72," or "W-203."

When several variations of a certain report are required, when a related series of reports will be issued, or when alterations of a prime specification are desired, a basic number can be assigned to designate the group, and a hyphened number to identify each individual report, such as "Report 203-1," "203-2," or "203-3."

MATCHED PARTS DESIGNATIONS

Parts which must be mated, and for which replacement as a matched set or pair is essential, shall be assigned a single number to designate each matched set or pair.

SYMMETRICALLY OPPOSITE PARTS

When feasible, symmetrically opposite parts may be described by showing one of the parts, in which case they shall be identified by adding a hyphened number after the drawing number. For example: "7654321-1 SHOWN" and "7654321-2 OPPOSITE" shall appear on the drawing, 7654321 being the basic drawing number. The use of odd hyphened numbers for the parts shown, and even hyphened numbers for the opposite parts, is preferred. As an alternative method, consecutive whole part numbers may be used and so indicated in the title block. Truly identical parts that are fully interchangeable shall carry only one part number.

INSEPARABLE ASSEMBLIES

When two or more parts are permanently fastened together by welding, riveting, brazing, cementing, bonding, etc., to form an inseparable assembly, the assembly shall be assigned an identifying number. The individual pieces may be assigned part or hyphened numbers.

FIND NUMBERS

A find number may be assigned to an item (part, assembly, etc.) on the field of a drawing for purposes of cross-referencing to items on a parts list, and as locators in lieu of using the item part number. The parts, assemblies, etc. so marked have other identifying numbers for purposes of procurement and marking that are cross-referenced to the find numbers in the integral or separate parts list, or in a table on the drawing.

IDENTIFICATION OF COMPUTER
PROGRAM TAPES

There are three basic kinds of computer program tapes: main, correction, and constants. They are prepared as either punched tapes or magnetic tapes. Tape identification application may be, as an example, through red or blue labels (red for in-house tapes and blue for all others). The identification is:

1. Typed on labels and spliced to the leading end for punched tapes.

2. Typed on labels affixed to the cassette used for magnetic tapes.

Main tape may include three areas of identification. The two basic areas include the design identification and the tape title and master or copy number information. The third element may involve the procuring agency information for deliverable tapes, when required. The design identification shall include the drawing number, reel identity, modification number, and correction number. The full identification, consisting of these four elements, represents the manufacturing part number. The drawing number may be prefixed with a "DP" (for punched tape) or "DM" (for magnetic tape).

With the advent of interactive graphics (IAG) and computer-aided design (CAD), punched and magnetic tapes are widely employed in graphics.

SUFFIX NUMBERING METHOD

This method is a variation of the semisignificant system. It consists of a significant base number combined with a nonsignificant (serial) suffix number, which may be a hyphen and a number or a letter and a number. This method is used for tabulating related details or assemblies on the same drawing. It is also used when detailed parts of an assembly or installation are dimensioned on the assembly or installation drawing. Here a number is assigned serially to the assembly or installation drawing. Each component part detailed on the assembly or installation drawing is in turn identified by a number consisting of the assembly drawing number followed by a serially assigned suffix number. In this letter application, the first portion of the part number becomes significant in that it identifies the assembly or installation which includes the component part. Examples using a (hyphened) suffix number:

1244	JACK ASSEMBLY
1244-1	BASE
1244-2	SCREW
1244-3	CAP
1244-4	ROD

Although the suffix number only may be shown in the body of the drawing and in the parts list, it is understood that the complete number includes both the basic portion and the suffix number—e.g., 1244-1 BASE.

Some organizations prefer to include (start with) the suffix number -1 on all initial drawings. This practice assures uniformity when a suffix number is added to an existing drawing. It merits attention.

Another method, using a plus or minus value as a suffix to indicate increments of variation from a basic design size, for the identification of parts for repair or service work, or for special applications of selected fits at assembly, has limited use. Parts in this category could be dowel pins, threaded fasteners, bushings, cylinders, or pistons. For example, a stud having a thread pitch diameter 0.004 larger than the basic design size would have the suffix +4, or a bushing having a bore 0.005 smaller than the basic design size would have the suffix −5 added to the basic part number. The letters P (plus) and M (minus) have also been used in preference to the plus and minus signs so as to avoid confusion with hyphened number applications. This method of identification should be carefully evaluated with respect to the possibility of errors and confusion with existing systems before adoption.

PATTERNS, CASTINGS, AND FORGINGS

Patterns, castings, and forgings are numbered the same as all other manufactured parts. The drawing number is the pattern, casting, and forging number. Appropriate prefixes may be used.

NUMBERING ASSIGNMENT AND CONTROL

Drawing and part numbers are essential. These identification numbers can be a source of endless confusion and excessive expense, though, if their assignments are not carefully controlled. The optimum practice is to provide a rapid means of obtaining numbers from a central point. A master drawing-number assignment record should be maintained on plain index cards, on punched cards, or in bound books. In addition, a disaster copy should be maintained and kept in a safe place.

Generally, the identification record includes the drawing number, title, and size. Other information usually included are the drafter's name, the date, the first application the drawing was used on, etc. With a computer input transmittal, the computer printout can include such information as follows:

1. Identification and description of parts, assemblies, and raw materials.
2. Design data: weights, configuration, auxiliary data, and special design data.
3. Reference data: customer documents, i.e., specifications, drawings, and standards.
4. Supplier (vendor) data: part name, drawing number, vendor's name, code identification, etc.
5. Next-assembly information.
6. Identification of item location on final product, such as a ship, airliner, or bridge.
7. Identification of item quantities by final product and by specific subassemblies.
8. Identification of effectiveness.
9. Rapid updating and revision.
10. High-speed automatic sorting capability (broken down by materials, purchased parts, standard parts, quantities, numerical list, indentured lists, multiple-usage lists, alphabetical lists, etc.).
11. Related data combined on a single document in a handy size.

An updated list of next-assembly numbers on drawings is not recommended, because it often is not dependable.

The benefits derived from automated systems are numerous, but they are not always understood. First of all, automated systems eliminate considerable manual drafting; consequently, design drafting costs are reduced. The time allotted to sorting of data is drastically reduced to a mere fraction of the time involved in a cumbersome, time-consuming manual operation, which may sometimes be virtually impossible. Also, automated systems provide an expanded range of data in an orderly and highly systematic manner.

VOIDED AND CANCELED NUMBERS

There are many occasions in which a drawing number is assigned but subsequent events make the drawing unnecessary or undesirable, and the drawing is not prepared. In other cases, a drawing may be prepared and released, then found unsuitable, with the result that it is no longer required. Numbers issued to these drawings should not be reassigned, but should be voided or canceled, depending on the circumstances. Attempts to reassign these drawing numbers are certain to result in difficulties arising from the possibility that the same number may be assigned to two drawings if the canceled drawing is later reinstated.

CHANGES REQUIRING NEW IDENTIFICATION

MIL-STD-100 provides clear direction on this element of numbering. Accordingly, the following list is based on this influential document.

1. When a part(s) has been submitted, a new drawing number or part number shall be assigned when a part or assembly is changed in such manner that any of the following conditions occur:

   Condition 1 Performance or durability is affected to such an extent that superseded parts must be discarded for reasons of safety or malfunctioning.

   Condition 2 Parts, subassemblies, or complete parts are changed to such an extent that the superseded and superseding parts are not interchangeable.

   Condition 3 When superseded parts are limited to use in specific articles or models of parts and the superseding parts are not so limited.

   Condition 4 When a part has been altered or selected, such as when a completed part is to be altered, or when an existing standard or vendor activity requires further selection, or there exists restriction of the part for fit, tolerance, performance, or reliability.

   Condition 5 When interchangeable repairable assemblies contain a noninterchangeable part, the part number reidentification of the noninterchangeable part, of its next assembly, and of all the progressively higher assemblies shall be changed up to and not including the assembly where interchangeability is reestablished.

2. When a part is changed in such a way that it necessitates a corresponding change to an operational, self-test or maintenance-test computer program, the part-number identification of the part and its next assembly and all progressively higher assemblies shall be changed up to and including the assembly where computer programs are affected.

CHANGES NOT REQUIRING NEW IDENTIFICATION

When a part or assembly is changed in such a manner that conditions defined in the preceding paragraph, Changes Requiring New Identification, do not occur, the part number shall not be changed. Under no condition shall the number be changed only because a new application is found for an existing part. However, when a design activity desires to create a tabulated listing or a standard because of a multiple application of an item, the foregoing need not apply. The new drawing will identify the document that superseded it.

GENERAL MOTORS (GM) NUMBERING EXPERIENCE

Several GM divisions in years past tried significant numbering systems. Invariably they confronted trouble either in their own organizations or in source and customer relations. Finally GM adopted as corporate practice a nonsignificant parts numbering system, centrally controlled insofar as allocation of number blocks is concerned, with divisional control of number assignments for functional parts and general office control of number assignments for standard parts. Fifty years of experience with that system has confirmed the belief of GM that nonsignificant part numbering is the simplest, the most flexible, and the only logical identification. An example of divisional block allocation is illustrated in the following partial listings of GM drawing numbers:

000001 to 049999	Division A
050001 to 100000	Division B
100001 to 230000	GM Standard Parts
230001 to 232000	Division C
232001 to 252500	Division D
252501 to 275000	Division E

Source identity was occasionally lost when responsibility for the manufacture of certain parts was moved from one division to another.

Fisher Body outlasted all other divisions in adherence to the significant numbering system. In composite body days, with drawing tied to patterns and templates, that system seemed to work out satisfactorily. Transition to metal bodies, with the problems encountered by the other metal fabricating divisions, soon showed up its deficiencies, and Fisher, like all the other divisions, found it expedient to convert to a nonsignificant numbering system.

SUMMARY

Nonsignificant part-numbering systems function independently of classifications, which may be revised or changed independently as circumstances indicate. The function of a part number is to identify the part. The identification should be complete, or the part number will fail in its purpose. Coding cannot be justified unless the code designation can stand alone without the aid of any supporting technical data such as drawings or descriptive lists. If reference to the supplementary information is necessary to establish the classification for coding or decoding, or if such reference is necessary to completely describe or identify the product, such coded designations compound difficulty, increase the chances of error, and slow up the whole reference process.

In many small businesses and in the organizational stages of large enterprises, an effectively devised code system seemingly justifies the time and effort expended on its development. It is very difficult and usually impractical in the long run to devise a coded identification system comprehensive enough for an extreme diversification of product or specialization of design. In such applications, the codes invariably become unwieldy, defeat their own purpose, emphasize rather than minimize the requirement for descriptive lists, and confuse rather than simplify the identification of parts. It is unfortunate that when such systems fail, the failure is usually attributed to the inadequacies of the particular code without questioning whether codes are worth using at all.

Index

About the Author

George E. Rowbotham, Editor in Chief, is engineering management editor of *Design Drafting & Reprographics* and former managing editor of *Graphic Science*. He served as chairman of General Motors Drafting Standards Committee and as Secretary of Ford Manufacturing Drafting Committee, and has been active on SAE, ANSI, Military, and similar standards committees. He has received international recognition for his work as manager, writer, consultant, and seminar chairman in engineering services support management, documentation, and graphics technology. He received a citation from the Department of Defense for his contribution to the development of Military Drafting Standards, and he has created or contributed to the development of more than 30 design and drafting manuals.